Studies in Systems, Decision and Control

Volume 323

Series Editor

Janusz Kacprzyk, Systems Research Institute, Polish Academy of Sciences, Warsaw, Poland

The series "Studies in Systems, Decision and Control" (SSDC) covers both new developments and advances, as well as the state of the art, in the various areas of broadly perceived systems, decision making and control–quickly, up to date and with a high quality. The intent is to cover the theory, applications, and perspectives on the state of the art and future developments relevant to systems, decision making, control, complex processes and related areas, as embedded in the fields of engineering, computer science, physics, economics, social and life sciences, as well as the paradigms and methodologies behind them. The series contains monographs, textbooks, lecture notes and edited volumes in systems, decision making and control spanning the areas of Cyber-Physical Systems, Autonomous Systems, Sensor Networks, Control Systems, Energy Systems, Automotive Systems, Biological Systems, Vehicular Networking and Connected Vehicles, Aerospace Systems, Automation, Manufacturing, Smart Grids, Nonlinear Systems, Power Systems, Robotics, Social Systems, Economic Systems and other. Of particular value to both the contributors and the readership are the short publication timeframe and the world-wide distribution and exposure which enable both a wide and rapid dissemination of research output.

** Indexing: The books of this series are submitted to ISI, SCOPUS, DBLP, Ulrichs, MathSciNet, Current Mathematical Publications, Mathematical Reviews, Zentralblatt Math: MetaPress and Springerlink.

More information about this series at http://www.springer.com/series/13304

Dongsheng Du · Shengyuan Xu ·
Vincent Cocquempot

Observer-Based Fault Diagnosis and Fault-Tolerant Control for Switched Systems

Springer

Dongsheng Du
Faculty of Automation
Huaiyin Institute of Technology
Huaian, Jiangsu, China

Shengyuan Xu
School of Automation
Nanjing University of Science
and Technology
Nanjing, Jiangsu, China

Vincent Cocquempot
UMR 9189 - CRIStAL
CNRS, Université de Lille
Centrale Lille
Lille, France

ISSN 2198-4182 ISSN 2198-4190 (electronic)
Studies in Systems, Decision and Control
ISBN 978-981-15-9075-7 ISBN 978-981-15-9073-3 (eBook)
https://doi.org/10.1007/978-981-15-9073-3

This Springer imprint is published by the registered company Springer Nature Singapore Pte Ltd.
The registered company address is: 152 Beach Road, #21-01/04 Gateway East, Singapore 189721, Singapore

Preface

The switched system is a system composed of a set of subsystems and of a switching law that indicates at each switching instant the active subsystem. The switching concept allows modeling a broad class of physical processes and practical systems. The study of the framework of switched systems can construct a bridge between the linear systems and the uncertain systems or nonlinear systems. Moreover, the study of switched systems provides additional insights into some long-standing and sophisticated problems, such as intelligent control, adaptive control, and robust analysis and control.

In practice, system faults are appear inevitably, which may influence the safe operation of the plant and cause system performance degradation, economic loss, and even disastrous consequences. Under this requirement, a lot of model-based fault diagnosis techniques, which have been proved to be efficient methods in dealing with faults, play an increasingly important role in fault-tolerant control (FTC). In general, observer-based technology has proved to be an efficient and feasible method in fault diagnosis and FTC. By constructing an observer or filer as a residual generator, the fault can be accurately diagnosed based on the obtained residual signal. Then an observer-based fault-tolerant controller can be designed.

This book devotes to present the observer-based fault diagnosis and FTC for switched systems. The focus is to address the problems of fault detection (FD), fault estimation (FE), and fault accommodation (FA) of actuator and sensor faults for switched systems by using adaptive observer, unknown input observer (UIO), reduced-order observer, proportional-integral observer (PIO) and filter. By using switched Lyapunov function, linear matrix inequality (LMI) and average dwell-time (ADT) techniques, a basic theoretical framework is formed toward the issues of fault diagnosis and FTC for switched systems. The proposed techniques provided efficient solutions in diagnosing actuator and sensor faults in the presence of switched systems.

In this research, the various methods proposed have been demonstrated by using numerical or practical examples. The simulation results show that the obtained results can achieve the prescribed performance requirements. This book can be used as a self-study book for upper division engineering students and researchers working with automatic control and switched systems.

Finally, I would like to thank my family, especially my parents, my wife, my son, and my daughter for their understanding and support in whatever I decided to do.

Huaian, Jiangsu, China Dongsheng Du
Nanjing, China Shengyuan Xu
Lille, France Vincent Cocquempot
May 2020

Acknowledgements The contents included in this book are an outgrowth of our academic research activities over the past several years. This book was partially supported by National Natural Science Foundation of China (No. 61873107), National Science Foundation for Post-doctoral Scientists of China (2016M601814), and Post-doctoral Foundation of Jiangsu Province (1601005B).

Contents

Acronyms

$\mathbb{R}$	Field of real numbers
$\mathbb{R}^n$	n-dimensional real Euclidean space
$\mathbb{R}^{n\times m}$	Space of $n \times m$ real matrices
I	Identity matrix
$I_{s\times s}$	Identity matrix of dimensions $s \times s$
O	Zero matrix
A	System matrix
$A_{m\times n}$	Matrix A of dimensions $m \times n$
A^{-1}	Inverse of matrix A
A^T	Transpose of matrix A
det(A)	Determinant of matrix A
rank(A)	Rank of matrix A
$\lambda(A)$	Eigenvalue of matrix A
$\lambda_{min}(A)$	Minimum eigenvalue of matrix A
$\lambda_{max}(A)$	Maximum eigenvalue of matrix A
$A \geq 0$	Symmetric positive semi-definite
$A > 0$	Symmetric positive definite
$A \leq 0$	Symmetric negative semi-definite
$A < 0$	Symmetric negative definite
$\|x\|$	Euclidean norm
$[a, b)$	The real number set $\{t \in \mathbb{R} : a \leq t < b\}$
$\forall$	For all
$\in$	Belong to
$Re(\cdot)$	Real parts of the arguments
$Im(\cdot)$	Imaginary parts of the arguments
lim	Limit
min	Minimum
max	Maximum
sup	Supremum
inf	Infimum

LMI	Linear matrix inequality
FD	Fault detection
FE	Fault estimation
FDI	Fault detection and identification
FA	Fault accommodation
FEA	Fault estimation and accommodation
FTC	Fault-tolerant control
DT	Dwell time
ADT	Average dwell time
PDT	Persistent dwell time
GUAS	Globally uniformly asymptotically stable
PIO	Proportional-integral observer
UIO	Unknown input observer
KYP	Kalman–Yakubovic–Popov
$diag\{X_1, X_2, \cdots, X_m\}$	Diagonal matrix with X_i as its ith diagonal element

Chapter 1
FD Filter Design for Switched Systems

1.1 Actuator FD Filter Design with Output Disturbances

In this section, the problem of actuator FD for discrete-time switched systems with output disturbance is investigated. By using the descriptor observer method, an H_∞ FD filter is designed to guarantee that the augmented system is admissible and satisfies a prescribed H_∞ performance index. By utilizing switched Lyapunov function approach, a sufficient condition for the admissible of the augmented system is obtained. Based on the obtained results, the desired FD filter can be designed. All the results are formulated in the form of LMIs. Finally, an example is proposed to illustrate the effectiveness of the developed method.

1.1.1 Introduction

Switched system belongs to hybrid systems, which is consisted of a number of subsystems and a switching signal specifying the switching between them [1]. In fact, many practical engineering systems can be modeled as switched systems, such as electrical engineering systems, traffic control systems, chemical process systems, and so on [2, 3]. Presently, the study of control and synthesis for switched systems is becoming a hot research topic, and many researchers have devoted themselves to investigating this challenging and meaningful issue. In the past few decades, a lot of achievements have been developed, readers can refer to [4–8] and the references therein.

It's well known that practical engineering systems are unavoidably affected by unexpected variations in external surroundings or sudden changes in signals. Due to these accidental reasons, different kinds of malfunctions or imperfect behaviors may appear during normal operations, and these phenomena are called faults. Generally

D. Du et al., *Observer-Based Fault Diagnosis and Fault-Tolerant Control for Switched Systems*, Studies in Systems, Decision and Control 323,
https://doi.org/10.1007/978-981-15-9073-3_1

speaking, faults can occur more easily in switched systems than those in non-switched systems for the reason of a large number of subsystems and the arbitrary switching between them. In reality, once an unexpected fault happens in one of the subsystems and is not detected on time, it might bring a big impact on the whole system and even cause potentially catastrophic damage [9]. Therefore, the investigation of FD problems for switched systems is of both theoretical and practical importance.

In this context, many theories and techniques have been developed in dealing with FD problem ([10–14] and references therein). Presently, several popular methods have been employed in dealing with this issue. For example, [15] proposed a model-based FD approach, while the parameter estimation approach was investigated to solve the problem in [16]. Reference [17] considered the FD problem by using generalized likelihood method. Among these methods, the model-based method is the most common way, which is to design an FD filter or observer generating a residual and compares it with a predefined threshold. When the residual evaluation function has a value larger than the threshold, an alarm is generated. For example, fault detection filter design for linear time-invariant systems was solved in [18]. FD issues for Markovian jump systems and uncertain fuzzy systems were separately studied in [19, 20].

In reviewing the development of these theories and techniques for different FD system designs, one of the commonly adopted ways in FD is to introduce a performance index and formulate the FD as an optimization problem. The H_∞ norm of the transfer-function matrix from unknown input to residual is accepted as a suitable and effective measure to estimate the influence of the unknown inputs; the H_∞ norm of transfer function from fault to residual has been proposed to evaluate the system sensitivity to the faults. In [21], an H_∞ filtering formulation of FD has also been presented to solve robust FD for uncertain systems. Different from the former scheme, the later one is to make the error between residual and fault (or, more generally, weighted fault) as small as possible, and provides an effective approach to uncertain system FD with the aid of an optimization tool [22].

In this section, the problem of actuator FD for discrete-time switched systems with output disturbance is investigated. The advantages can be summarized in the following aspects: firstly, for control input, actuator fault and unknown output disturbance in the discrete-time switched system, a descriptor-based FD filter is constructed such that the residual system is admissible with H_∞ performance index; secondly, a sufficient condition for the admissibility of the residual system is obtained; thirdly, compared with other FD schemes, descriptor observer approach can not only lead to a simple design procedure but also can observe disturbance signal. Therefore, the descriptor approach doesn't need any prior knowledge of disturbance signal.

1.1.2 Problem Statements and Preliminaries

Consider the following discrete-time switched system:

$$x(t+1) = \sum_{i=1}^{N} \delta_i(t)(A_i x(t) + B_i u(t) + F_i f(t)) \tag{1.1}$$

$$y(t) = \sum_{i=1}^{N} \delta_i(t)(C_i x(t) + D_i d(t)), \tag{1.2}$$

where $x(t) \in \mathbb{R}^n$ is the state vector, $u(t) \in \mathbb{R}^m$ is the control input vector, $d(t) \in \mathbb{R}^d$ is the output disturbance vector, $f(t) \in \mathbb{R}^f$ is the fault vector, and $y(t) \in \mathbb{R}^p$ is the measurable output vector. The function $\delta_i(t)$ is the switching signal and assumed be a priori unknown, but its instantaneous value is available in real time, where

$$\delta_i(t) : Z^+ = \{0, 1, 2\ldots\} \to \{0, 1\}, \quad \sum_{i=1}^{N} \delta_i(t) = 1, \quad \forall t \in Z^+.$$

A_i, B_i, C_i, D_i, and F_i are constant matrices with appropriate dimensions.

For the purpose of this section, the following assumptions are given:

- (A_i, C_i) is observable for $i = 1, \ldots, N$.
- Matrices D_i $(i = 1, \ldots, N)$ is full column rank.

Define the following augmented state and matrices:

$$\bar{x}(t) = \begin{bmatrix} x(t) \\ d(t) \end{bmatrix}, \bar{E} = \begin{bmatrix} I & 0 \\ 0 & 0 \end{bmatrix}, \bar{A}_i = \begin{bmatrix} A_i & 0 \\ 0 & -D_i \end{bmatrix},$$
$$\bar{B}_i = \begin{bmatrix} B_i \\ 0 \end{bmatrix}, \bar{D}_i = \begin{bmatrix} 0 \\ D_i \end{bmatrix}, \bar{F}_i = \begin{bmatrix} F_i \\ 0 \end{bmatrix}, \bar{C}_i = \begin{bmatrix} C_i & D_i \end{bmatrix}. \tag{1.3}$$

Then the system (1.1)–(1.2) can be expressed as the following descriptor system:

$$\bar{E}\bar{x}(t+1) = \sum_{i=1}^{N} \delta_i(t)(\bar{A}_i \bar{x}(t) + \bar{B}_i u(t) + \bar{D}_i d(t) + \bar{F}_i f(t)) \tag{1.4}$$

$$y(t) = \sum_{i=1}^{N} \delta_i(t)\bar{C}_i \bar{x}(t). \tag{1.5}$$

For system (1.4)–(1.5), we employ the following filter as residual generator:

$$\hat{\tilde{x}}(t+1) = \sum_{i=1}^{N} \delta_i(t)(A_{fi}\hat{\tilde{x}}(t) + B_{fi}y(t)), \tag{1.6}$$

$$r(t) = \sum_{i=1}^{N} \delta_i(t)(C_{fi}\hat{\tilde{x}}(t) + D_{fi}y(t)), \tag{1.7}$$

where $\hat{\tilde{x}}(t) \in \mathbb{R}^{n+d}$ is the filter's state, $r(t) \in \mathbb{R}^p$ is the so-called residual signal. The matrices A_{fi}, B_{fi}, C_{fi}, and D_{fi} are the filter parameters to be determined.

Remark 1.1 From the formation of the filter (1.6)–(1.7), one can get that the magnitude of the original system's state $x(t)$ and the output disturbance $d(t)$ of system (1.1)–(1.2) can be simultaneously estimated.

In order to detect the actuator fault $f(t)$ more efficient and fast, sometimes one is more interested in the fault signal within a certain frequency interval, which can be formulated as the weighted fault $\bar{f}(s) = W_f(s)f(s)$ with $W_f(s)$ being a given stable weighting matrix. One minimal state-space realization of $\bar{f}(s) = W_f(s)f(s)$ is supposed to be

$$\tilde{x}(t+1) = A_W\tilde{x}(t) + B_W f(t), \tag{1.8}$$

$$\bar{f}(t) = C_W\tilde{x}(t) + D_W f(t), \tag{1.9}$$

where $\tilde{x}(t) \in \mathbb{R}^{n_W}$ is the state of the weighted fault, $\bar{f}(t) \in \mathbb{R}^f$ is the weighted fault. A_W, B_W, C_W, and D_W are known constant matrices with appropriate dimensions.

Remark 1.2 As the technique developed in [23], a suitable weighting matrix $W_f(s)$ is introduced to limit the frequency interval of the interested fault, which can definitely improve the FD performance of the systems. From another aspect to see, the use of weighted fault $\bar{f}(t)$ is more general than using the original fault $f(t)$, because if we impose $W_f(s) = I$, we can obtain $\bar{f}(t) = f(t)$.

Denoting

$$e(t) = \left[\bar{x}^T(t)\ \hat{\tilde{x}}^T(t)\ \tilde{x}^T(t)\right]^T, \tag{1.10}$$

$$\omega(t) = \left[u^T(t)\ d^T(t)\ f^T(t)\right]^T, \tag{1.11}$$

$$r_e(t) = r(t) - \bar{f}(t). \tag{1.12}$$

Augmenting the model of system (1.4)–(1.5) to include the states of (1.6)–(1.7) and (1.8)–(1.9), the following augmented system can be obtained:

$$\tilde{E}e(t+1) = \sum_{i=1}^{N} \delta_i(t)(A_{ei}e(t) + B_{ei}\omega(t)), \tag{1.13}$$

$$r_e(t) = \sum_{i=1}^{N} \delta_i(t)(C_{ei}e(t) + D_{ei}\omega(t)), \tag{1.14}$$

where

$$\tilde{E} = \begin{bmatrix} \bar{E} & 0 & 0 \\ 0 & I & 0 \\ 0 & 0 & I \end{bmatrix}, A_{ei} = \begin{bmatrix} \bar{A}_i & 0 & 0 \\ B_{fi}\bar{C}_i & A_{fi} & 0 \\ 0 & 0 & A_W \end{bmatrix}, B_{ei} = \begin{bmatrix} \bar{B}_i & \bar{D}_i & \bar{F}_i \\ 0 & 0 & 0 \\ 0 & 0 & B_W \end{bmatrix},$$
$$D_{ei} = \begin{bmatrix} 0 & 0 & -D_W \end{bmatrix}, C_{ei} = \begin{bmatrix} D_{fi}\bar{C}_i & C_{fi} & -C_W \end{bmatrix}. \tag{1.15}$$

After designing the residual generator (1.6)–(1.7), the last step to a successful FD is the residual evaluation stage, which needs to construct an residual evaluation function and a threshold. In this section, the residual evaluation function $J_L(r(t))$ and the threshold J_{th} are selected as

$$J_L(r(t)) = \|r(t)\|_2 = \left(\sum_{k=k_0}^{k_0+L} r^T(t)r(t) \right)^{\frac{1}{2}}, \tag{1.16}$$

$$J_{th} = \sup_{u(t)\in l_2, d(t)\in l_2, f(t)=0} \|r(t)\|_2, \tag{1.17}$$

where k_0 denotes the initial evaluation time instant, L is the evaluation time steps. Based on this, the occurrence of the fault can be successfully detected by comparing $J_L(r(t))$ and J_{th} according to the following regulation:

$$J_L(r(t)) > J_{th} \Longrightarrow with\ \ faults \Longrightarrow alarm. \tag{1.18}$$
$$J_L(r(t)) \leq J_{th} \Longrightarrow no\ \ faults. \tag{1.19}$$

In order to minimize the effect of the disturbance and improve the sensitivity of the residual to the fault, the FD filter design can be formulated as an H_∞ filter problem. i.e., the problem to be addressed in this work is expressed as follows: to develop the filter (1.6)–(1.7) for the system (1.4)–(1.5) such that the augmented system (1.13)–(1.14)

- is asymptotically stable when $\omega(t) = 0$;
- under zero-initial condition, the minimum of γ is made small in the feasibility of

$$\sup_{\|\omega(t)\|_2 \neq 0} \frac{\|r_e(t)\|_2}{\|\omega(t)\|_2} < \gamma, \quad \gamma > 0. \tag{1.20}$$

For the purpose of this section, we give the following definitions.

Definition 1.1 ([24]) Consider the switched system $Ex(t+1) = \sum_{i=1}^{N} \delta_i(t)A_i x(t)$,

- the pair (E, A_i) is said to be regular if $det(sE - A_i)$ is not identically zero for all $i = 1, \ldots, N$.
- the pair (E, A_i) is said to be causal if $deg(det(sE - A_i)) = rankE$ for all $i = 1, \ldots, N$.

- the pair (E, A_i) is said to be regular and causal if every pair (E, A_i) is regular and causal for all $i = 1, \ldots, N$.

Definition 1.2 ([25]) System (1.1)–(1.2) with $u(t) = 0$, $d(t) = 0$, and $f(t) = 0$, is said to be

- regular, causal, if the pair (E, A_i) is regular, causal;
- uniformly asymptotically stable, if for any $\varepsilon > 0$, there is a $\vartheta(\varepsilon) > 0$ such that for arbitrary switching signal $\delta_i(t)$, $sup\|\phi(s)\| < \vartheta$ implies $\|x(t)\| < \varepsilon$ for all $t \geq 0$, and there is a $\vartheta'(\varepsilon) > 0$ such that for arbitrary switching signal $\delta_i(t)$, $sup\|\phi(s)\| < \vartheta'$ implies $x(t) \to 0$ as $t \to \infty$.

Definition 1.3 System (1.13)–(1.14) with $\omega(t) = 0$ is said to be uniformly asymptotically stable with γ-disturbance attenuation if system (1.13)–(1.14) with $\omega(t) = 0$ is regular, causal, uniformly asymptotically stable, and for a given scalar $\gamma > 0$, for any disturbance $\omega(t) \in l_2[0, \infty)$, the following H_∞ performance is satisfied:

$$\sum_{t=0}^{\infty} r_e^T(t) r_e(t) \leq \sum_{t=0}^{\infty} \gamma^2 \omega^T(t)\omega(t). \tag{1.21}$$

Remark 1.3 Regularity and causality of the switched system (1.13) with $\omega(t) = 0$ ensure that for the arbitrary switching signal $\delta_i(t)$, the solution to this system exists and is unique for any compatible initial conditions. However, even if the switched system (1.13) is regular and causal, it still has finite jumps because of the incompatible initial conditions caused by subsystems switching. These jumps are generally unavoidable [26]. In this section, for simplicity, we assume that such finite jumps cannot destroy the stability of the switched system.

1.1.3 FD Filter Design

In this section, the FD filter design problem will be solved for the descriptor discrete-time switched systems (1.4)–(1.5). The fault detection analysis problem is firstly solved, and then based on this, a full-rank FD filter is designed.

1.1.3.1 FD Analysis

Theorem 1.1 *For a prescribed scalar $\gamma > 0$, the residual system (1.13)–(1.14) is admissible with H_∞ performance γ if there exist positive definite matrices P_i, P_j, and matrices R and Q_i for all $i, j = 1, \ldots, N$ such that*

$$\begin{bmatrix} \Lambda_i & 0 & C_{ei}^T & A_{ei}^T \\ * & -\gamma^2 I & D_{ei}^T & B_{ei}^T \\ * & * & -I & 0 \\ * & * & * & -P_j^{-1} \end{bmatrix} < 0 \tag{1.22}$$

where $\Lambda_i = -\tilde{E}^T P_i \tilde{E} + A_{ei}^T R Q_i^T + Q_i R^T A_{ei}$, *and* R *is any matrix with full column which satisfies* $\tilde{E}^T R = 0$.

Proof Firstly, we prove the system (1.13)–(1.14) with $\omega(t) = 0$ is admissible. When $\omega(t) = 0$, the system (1.13)–(1.14) is written as follows:

$$\tilde{E}e(t+1) = \sum_{i=1}^{N} \delta_i(t) A_{ei} e(t), \tag{1.23}$$

$$r_e(t) = \sum_{i=1}^{N} \delta_i(t) C_{ei} e(t). \tag{1.24}$$

Suppose that there exist nonsingular matrices G and H such that

$$G\tilde{E}H = \begin{bmatrix} I & 0 \\ 0 & 0 \end{bmatrix}, GA_{ei}H = \begin{bmatrix} A_{11ei} & A_{12ei} \\ A_{21ei} & A_{22ei} \end{bmatrix}, G^{-T} P_i G^{-1} = \begin{bmatrix} P_{11i} & P_{12i} \\ P_{21i} & P_{22i} \end{bmatrix}. \tag{1.25}$$

Then we denote

$$R = G^T \begin{bmatrix} 0 \\ I \end{bmatrix} N, \quad Q_i = H^{-T} \begin{bmatrix} Q_{1i} \\ Q_{2i} \end{bmatrix} N^{-T}. \tag{1.26}$$

From inequality (1.22), we have

$$A_{ei}^T P_j A_{ei} - \tilde{E}^T P_i \tilde{E} + A_{ei}^T R Q_i^T + Q_i R^T A_{ei} < 0. \tag{1.27}$$

Pre- and post-multiplying inequality (1.27) by H^T and H, respectively, we have

$$\begin{bmatrix} \star & \star \\ \star & \Gamma_i \end{bmatrix} < 0, \tag{1.28}$$

where $\star$ denotes submatrix which is not relevant to our discussion and the form of Γ_i is shown as follows:

$$\begin{aligned} \Gamma_i = {} & A_{12ei}^T P_{11j} A_{12ei} + A_{22ei}^T P_{21j} A_{12ei} + A_{12ei}^T P_{12j} A_{22ei} \\ & + A_{22ei}^T P_{22j} A_{22ei} + A_{22ei}^T Q_{2i}^T + Q_{2i} A_{22ei} - P_{22i} < 0, \end{aligned} \tag{1.29}$$

which implies that matrices A_{22ei} are nonsingular. Therefore, under the condition that $\omega(t) = 0$, one can conclude that system (1.13)–(1.14) is regular and causal from Definition 1.1.

Consider the following switched Lyapunov function:

$$V(e(t)) = e^T(t)\tilde{E}^T\left(\sum_{i=1}^{N}\delta_i(t)P_i\right)\tilde{E}e(t). \tag{1.30}$$

Define $\varDelta V(t) = V(t+1) - V(t)$, then along the solution of (1.13), we have

$$\begin{aligned}
\varDelta V(t) &= e^T(t+1)\tilde{E}^T\left(\sum_{i=1}^{N}\delta_i(t+1)P_i\right)\tilde{E}e(t+1) - e^T(t)\tilde{E}^T\left(\sum_{i=1}^{N}\delta_i(t)P_i\right)\tilde{E}e(t)\\
&= e^T(t)\left[\sum_{i=1}^{N}\delta_i(t)A_{ei}^T\left(\sum_{i=1}^{N}\delta_i(t+1)P_i\right)\sum_{i=1}^{N}\delta_i(t)A_{ei}\right.\\
&\quad\left.-\tilde{E}^T\left(\sum_{i=1}^{N}\delta_i(t)P_i\right)\tilde{E}\right]e(t)
\end{aligned} \tag{1.31}$$

For any nonzero state vector $e(t)$ and the particular case $\delta_i(t) = 1$, $\delta_{l\neq i}(t) = 0$, and $\delta_j(t+1) = 1$, $\delta_{l\neq j}(t+1) = 0$, one can get

$$\varDelta V(t) = e^T(t)\left(A_{ei}^T P_j A_{ei} - \tilde{E}^T P_i \tilde{E}\right)e(t). \tag{1.32}$$

Notice that $\tilde{E}^T R = 0$, then

$$\begin{aligned}
0 &= 2e^T(t+1)\tilde{E}^T R\left(\sum_{i=1}^{N}\delta_i(t)Q_i\right)\tilde{E}e(t+1)\\
&= e^T(t)\left[\left(\sum_{i=1}^{N}\delta_i(t)A_{ei}^T\right)R\left(\sum_{i=1}^{N}\delta_i(t)Q_i^T\right)\right.\\
&\quad\left.+\left(\sum_{i=1}^{N}\delta_i(t)Q_i\right)R^T\left(\sum_{i=1}^{N}\delta_i(t)A_{ei}\right)\right]e(t).
\end{aligned} \tag{1.33}$$

Combining the conditions in (1.32) and (1.33), one can get

$$\varDelta V(t) = e^T(t)\left(A_{ei}^T P_j A_{ei} - \tilde{E}^T P_i \tilde{E} + A_{ei}^T R Q_i^T + Q_i R^T A_{ei}\right)e(t). \tag{1.34}$$

Therefore, from the condition in (1.22), one can obtain that $\varDelta V(t) < 0$. Then we can conclude that the system (1.13)–(1.14) is asymptotically stable.

Based on the above discussions, one can get that the system (1.13)–(1.14) is admissible. In the following, the γ performance index will be proved.

Define the following performance index:

$$J = \sum_{t=0}^{K-1} [r_e^T(t) r_e(t) - \gamma^2 \omega^T(t) \omega(t)]. \tag{1.35}$$

Under the zero initial condition, one has

$$\begin{aligned} J &= \sum_{t=0}^{K-1} [r_e^T(t) r_e(t) - \gamma^2 \omega^T(t) \omega(t) + \Delta V(t)] - V(K) \\ &\le \sum_{t=0}^{K-1} [r_e^T(t) r_e(t) - \gamma^2 \omega^T(t) \omega(t) + \Delta V(t)]. \end{aligned} \tag{1.36}$$

For any nonzero state vector $e(t)$ and the particular case $\delta_i(t) = 1$, $\delta_{l \neq i}(t) = 0$, and $\delta_j(t+1) = 1$, $\delta_{l \neq j}(t+1) = 0$, one can get

$$\Delta V(t) \le \upsilon^T(t) \Theta_{ij} \upsilon(t), \upsilon(t) = \begin{bmatrix} e(t) \\ \omega(t) \end{bmatrix}, \tag{1.37}$$

where

$$\begin{aligned} \Theta_{ij} &= \begin{bmatrix} \Theta_{11ij} & A_{ei}^T P_j B_{ei} + C_{ei}^T D_{ei} \\ * & B_{ei}^T P_j B_{ei} - \gamma^2 I + D_{ei}^T D_{ei} \end{bmatrix}, \\ \Theta_{11ij} &= A_{ei}^T P_j A_{ei} - \tilde{E}^T P_i \tilde{E} + A_{ei}^T R Q_i^T + Q_i R^T A_{ei} + C_{ei}^T C_{ei}. \end{aligned} \tag{1.38}$$

According to Schur complement Lemma and the condition in (1.22), one can get that $\Theta_{ij} < 0$, which implies that $J < 0$. Therefore, we obtain $\|r_e(t)\|_2 < \gamma \|\omega(t)\|_2$ for any nonzero $\omega(t) \in l_2[0, \infty)$. Thus the proof is completed. □

In order to design the FD filter more conveniently, we extend Theorem 1.1 to the following corollary.

Corollary 1.1 *For a prescribed scalar $\gamma > 0$, the residual system (1.13)–(1.14) is admissible with H_∞ performance γ if there exist positive definite matrices P_i, P_j, and matrices R, Ω and Q_i for all $i, j = 1, \ldots, N$ such that*

$$\begin{bmatrix} \Lambda_i & 0 & C_{ei}^T & A_{ei}^T \Omega \\ * & -\gamma^2 I & D_{ei}^T & B_{ei}^T \Omega \\ * & * & -I & 0 \\ * & * & * & P_j - (\Omega + \Omega^T) \end{bmatrix} < 0, \tag{1.39}$$

where $\Lambda_i = -\tilde{E}^T P_i \tilde{E} + A_{ei}^T R Q_i^T + Q_i R^T A_{ei}$, and R is any matrix with full column which satisfies $\tilde{E}^T R = 0$.

Proof Suppose the LMIs (1.39) hold. Noting the inequality

$$(P_i - \Omega)^T P_i^{-1}(P_i - \Omega) \geq 0 \tag{1.40}$$

implies that

$$P_i - (\Omega + \Omega^T) \geq -\Omega^T P_i^{-1} \Omega \tag{1.41}$$

which together with (1.39) yield

$$\begin{bmatrix} \Lambda & 0 & C_{ei}^T & A_{ei}^T \Omega \\ * & -\gamma^2 I & D_{ei}^T & B_{ei}^T \Omega \\ * & * & -I & 0 \\ * & * & * & -\Omega^T P_j^{-1} \Omega \end{bmatrix} < 0, \tag{1.42}$$

which pre- and post-multiplying $diag\{I, I, I, \Omega^{-T}\}$ and $diag\{I, I, I, \Omega^{-1}\}$ yields (1.22), then we conclude the proof. □

Remark 1.4 With the introduction of a new additional matrix Ω, we obtain a sufficient condition in which the matrices P_i and P_j are not involved in any product with matrices A_{ei}, B_{ei}, C_{ei}, and D_{ei}, which makes an FD filter design feasible.

1.1.3.2 FD Filter Design

In this subsection, the FD filter design problem is investigated based on Corollary 1.1, i.e., a method will be developed to determine the FD filter matrices expressed in (1.6) and (1.7), such that the residual system (1.13)–(1.14) is asymptotically stable and the performance defined in (1.20) is guaranteed.

Theorem 1.2 *For prescribed scalars $\gamma > 0$ and $\eta > 0$, if there exist positive definite matrices P_{11i}, P_{22i}, P_{33i}, P_{11j}, P_{22j}, P_{33j}, and matrices P_{12i}, P_{13i}, P_{23i}, P_{12j}, P_{13j}, P_{23j}, U, L_1, L_2, Q_{1i}, Q_{3i}, $\hat{A}_{fi}$, $\hat{B}_{fi}$, $\hat{C}_{fi}$, $\hat{D}_{fi}$ for any $i, j = 1, \ldots, N$ satisfying the following LMIs:*

$$\Upsilon_{ij} = \begin{bmatrix} \Upsilon_{11ij} & 0 & \Upsilon_{13ij} & \Upsilon_{14ij} \\ * & -\gamma^2 I & \Upsilon_{23ij} & \Upsilon_{24ij} \\ * & * & -I & 0 \\ * & * & * & \Upsilon_{44ij} \end{bmatrix} < 0, \tag{1.43}$$

where

$$\Upsilon_{11ij} = \begin{bmatrix} (1,1) & -\bar{E}^T P_{12i} & -\bar{E}^T P_{13i} + \bar{A}_i^T R_1 Q_{3i}^T \\ * & -P_{22i} & -P_{23i} \\ * & * & -P_{33i} \end{bmatrix},$$

$$(1,1) = -\bar{E}^T P_{11i} \bar{E} + \bar{A}_i^T R_1 Q_{1i}^T + Q_{1i} R_1^T \bar{A}_i,$$

$$\Upsilon_{13ij} = \begin{bmatrix} \bar{C}_i^T \hat{D}_{fi}^T \\ \hat{C}_{fi}^T \\ -C_W^T \end{bmatrix}, \Upsilon_{23ij} = \begin{bmatrix} 0 \\ 0 \\ -D_W^T \end{bmatrix},$$

$$\Upsilon_{14ij} = \begin{bmatrix} \bar{A}_i^T U^T + \bar{C}_i^T \hat{B}_{fi}^T & \bar{A}_i^T L_1^T + \bar{C}_i^T \hat{B}_{fi}^T & 0 \\ \hat{A}_{fi}^T & \hat{A}_{fi}^T & 0 \\ 0 & 0 & \eta A_W^T \end{bmatrix},$$

$$\Upsilon_{24ij} = \begin{bmatrix} \bar{B}_i^T U^T & \bar{B}_i^T L_1^T & 0 \\ \bar{D}_i^T U^T & \bar{D}_i^T L_1^T & 0 \\ \bar{F}_i^T U^T & \bar{F}_i^T L_1^T & 0 \end{bmatrix},$$

$$\Upsilon_{44ij} = \begin{bmatrix} (2,2) & P_{12j} - (L_2 + L_1^T) & P_{13j} \\ * & P_{22j} - (L_2 + L_2^T) & P_{23j} \\ * & * & P_{33j} - 2\eta I \end{bmatrix},$$

$$(2,2) = P_{11j} - (U + U^T).$$

R_1 is any matrix with full rank and satisfies $\bar{E}^T R_1 = 0$. Then there exists an FD filter in the form of (1.6)–(1.7), such that the residual signal system in (1.13)–(1.14) is asymptotically stable with H_∞ performance γ. Moreover, if the aforementioned conditions are satisfied, the parameters for the FD filter in (1.6)–(1.7) can be given by

$$\begin{bmatrix} A_{fi} & B_{fi} \\ C_{fi} & D_{fi} \end{bmatrix} = \begin{bmatrix} X^{-1} & 0 \\ 0 & I \end{bmatrix} \begin{bmatrix} \hat{A}_{fi} & \hat{B}_{fi} \\ \hat{C}_{fi} & \hat{D}_{fi} \end{bmatrix} \begin{bmatrix} X^{-T} Y^T & 0 \\ 0 & I \end{bmatrix}, \tag{1.44}$$

where matrices X and Y are any nonsingular matrices satisfying $XY^{-T}X^T = L_2$.

Proof Based on the condition $\tilde{E}^T R = 0$ in Theorem 1.1, one can conclude that the matrix R has the following form:

$$R = \left[R_1^T \; 0 \; 0 \right]^T. \tag{1.45}$$

Define the following matrices:

$$\Omega = \begin{bmatrix} \Omega_1 & \Omega_2 & 0 \\ \Omega_3 & \Omega_4 & 0 \\ 0 & 0 & \eta I \end{bmatrix}^T, Q_i = \begin{bmatrix} Q_{1i} \\ 0 \\ Q_{3i} \end{bmatrix}, \Phi = \begin{bmatrix} I & 0 & 0 \\ 0 & \Omega_4^{-T} \Omega_2^T & 0 \\ 0 & 0 & I \end{bmatrix}, \tag{1.46}$$

where Ω_2 and Ω_4 are nonsingular matrices. Therefore, matrix Φ is invertible.

Performing a congruence transformation to inequality (1.39) via $diag\{\Phi, I, I, \Phi\}$, and define

$$\hat{A}_{fi} = \Omega_2 A_{fi} \Omega_4^{-T} \Omega_2^T, \hat{B}_{fi} = \Omega_2 B_{fi}, \hat{C}_{fi} = C_{fi} \Omega_4^{-T} \Omega_2^T, \hat{D}_{fi} = D_{fi},$$
$$L_1 = \Omega_2 \Omega_4^{-1} \Omega_3, L_2 = \Omega_2 \Omega_4^{-T} \Omega_2^T, U = \Omega_1, X = \Omega_2, Y = \Omega_4,$$
$$\Phi^T P_i \Phi = \begin{bmatrix} P_{11i} & P_{12i} & P_{13i} \\ * & P_{22i} & P_{23i} \\ * & * & P_{33i} \end{bmatrix}, \Phi^T P_j \Phi = \begin{bmatrix} P_{11j} & P_{12j} & P_{13j} \\ * & P_{22j} & P_{23j} \\ * & * & P_{33j} \end{bmatrix}. \quad (1.47)$$

Then, combining the conditions in (1.46) and (1.47), one can conclude that the inequality in (1.43) is equivalent to $\Phi^T \Upsilon_{ij} \Phi < 0$, which clearly guarantees the inequality in (1.39). Thus, the proof is completed. □

1.1.4 A Numerical Example

In this section, a numerical example is given to illustrate the developed method is feasible. Consider the switched system (1.1)–(1.2) with parametric matrices as follows:

$$A_1 = \begin{bmatrix} 0.8 & -0.3 \\ -0.1 & 0.2 \end{bmatrix}, B_1 = \begin{bmatrix} 0.1 \\ -0.2 \end{bmatrix}, F_1 = \begin{bmatrix} -0.1 \\ 0.09 \end{bmatrix}, C_1 = \begin{bmatrix} 0.1 & -0.7 \end{bmatrix}, D_1 = 0.2,$$
$$A_2 = \begin{bmatrix} 0.2 & -0.4 \\ -0.1 & 0.5 \end{bmatrix}, B_2 = \begin{bmatrix} 0.01 \\ -0.05 \end{bmatrix}, F_2 = \begin{bmatrix} -0.1 \\ 0.7 \end{bmatrix}, C_2 = \begin{bmatrix} 1 & -2 \end{bmatrix}, D_2 = 0.2.$$

Suppose the fault weighting matrix $A_W = 0.5$, $B_W = 0.25$, $C_W = 1.0$, and $D_W = 0.5$. If we choose the matrix $R_1 = \begin{bmatrix} 0 & 0 & 1 \end{bmatrix}^T$ and scalar $\eta = 2$ and solve the conditions (1.43) and (1.44) in Theorem 1.2, we can get when the minimum performance index $\gamma = 0.71$, the parametric matrices of the desired FD filter parameters are obtained as follows:

$$A_{f1} = \begin{bmatrix} 0.0039 & 0.0010 & 0.0000 \\ 0.0003 & 0.0001 & 0.0000 \\ -0.0000 & -0.0000 & -0.0000 \end{bmatrix}, B_{f1} = \begin{bmatrix} -1.3687 \\ -0.1102 \\ -0.0000 \end{bmatrix},$$
$$C_{f1} = \begin{bmatrix} -0.0024 & -0.0006 & -0.0000 \end{bmatrix}, D_{f1} = -0.0221.$$
$$A_{f2} = 10^3 * \begin{bmatrix} 0.7411 & 0.2789 & 0.0000 \\ 0.0312 & -0.3112 & -0.0000 \\ -0.0000 & 0.0000 & 0.0000 \end{bmatrix}, B_{f2} = \begin{bmatrix} -0.4493 \\ 0.0223 \\ -0.0000 \end{bmatrix},$$
$$C_{f2} = \begin{bmatrix} -0.0024 & -0.0006 & -0.0000 \end{bmatrix}, D_{f2} = -0.0335.$$

In this section, the main purpose is to realize FD for discrete-time switched system with sensor output disturbance. For the simulation, the sensor unknown output distur-

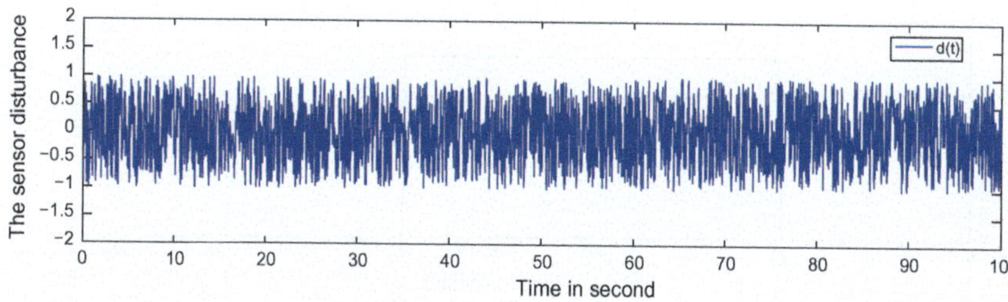

Fig. 1.1 Unknown output disturbance $d(t)$

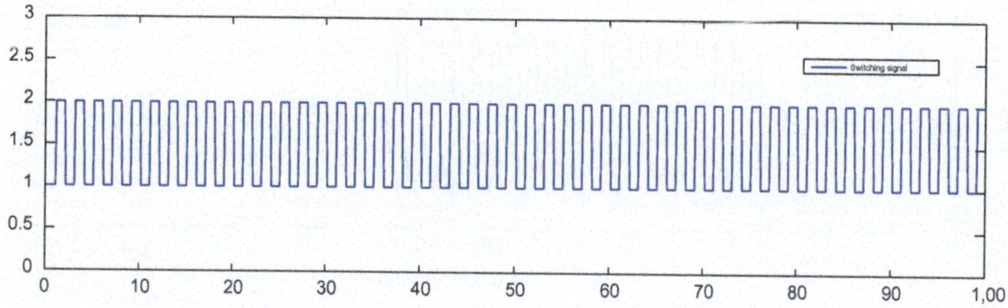

Fig. 1.2 Switched signal $\delta_i(t)$

bance is assumed to be random noise with sample time 0.1, as shown in Fig. 1.1. The arbitrary switched signal is shown in Fig. 1.2. The control input is $u(t) = 0.5cos(t)$. The fault signal is set up as

$$f(t) = \begin{cases} 5, & 20 \leq t \leq 60 \\ 0, & others \end{cases}$$

then the weighted fault signal $\bar{f}(t)$ is described in Fig. 1.3. We set the initial state is $x(0) = \begin{bmatrix} 0.3 & -0.2 \end{bmatrix}^T$, Figs. 1.4, 1.5, 1.6 separately depict the trajectories of the state $x(t)$, the state estimation $\hat{x}(t)$, and the measurement output $y(t)$. Figures 1.7 and 1.8 show the response of residual signal $r(t)$ without fault $f(t)$ and under the constant fault $f(t)$. Figure 1.9 presents the residual evaluation function $J_L(r(t))$ for both the faulty case (solid line) and fault-free case (dashed line). With a selected threshold $J_{th} = 2.2656$, the simulation results show that $J_L(21.5) = 2.7024 > J_{th}$ for $t = 21.5$, which means that the fault $f(t)$ can be successfully detected two time steps after its occurrence.

1.1.5 Summary

In this section, FD filter design for discrete-time switched systems under actuator fault and sensor disturbance is considered. By constructing a descriptor system,

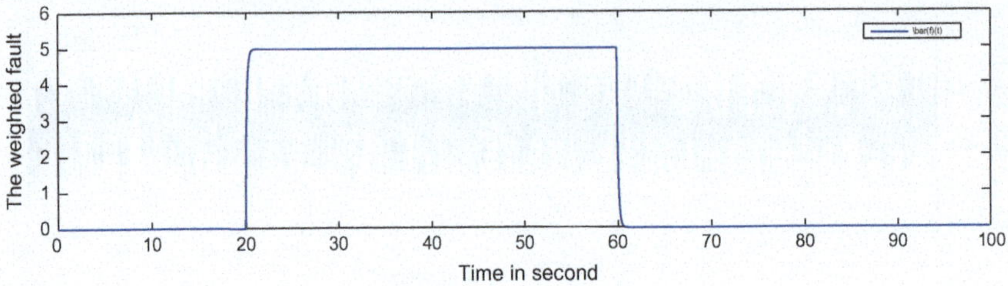

Fig. 1.3 The weighted fault signal $\bar{f}(t)$

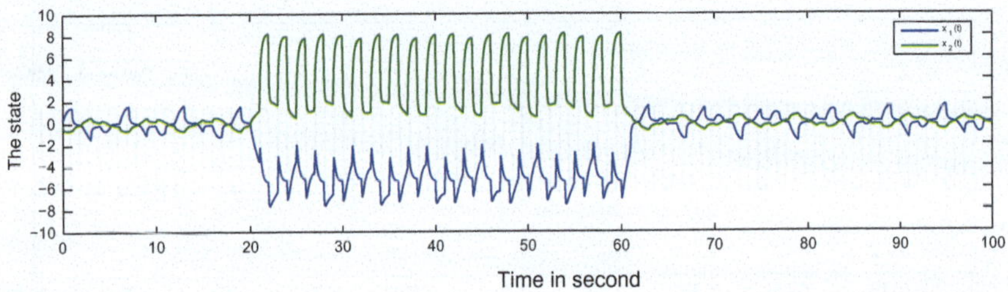

Fig. 1.4 State vectors $x(t)$

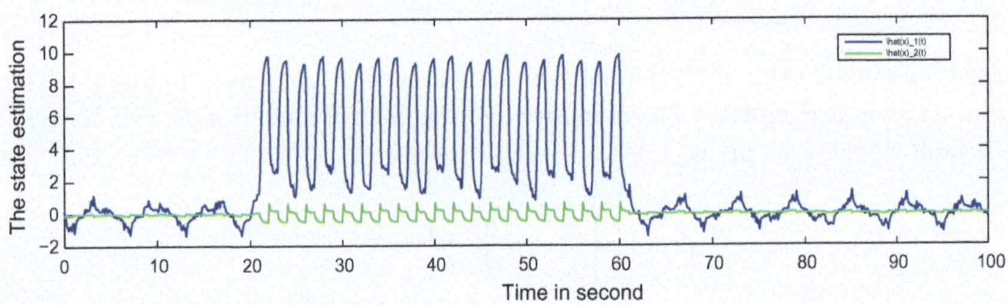

Fig. 1.5 The estimation of state vectors $\hat{x}(t)$

an efficient condition is obtained to guarantee the existence of an FD filter and such that the augmented systems are asymptotically stable with H_∞ performance. Finally, simulation results have illustrated the proposed methodology and shown the effectiveness of the results.

1.2 Finite-Frequency Actuator FD Filter Design

This section is concerned with the problem of the FD filter design for continuous-time switched systems with actuator faults. The actuator faults and the unknown disturbances are considered to be in finite-frequency domain. By using the switched Lyapunov function and the ADT techniques, efficient conditions are obtained, which

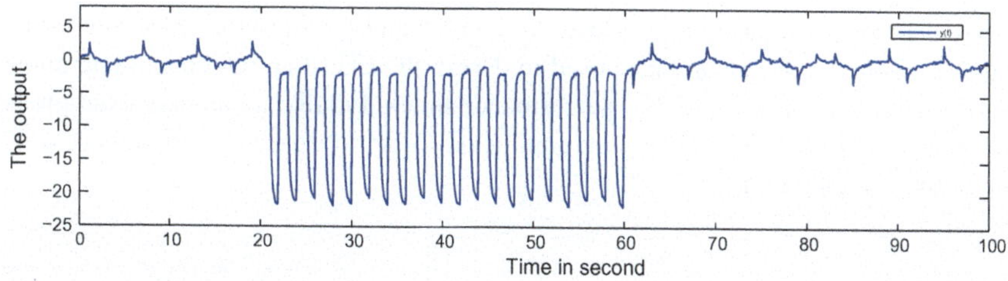

Fig. 1.6 The output vector $y(t)$

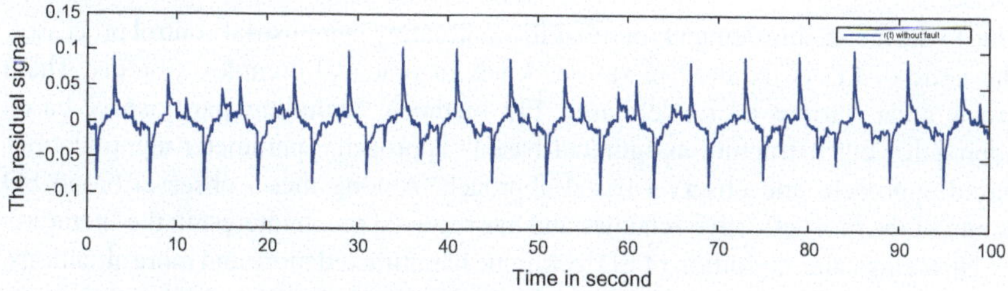

Fig. 1.7 Residual signal $r(t)$ without fault

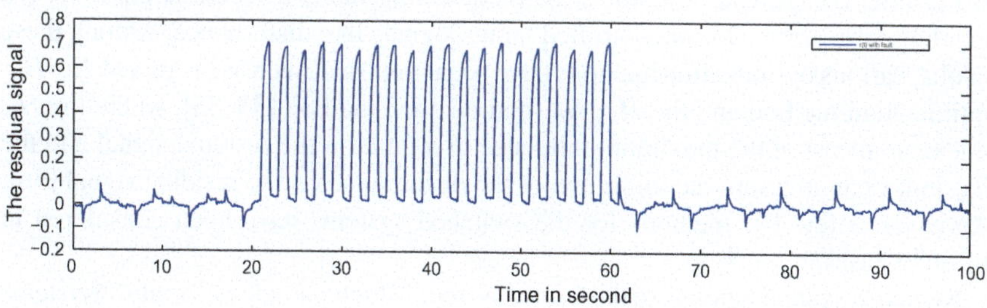

Fig. 1.8 Residual signal $r(t)$ with fault

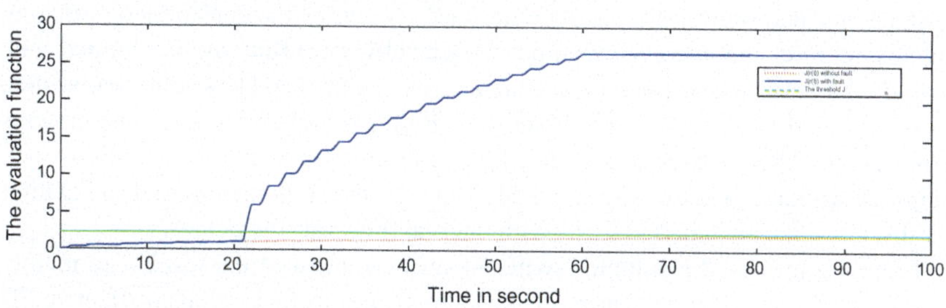

Fig. 1.9 Evaluation function $J_L(r(t))$

can realize the residual signal sensitive to the fault and robust to the unknown disturbances. LMIs conditions are proposed to design the FD filter, which can guarantee the finite-frequency H_- and H_∞ performance index. Finally, a practical example is provided and simulation results are conducted to demonstrate the effectiveness of the proposed approach.

1.2.1 Introduction

Due to the increasing demand for reliability and safety in industrial control processes, the issue of FD is required in various kinds of practical complex systems. There are a great variety of model-based FD methods, including parity-space-based approach, eigenstructure-assignment-based approach, parameter-identification-based approach, and observer-based approach. Among these, observer-based FD is one of the most effective schemes and has received much interest in the literature.

Nowadays, the exploition of FD technique has attracted more and more attentions. Especially, observer-based FD method have been proposed for different systems, designing FD observer or filter to monitor systems; see, e.g., [27–32]. The objectives of FD filter design can be defined as: (i) detecting faults as soon as possible, (ii) avoiding false alarms for uncontrolled inputs signals like disturbance. Among these results, various optimization methods have been proposed, such as the mixed H_2/H_∞ optimization method and the H_-/H_∞ optimization method [33–35], where the H_- index can measure the maximum influence of a fault on the residual signal and the H_∞ index can measure the minimum of the disturbance on the residual signal [36]. Moreover, a few FD methods for the switched systems have been considered in different cases.

Switched system belongs to the hybrid system, which includes several subsystems and a rule which orchestrates the switching among them. Many real engineering systems can be modeled as such systems, such as circuit system [37], aerospace systems [38], chemical systems [39], etc. ADT technology, one of the constrained switching signals, requires that the average time of each subsystem running in a limited time period is larger than a constant [40], which is more general and less conservative than classical dwell time. In [41], FD based on UIO for switched discrete-time systems was studied. The problem of actuator FD for discrete-time switched systems with output disturbance was investigated in [42]. In [43], the authors proposed an FD filter for discrete-time switched systems with interval time-varying delays.

From the proposed results for switched systems, most of the results are mainly considered for the discrete-time case, which motivates us to develop an effective FD filter for continuous-time switched systems. Presently, there have some technologies proposed to deal with the above problems. For example, the FD filter design problems for continuous-time switched system were studied in [44–46]. However, both of the papers mainly considered the robust performance and the fault sensitivity was not given. In [47], robust control and FD for continuous-time switched systems subject to a dwell time constraint were studied, while the authors followed the rule of classical dwell-time switching signal to detect faults, which is more conservative and less general than ADT. Therefore, motivated by the problems existing in the obtained achievements, we study an effective FD filter for continuous-time switched systems, and followed the ADT switching rule, the H_∞ and H_- performance indices are guaranteed by the design procedure.

In this section, the problem of FD for a class of switched systems in the finite-frequency domain is addressed. We consider the finite-frequency H_- performance and H_∞ performance, which can better detect the fault of the control systems and reduce the effect of disturbance by using the finite-frequency domain approach. Then a new bounded real lemma is obtained. Based on the new bounded real lemma, the sufficient conditions for designing the filter which can guarantee the H_- and H_∞ performances are given in terms of solving a set of LMIs. Compared to the existing FD method in the full frequency domain [48], the designed filter is more sensitive to the fault signal and more robust against disturbances due to the introduced additional slack matrix variable. Finally, a simulation example is given to show the effectiveness of the proposed finite-frequency FD.

1.2.2 Preliminaries and Problem Statement

1.2.2.1 System Description

Consider the following switched system:

$$\dot{x}(t) = \sum_{i=1}^{N} \sigma_i(t)(A_i x(t) + B_i d(t) + F_i f(t)) \tag{1.48}$$

$$y(t) = \sum_{i=1}^{N} \sigma_i(t)(C_i x(t) + D_i d(t)), \tag{1.49}$$

Where $x(t) \in \mathbb{R}^n$, $y(t) \in \mathbb{R}^q$, $d(t) \in \mathbb{R}^d$ and $f(t) \in \mathbb{R}^f$ are, respectively, the state variable, the measured output, the disturbance input, and the actuator fault signal. A_i, B_i, C_i, D_i and F_i are coefficient matrices with appropriate dimensions.

The piecewise constant function

$$\sigma_i(t) : [0, \infty) \to \{0, 1\}(i \in \mathbb{N}), \quad \sum_{i=1}^{N} \sigma_i(t) = 1 \tag{1.50}$$

is the switching signal, which specifies which subsystem is activated at the switching instant.

The frequencies of $d(t)$ and $f(t)$ reside in a known finite-frequency set Θ, which is defined as

$$\Theta := \begin{cases} \{\omega \in \mathbb{R} :\mid \omega \mid \leq \varpi_l, \varpi_l \geq 0\}, & LF \\ \{\omega \in \mathbb{R} : \varpi_1 \leq \omega \leq \varpi_2, \varpi_1 \leq \varpi_2\}, & MF \\ \{\omega \in \mathbb{R} :\mid \omega \mid \geq \varpi_h, \varpi_h \geq 0\}, & HF \end{cases} \tag{1.51}$$

Remark 1.5 Switched nonlinear systems can be found in various domains, such as mobile robots, network control systems, automotive, dc converters, and so on. Recently, some results have been proposed for switched nonlinear systems.

Assumption 1 There exists a switching function $\sigma_i(t)$ such that the system (1.48) with $d(t) = 0$ and $f(t) = 0$ is admissible.

Definition 1.4 ([49]) For any switching signal $\sigma_i(t)$ and any $t_2 > t_1 > 0$, let $N_{\sigma_i(t)}$ (t_1, t_2) denote thc number of switchings $\sigma_i(t)$ on an interval (t_1, t_2). If

$$N_{\sigma_i(t)}(t_1, t_2) \leq N_0 + \frac{t_2 - t_1}{\tau_a}$$

holds for a given $N_0 \geq 0$ and $\tau_a > 0$, then the constant τ_a is called the ADT and N_0 is the chattering bound.

1.2.2.2 FD Filter

In order to detect the actuator fault, the following FD filter is introduced:

$$\dot{x}_f(t) = \sum_{i=1}^{N} \sigma_i(t)(A_{fi}x_f(t) + B_{fi}y(t)) \tag{1.52}$$

$$r(t) = \sum_{i=1}^{N} \sigma_i(t)C_{fi}x_f(t) \quad , \tag{1.53}$$

where $x_f(t) \in \mathbb{R}^n$ is the filter state, $r(t) \in \mathbb{R}^p$ is the filter output.

Augmenting the model of system (1.48) to include the states of (1.52), we obtain the following system:

$$\dot{\eta}(t) = \sum_{i=1}^{N} \sigma_i(t)(\bar{A}_i\eta(t) + \bar{B}_i d(t) + \bar{F}_i f(t))r(t) = \sum_{i=1}^{N} \sigma_i(t)(\bar{C}_i\eta(t)), s \tag{1.54}$$

where $\eta(t) = \left[x^T(t)\ x_f^T(t)\right]^T$, and

$$\bar{A}_i = \begin{bmatrix} A_i & 0 \\ B_{fi}C_i & A_{fi} \end{bmatrix}, \ \bar{B}_i = \begin{bmatrix} B_i \\ B_{fi}D_i \end{bmatrix}, \ \bar{F}_i = \begin{bmatrix} F_i \\ 0 \end{bmatrix}, \ \bar{C}_i = \begin{bmatrix} 0 \\ C_{fi}^T \end{bmatrix}^T. \tag{1.55}$$

1.2.2.3 Problem Statement

In practice, for some systems, the main effects of disturbances and faults occupy different frequency domains. Particularly, fault signals usually emerge in the low-frequency domain, for example, the constant struck fault just belongs to the low-frequency domain [50]. Meanwhile, the disturbances are always in the high frequency such as high-frequency noise.

As mentioned earlier, in this paper, we assume that faults are low-frequency signals, while disturbances reside in the high-frequency domain. Consider the following two finite-frequency ranges for frequency ω in disturbance $d(t)$ and fault $f(t)$:

$$\Theta_d := \{\omega \in \mathbb{R} :\mid \omega \mid \geq \varpi_h, \varpi_h \geq 0\}, \tag{1.56}$$

$$\Theta_f := \{\omega \in \mathbb{R} :\mid \omega \mid \leq \varpi_l, \varpi_l \geq 0\}. \tag{1.57}$$

Remark 1.6 For an incipient signal, the fault information is always contained within a low-frequency band as the fault development is slow. For the type fault signal, an FD filter design typically requires a high fault sensitivity in a low-frequency range.

Definition 1.5 The system (1.54) has the finite-frequency H_∞ index bound γ, if the inequality

$$\int_0^\infty e^{-\alpha t} r^T(t) r(t) dt \leq \gamma^2 \int_0^\infty d^T(t) d(t) dt \tag{1.58}$$

holds for all solutions of (1.54) with $d(t) \in L_2$ such that

$$\int_0^\infty \tau(w + \varpi_h)(w - \varpi_h)\eta(\omega)\dot{\eta}(\omega) d\omega > 0 \tag{1.59}$$

under the zero initial condition.

Definition 1.6 The system (1.54) has the finite-frequency H_- index bound β, if the inequality

$$\int_0^\infty r^T(t) r(t) dt \geq \beta^2 \int_0^\infty e^{-\alpha t} f^T(t) f(t) dt \tag{1.60}$$

holds for all solutions of (1.54) with $d(t) \in L_2$ such that

$$\int_0^\infty -\tau(w + \varpi_l)(w - \varpi_l)\eta(\omega)\dot{\eta}(\omega) d\omega > 0 \tag{1.61}$$

under the zero initial condition.

In this paper, an H_-/H_∞ FD filter is designed to make the residual signal sensitive to the fault and robust to the unknown disturbances, and to guarantee that the augmented model (1.54) is asymptotically stable. To this end, considering the frequency characterizations of $d(t)$ and $f(t)$, the finite-frequency FD problem can be formulated as follows:

Given three scalars α, β, andγ, under the condition that the filter (1.54) is stable, to solve the following optimization problem:

$$\begin{aligned} &max\ \beta \\ &s.t.(1.57), (1.59). \end{aligned} \tag{1.62}$$

Remark 1.7 In the optimization problem (1.62), condition (1.58) is a H_∞ performance constraint where γ denotes the worst-case criterion for the effects of the disturbance $d(t)$ on the residual $r(t)$ and condition (1.60) is a H_- performance constraint where β is a measurement of the fault sensitivity in the worst case from faults $f(t)$ to the residual $r(t)$.

1.2.2.4 Preliminaries

In the subsequent proof process, the following lemmas are provided.

Lemma 1.1 ([51]) *Let* $\xi \in \mathbb{R}^n$, $Q \in \mathbb{R}^{n\times n}$, *and* $\aleph \in \mathbb{R}^{n\times m}$. *The following statements are equivalent:*
(1) $\xi^T Q\xi < 0,\ \forall \aleph^T \xi = 0,\ \xi \neq 0,$
(2) $\exists \mathscr{Y} \in \mathbb{R}^{m\times n} : Q + \aleph\mathscr{Y} + \mathscr{Y}^T\aleph^T < 0.$

Lemma 1.2 ([52]) *Consider the continuous-time switched system* $\dot{x}(t) = f_{\sigma_i(t)}(x(t))$, *and let* $\alpha > 0$, $\mu > 1$ *be given constants. Suppose that there exist* C^1 *functions* $V_{\sigma_i(t)} : \mathbb{R}^n \to \mathbb{R}$ *and two class* K_∞ *functions* K_1 *and* K_2 *such that*

$$K_1(|x(t)|) \leq V_{\sigma_i(t)}(x(t)) \leq K_2(|x(t)|) \tag{1.63}$$

$$\dot{V}_{\sigma_i(t)}(x(t)) \leq -\alpha V_{\sigma_i(t)}(x(t)) \tag{1.64}$$

and $\forall(\sigma_i(t_l) = i, \sigma_i(t_l^-) = j) \in \mathcal{N} \times \mathcal{N}, i \neq j,$

$$\dot{V}_i(x(t)) \leq \mu \dot{V}_j(x(t)) \tag{1.65}$$

then the switched system $\dot{x}(t) = f_{\sigma_i}(t)(x(t))$ *is globally uniformly asymptotically stable (GUAS) for any switching signal with average dwell time*

$$T > T^T = \frac{\ln \mu}{\alpha}. \tag{1.66}$$

Lemma 1.3 ([53]) *If there exist functions* $f(t)$ *and* $g(t)$ *satisfying*

$$\dot{f}(t) \leq -\alpha f(t) + mg(t),$$

where α and m are constant, then

$$f(t) \leq e^{-\alpha(t-t_0)} f(t_0) + m \int_{t_0}^{t} e^{-\alpha(t-s)} g(s) ds.$$

Lemma 1.4 ([54]) *For any signal $\eta(t)$ residing in a known finite-frequency set Θ defined in (1.51), if there exist a symmetric matrix P and a symmetric positive definite matrix Q, then*

$$\int_0^\infty \vartheta^T(t) \Upsilon_d \vartheta(t) dt = \int_0^\infty \vartheta^T(t)(\Phi \otimes P + \Psi \otimes Q)\vartheta(t) dt \geq 0, \tag{1.67}$$

where $\vartheta(t) = \begin{bmatrix} \dot{\eta}(t) \\ \eta(t) \end{bmatrix}$, $\Phi = \begin{bmatrix} 0 & 1 \\ 1 & 0 \end{bmatrix}$ *and Ψ is defined as follows:*

$$\Psi := \begin{cases} \begin{bmatrix} -1 & 0 \\ 0 & \varpi_l^2 \end{bmatrix}, \mid \omega \mid \leq \varpi_l, & LF, \\ \begin{bmatrix} -1 & \frac{\varpi_1+\varpi_2}{2} \\ \frac{\varpi_1+\varpi_2}{2} & -\varpi_1\varpi_2 \end{bmatrix}, \varpi_1 \leq \omega \leq \varpi_2, & MF, \\ \begin{bmatrix} 1 & 0 \\ 0 & -\varpi_h^2 \end{bmatrix}, \mid \omega \mid \geq \varpi_h, & HF \end{cases} \tag{1.68}$$

1.2.3 Main Results

In this section, a finite-frequency FD filter design for continuous-time switched systems is formulated, based on an LMI approach will be solved.

1.2.3.1 Finite-Frequency Performance Analysis

In this subsection, we assume that the FD filter parameters in (1.52) are known, and the proposed the sufficient condition such that the filter error system (1.54) satisfies these two specifications (1.58) and (1.60). Then we propose the theorems to obtain the filter which guarantee (1.58) and (1.60) based on the previous theorems.

- Inequality condition for finite-frequency H_∞ performance

It is well known that the disturbances usually are high-frequency signals in practice. So, in this paper the disturbances are assumed to belong to the following high-frequency domain:

$$\Theta = \{\omega \in \mathbb{R} : \mid \omega \mid \geq \varpi_h, \varpi_h \geq 0\}. \tag{1.69}$$

Set $f(t) = 0$, the filter error system (1.54) becomes

$$\dot{\eta}(t) = \sum_{i=1}^{N} \sigma_i(t)(\bar{A}_i \eta(t) + \bar{B}_i d(t)) r(t) = \sum_{i=1}^{N} \sigma_i(t) \bar{C}_i \eta(t). \tag{1.70}$$

In the following, the theorem is given to ensure that the error system (1.70) is GUAS and has H_∞ performance.

Theorem 1.3 *For a given scalar $\alpha > 0, \gamma > 0$, the high-frequency H_∞ performance (1.58) is guaranteed for system (1.70), if there exist symmetric positive-definite matrices P_i^d, Q_d, symmetric matrices P_d, a matrix V_{di}, $i = 1, \ldots, r$, and a scalar μ such that for a switching signal with average dwell time satisfying $T > T^T = \frac{ln\mu}{\alpha}$*

$$\Lambda_{di} + \Lambda_{d1i} + \Lambda_{d2i} < 0, \tag{1.71}$$

$$\Lambda_{di} > 0, \tag{1.72}$$

$$P_i^d \leq \mu P_j^d, \quad i \neq j, \tag{1.73}$$

where

$$\Lambda_{di} = \begin{bmatrix} \bar{A}_i & \bar{B}_i \\ I & 0 \end{bmatrix}^T (\Phi \otimes P_d + \Psi \otimes Q_d) \begin{bmatrix} \bar{A}_i & \bar{B}_i \\ I & 0 \end{bmatrix},$$

$$\Lambda_{d1i} = \begin{bmatrix} \bar{C}_i & 0 \\ 0 & I \end{bmatrix}^T \begin{bmatrix} I & 0 \\ 0 & -\gamma^2 I \end{bmatrix} \begin{bmatrix} \bar{C}_i & 0 \\ 0 & I \end{bmatrix},$$

$$\Lambda_{d1i} = \begin{bmatrix} \bar{A}_i & \bar{B}_i \\ I & 0 \end{bmatrix}^T \begin{bmatrix} 0 & P_i^d \\ P_i^d & \alpha P_i^d \end{bmatrix} \begin{bmatrix} \bar{A}_i & \bar{B}_i \\ I & 0 \end{bmatrix}.$$

Proof For the system (1.70), we choose a Lyapunov functional candidate of the form

$$\begin{aligned} V(t) = \eta^*(t) P_i^d \eta(t) &= \begin{bmatrix} \eta(t) \\ d(t) \end{bmatrix}^T \begin{bmatrix} P_i^d & 0 \\ 0 & 0 \end{bmatrix} \begin{bmatrix} \eta(t) \\ d(t) \end{bmatrix} \\ &= \xi^T(t) \begin{bmatrix} \bar{A}_i & \bar{B}_i \\ I & 0 \end{bmatrix}^T \begin{bmatrix} 0 & 0 \\ 0 & P_i^d \end{bmatrix} \begin{bmatrix} \bar{A}_i & \bar{B}_i \\ I & 0 \end{bmatrix} \xi(t) \end{aligned} \tag{1.74}$$

then the derivative of $V(t)$ along (1.70) satisfies

$$\begin{aligned} \dot{V}(t) &= \dot{\eta}^T(t) P_i^d \eta(t) + \eta^T(t) P_i^d \dot{\eta}(t) \\ &= \eta^T(t)(\bar{A}_i^T P_i^d + P_i^d \bar{A}_i)\eta(t) + d^T(t) \bar{B}_i^T P_i^d \eta(t) + \eta^T(t) P_i^d \bar{B}_i d(t) \\ &= \xi^T(t) \begin{bmatrix} \bar{A}_i & \bar{B}_i \\ I & 0 \end{bmatrix}^T \begin{bmatrix} 0 & P_i^d \\ P_i^d & 0 \end{bmatrix} \begin{bmatrix} \bar{A}_i & \bar{B}_i \\ I & 0 \end{bmatrix} \xi(t). \end{aligned} \tag{1.75}$$

From system (1.70), we have

$$
\begin{aligned}
& r^T(t)r(t) - \gamma^2 d^T(t)d(t) \\
&= \xi^T(t)\begin{bmatrix} C_i^T C_i & 0 \\ 0 & -\gamma^2 I \end{bmatrix}\xi(t) \\
&= \xi^T(t)\begin{bmatrix} \bar{C}_i & 0 \\ 0 & I \end{bmatrix}^T \begin{bmatrix} I & 0 \\ 0 & -\gamma^2 I \end{bmatrix}\begin{bmatrix} \bar{C}_i & 0 \\ 0 & I \end{bmatrix}\xi(t).
\end{aligned} \tag{1.76}
$$

Let $0 = t_0 < t_1 < \cdots < t_k < \cdots, k = 1, \ldots,$ denote the switching point, and we suppose that the i_kth subsystem is active when $i \in [t_k, t_{k+i})$. From (1.71) we have

$$
\begin{aligned}
& \xi^T(t)(\Lambda_{di_k} + \Lambda_{d1i_k} + \Lambda_{d2i_k})\xi(t) \\
&= \xi^T(t)\Big(\begin{bmatrix} \bar{A}_i & \bar{B}_i \\ I & 0 \end{bmatrix}^T \begin{bmatrix} 0 & P_i^d \\ P_i^d & 0 \end{bmatrix}\begin{bmatrix} \bar{A}_i & \bar{B}_i \\ I & 0 \end{bmatrix} + \begin{bmatrix} \bar{A}_i & \bar{B}_i \\ I & 0 \end{bmatrix}^T \begin{bmatrix} 0 & 0 \\ 0 & \alpha P_i^d \end{bmatrix}\begin{bmatrix} \bar{A}_i & \bar{B}_i \\ I & 0 \end{bmatrix} \\
&\quad + \begin{bmatrix} \bar{C}_i & 0 \\ 0 & I \end{bmatrix}^T \begin{bmatrix} I & 0 \\ 0 & -\gamma^2 I \end{bmatrix}\begin{bmatrix} \bar{C}_i & 0 \\ 0 & I \end{bmatrix}\Big)\xi(t) \\
&\quad + \xi^T(t)\Big(\begin{bmatrix} \bar{A}_i & \bar{B}_i \\ I & 0 \end{bmatrix}^T (\Phi \otimes P_d + \Psi \otimes Q_d)\begin{bmatrix} \bar{A}_i & \bar{B}_i \\ I & 0 \end{bmatrix}\Big)\xi(t) \\
&= \dot{V}_{i_k}(t) + \alpha V_{i_k}(t) + \Sigma_{i_k} + e^{N(0,t)\ln\mu}\vartheta_{i_k}^T \Upsilon_d \vartheta_{i_k} < 0,
\end{aligned} \tag{1.77}
$$

where

$$
\vartheta_{i_k} = e^{-N(0,t)ln\mu/2}\begin{bmatrix} \dot{\eta}(t) \\ \eta(t) \end{bmatrix} = \begin{bmatrix} \dot{x_{i_k}}(t) \\ x_{i_k}(t) \end{bmatrix}, \tag{1.78}
$$

$$
\Lambda_{d2i_k} = \Sigma_{i_k} = r_{i_k}^T(t)r_{i_k}(t) - \gamma^2 d^T(t)d(t), \tag{1.79}
$$

$$
\Lambda_{di_k} = \eta(t)^T(\Phi \otimes P_{di} + \Psi \otimes Q_d)\eta(t), \tag{1.80}
$$

$$
\Upsilon_d = \Phi \otimes P_{di} + \Psi \otimes Q_d. \tag{1.81}
$$

Set $\eta(t_{k+1}) = \eta_{i_{k+1}}(t_{k+1})$, the state of the switched system can be expressed as

$$
\eta(t) = \sum_{i=1}^{N}\Big(\eta_{i_k}(t)[u(t - t_k) - u(t - t_{k+1})]\Big), \qquad t \in [0, \infty), \tag{1.82}
$$

We have $\vartheta(t) = \begin{bmatrix} \dot{\chi}(t) \\ \chi(t) \end{bmatrix}$. Considering (1.82),

$$
\chi(t) = \sum_{i=1}^{N}\Big(e^{-\frac{k\ln\mu}{2}}\eta_{i_k}(t)[u(t - t_k) - u(t - t_{k+1})]\Big), \qquad t \in [0, \infty). \tag{1.83}
$$

Obviously, $\chi(t)$ is also in the frequency range Θ_d and asymptotically stable. In addition, from (1.83)

$$\dot{\chi}(t) = \sum_{i=1}^{N} \Big(e^{-\frac{k\ln\mu}{2}} \dot{\eta}_{i_k}(t)[u(t-t_k) - u(t-t_{k+1})]$$
$$+ e^{-\frac{k\ln\mu}{2}} \eta_{i_k}(t)[\delta(t-t_k) - \delta(t-t_{k+1})]\Big), \qquad t \in [0, \infty), \tag{1.84}$$

where $\delta(t)$ is the unit pulse signal. Therefore, by Lemma 1.4, we have

$$\int_0^\infty \vartheta^T(s)\Upsilon_d \vartheta(s) ds \geq 0. \tag{1.85}$$

From Lemma 1.3, we have

$$V_{i_k}(t) - e^{-\alpha(t-t_k)} V_{i_k}(t_k) - \int_{t_k}^{t} e^{-\alpha(t-s)}(\Sigma_{i_k}(s) + e^{N(0,t)\ln\mu} \vartheta_{i_k}^T \Upsilon_d \vartheta_{i_k}) ds < 0. \tag{1.86}$$

Besides, from (1.73), we have

$$V_{i_k}(t_k) \leq \mu V_{i_{k-1}}(t_k^-). \tag{1.87}$$

Then we can obtain the following formula with (1.86), (1.87):

$$\begin{aligned}
V_{i_k}(t) &\leq \mu e^{-\alpha(t-t_k)} V_{i_{k-1}}(t_k^-) - \int_{t_k}^{t} e^{-\alpha(t-s)}(\Sigma_{i_k}(s) + e^{N(0,t)\ln\mu} \vartheta_{i_k}^T \Upsilon_d \vartheta_{i_k}) ds \\
&\leq \mu^{N(0,t)} e^{-\alpha t} V_{i_0}(0) - \mu^{N(0,t)} \int_{0}^{t_1} e^{-\alpha(t-s)}(\Sigma_{i_k}(s) + e^{N(0,t)\ln\mu} \vartheta_{i_k}^T \Upsilon_d \vartheta_{i_k}) ds \\
&\quad - \cdots - \mu^0 \int_{t_k}^{t} e^{-\alpha(t-s)}(\Sigma_{i_k}(s) + e^{N(0,t)\ln\mu} \vartheta_{i_k}^T \Upsilon_d \vartheta_{i_k}) ds \\
&= e^{-\alpha(t-t_k)+N(0,t)\ln\mu} V_{i_0}(0) \\
&\quad - \int_0^t e^{-\alpha(t-s)+N(s,t)\ln\mu}(\Sigma_{i_k}(s) + e^{N(0,t)\ln\mu} \vartheta_{i_k}^T \Upsilon_d \vartheta_{i_k}) ds.
\end{aligned} \tag{1.88}$$

Considering the zero initial condition and $V_{i_k}(t) > 0$, (1.88) implies

$$\int_0^t e^{-\alpha(t-s)+N(s,t)\ln\mu}(\Sigma_{i_k}(s) + e^{N(0,t)\ln\mu} \vartheta_{i_k}^T \Upsilon_d \vartheta_{i_k}) ds \leq 0 \tag{1.89}$$

and multiplying both sides of (1.89) by $e^{-N(0,t)\ln\mu}$ we have

$$\begin{aligned}
&\int_0^t e^{-\alpha(t-s)-N(0,s)\ln\mu} r_{i_k}^T(s) r_{i_k}(s) ds \\
&\leq \int_0^t e^{-\alpha(t-s)-N(0,s)\ln\mu}(\gamma^2 d^T(s)d(s) - e^{N(0,t)\ln\mu} \vartheta_{i_k}^T \Upsilon_d \vartheta_{i_k}) ds.
\end{aligned} \tag{1.90}$$

Because of the fact that $N(0, s) \le \frac{s}{T} < \alpha = \frac{s}{T^T}$ and $T^T = \frac{\ln \mu}{\alpha}$, we have $N(0, s) \ln \mu < \alpha s$ and (1.90) implies

$$\int_0^t e^{-\alpha(t-s)-\alpha s} r_{i_k}^T(s) r_{i_k}(s) ds \le \int_0^t e^{-\alpha(t-s)} (\gamma^2 d^T(s) d(s) - \vartheta_{i_k}^T \Upsilon_d \vartheta_{i_k}) ds. \quad (1.91)$$

Integrating (1.91) from $t = 0$ to $t = \infty$

$$\begin{aligned} &\int_0^\infty \Big(\int_0^t e^{-\alpha(t-s)-\alpha s} r_{i_k}^T(s) r_{i_k}(s) ds \Big) dt \\ &\le \int_0^\infty \Big(\int_0^t e^{-\alpha(t-s)} (\gamma^2 d^T(s) d(s) - \vartheta_{i_k}^T \Upsilon_d \vartheta_{i_k}) ds \Big) dt. \end{aligned} \quad (1.92)$$

Then,

$$\begin{aligned} &\Big(\int_0^\infty e^{-\alpha s} r_{i_k}^T(s) r_{i_k}(s) ds \Big) \Big(\int_s^\infty e^{-\alpha(t-s)} dt \Big) \\ &\le \Big(\int_0^\infty (\gamma^2 d^T(s) d(s) - \vartheta_{i_k}^T \Upsilon_d \vartheta_{i_k}) ds \Big) \Big(\int_s^\infty e^{-\alpha(t-s)} dt \Big). \end{aligned} \quad (1.93)$$

Eq. (1.93) implies

$$\int_0^\infty e^{-\alpha t} r^T(t) r(t) dt \le \gamma^2 \int_0^\infty d^T(t) d(t) dt - \int_0^\infty \vartheta_{i_k}^T \Upsilon_d \vartheta_{i_k} dt. \quad (1.94)$$

Because of the premise lemma, Lemma 4, we have

$$\int_0^\infty \vartheta_{i_k}^T \Upsilon_d \vartheta_{i_k} dt \ge 0. \quad (1.95)$$

So, we obtain

$$\int_0^\infty e^{-\alpha t} r^T(t) r(t) dt \le \gamma^2 \int_0^\infty d^T(t) d(t) dt. \quad (1.96)$$

The proof is completed. □

- Inequality condition for finite-frequency H_- performance

Just as mentioned previously, on the contrary, the fault signals usually emerge in the low-frequency domain where disturbances usually are high-frequency signals. Therefore, in this paper, we assume that the fault signals belong to the following low-frequency range:

$$\Theta = \{\omega \in \mathbb{R} : \mid \omega \mid \le \varpi_l, \varpi_l \ge 0\}. \quad (1.97)$$

Set $d(t) = 0$, the filter error system (1.54) becomes

$$\dot{\eta}(t) = \sum_{i=1}^{N} \sigma_i(t)(\bar{A}_i\eta(t) + \bar{F}_i f(t))r(t) = \sum_{i=1}^{N} \sigma_i(t)(\bar{C}_i\eta(t)).s \tag{1.98}$$

In the following, the theorem is given to ensure that the error system (1.98) is GUAS and has H_∞ performance.

Theorem 1.4 *For a given scalar $\alpha > 0$, $\beta > 0$, the low-frequency H_- performance (1.58) is guaranteed for system (1.98), if there exist symmetric positive definite matrices P_i^f, Q_f, symmetric matrices P_f, a matrix V_{fi}, $i = 1, \ldots, r$, and a scalar μ such that for a switching signal with average dwell time satisfying $T > T^* = \frac{ln\mu}{\alpha}$*

$$\Lambda_{fi} + \Lambda_{f1i} + \Lambda_{f2i} < 0, \tag{1.99}$$

$$\Lambda_{fi} > 0, \tag{1.100}$$

$$P_i^f \le \mu P_j^f, \quad i \neq j, \tag{1.101}$$

where

$$\Lambda_{fi} = \begin{bmatrix} \bar{A}_i & \bar{B}_i \\ I & 0 \end{bmatrix}^T (\Phi \otimes P_f + \Psi \otimes Q_f) \begin{bmatrix} \bar{A}_i & \bar{B}_i \\ I & 0 \end{bmatrix},$$

$$\Lambda_{f1i} = \begin{bmatrix} \bar{C}_i & 0 \\ 0 & I \end{bmatrix}^T \begin{bmatrix} I & 0 \\ 0 & -\gamma^2 I \end{bmatrix} \begin{bmatrix} \bar{C}_i & 0 \\ 0 & I \end{bmatrix},$$

$$\Lambda_{f1i} = \begin{bmatrix} \bar{A}_i & \bar{B}_i \\ I & 0 \end{bmatrix}^T \begin{bmatrix} 0 & P_i^f \\ P_i^f & \alpha P_i^f \end{bmatrix} \begin{bmatrix} \bar{A}_i & \bar{B}_i \\ I & 0 \end{bmatrix}.$$

Proof Theorem 1.4 is proposed by using a similar method to Theorem 1.3. □

1.2.4 FD Filter Design

In the preliminaries, Theorems 1.3 and 1.4 provide finite-frequency performance analysis conditions when the filter parameters are known. However, the parameters of the filter (1.52) are unknown. In the following, we will develop some decoupling techniques to transform the conditions in Theorems 1.3 and 1.4 into LMIs conditions to obtain the FD filter gains.

3.2.1 FD filter design for low-frequency H_- performance

The next theorem provides FD filter design conditions under which the finite-frequency performance (1.60) can be guaranteed based on Theorem 1.4.

Theorem 1.5 *Consider the continuous-time switched fuzzy system (1.48). For a given positive scalar α, β, there exists an FD filter in the form of (1.52), such that* *the filter error system (1.54) with $d(t) = 0$ satisfies the finite-frequency performance (1.60) when ω belongs to (1.97) if there exist matrices*

$$P_f = \begin{bmatrix} P_f^{11} & 0 \\ 0 & P_f^{22} \end{bmatrix} > 0, \qquad Q_f = \begin{bmatrix} Q_f^{11} & Q_f^{12} \\ * & Q_f^{22} \end{bmatrix} > 0, \tag{1.102}$$

$$P_i^f = \begin{bmatrix} P_i^{f11} & 0 \\ 0 & P_i^{f22} \end{bmatrix} > 0, \qquad i = 1, \ldots, r, \tag{1.103}$$

the matrices $\hat{A}_{fi}, \hat{B}_{fi}, C_{fi}, M_{f1i}, M_{f2i}, M_{f3i}, Z_{f1i}, Z_{f2i}, Z_{f3i}$, and $N_i, i = 1, \ldots, r$ and scalar $\mu > 1$, ς such that the following inequalities hold:

$$\varsigma^2 - C_{fi} T < 0, \quad i = 1, \ldots, r, \tag{1.104}$$

$$P_i^f \leq \mu P_j^f, \quad i \neq j, \tag{1.105}$$

$$\begin{bmatrix} \Upsilon_{11i} & * & * \\ \Upsilon_{21i} & \Upsilon_{22i} & * \\ \Upsilon_{31i} & \Upsilon_{32i} & \Upsilon_{33i} \end{bmatrix} < 0, \tag{1.106}$$

where

$$T \in \mathbb{R}^n, \qquad \|T\|_2 = \varsigma,$$

$$\Upsilon_{11i} = \begin{bmatrix} -Q_f^{11} - He(Z_{f1i}) & * \\ -Q_f^{21} - N_i - M_{f1i}^T & -Q_f^{22} - He(N_i) \end{bmatrix},$$

$$\Upsilon_{21i} = \begin{bmatrix} P_i^{f11} + P_f^{11} + Z_{f1i}A_i + \hat{B}_{fi}C_i - Z_{f2i}^T & \hat{A}_{fi} - M_{f2i}^T \\ M_{f1i}A_i + \hat{B}_{fi}C_i - N_i^T & P_i^{f22} + P_f^{22} + \hat{A}_{fi} - N_i^T \end{bmatrix},$$

$$\Upsilon_{31i} = \begin{bmatrix} Z_{f1i}B_i + Z_{f3i}^T B_i \\ M_{f1i}B_i + \hat{B}_i \end{bmatrix},$$

$$\Upsilon_{22i} = \begin{bmatrix} \varpi_l^2 Q_f^{11} + He(Z_{f2i}A_i + \hat{B}_{fi}C_i) & * \\ \varpi_l^2 Q_f^{12} + \hat{A}_{fi} + A_i^T M_{f2i}^T + C_i^T B_{fi}^T & \varpi_l^2 Q_f^{22} + He(\hat{A}_{fi}) - \varsigma^2 I \end{bmatrix},$$

$$\Upsilon_{32i} = \begin{bmatrix} Z_{f2i}B_i + A_i^T Z_{f3i}^T B_i - C_i^T \hat{B}_{fi}^T B_i \\ M_{f2i}B_i - \hat{A}_{fi}^T B_i \end{bmatrix},$$

$$\Upsilon_{33i} = \beta^2 I - He(B_i^* Z_{f3i} B_i).$$

Then $A_{fi} = N_i^{-1}\hat{A}_{fi}$, $B_{fi} = N_i^{-1}\hat{B}_{fi}$, and C_{fi} are the parameters of the filter (1.52).

Proof Considering Theorem 1.4, from (1.99), we have

$$\begin{bmatrix} \bar{A}_i & \bar{B}_i \\ I & 0 \end{bmatrix}^T \begin{bmatrix} -Q_f & P_i^f + P_f \\ * & \alpha P_i^f + \varpi_l^2 Q_f \end{bmatrix} \begin{bmatrix} \bar{A}_i & \bar{B}_i \\ I & 0 \end{bmatrix} + \begin{bmatrix} \bar{C}_i & 0 \\ 0 & I \end{bmatrix}^T \begin{bmatrix} I & 0 \\ 0 & -\beta^2 I \end{bmatrix} \begin{bmatrix} \bar{C}_i & 0 \\ 0 & I \end{bmatrix} < 0. \tag{1.107}$$

By setting

$$\Pi = \begin{bmatrix} -Q_f & * & * \\ P_i^f + P_f & \alpha P_i^f + \varpi_l^2 Q_f - \bar{C}_{fi}^T \bar{C}_{fi} & * \\ 0 & 0 & \beta^2 I \end{bmatrix}. \tag{1.108}$$

Eq. (1.107) can be written as

$$\begin{bmatrix} \bar{A}_i & \bar{B}_i \\ I & 0 \\ 0 & I \end{bmatrix}^T \Pi \begin{bmatrix} \bar{A}_i & \bar{B}_i \\ I & 0 \\ 0 & I \end{bmatrix} = S_{1i}^T \Pi S_{1i} < 0. \tag{1.109}$$

Then applying Lemma 1.1, (1.109) is equivalent to

$$\Pi + He(Y_{fi} S_{2i}) < 0, \tag{1.110}$$

where $S_{2i} = \begin{bmatrix} -I & \bar{A}_i & \bar{B}_i \end{bmatrix}$, $S_{2i} S_{1i} = 0$, and Y_{fi} can be set as $\begin{bmatrix} Y_{f1i}^T & Y_{f2i}^T & -Y_{f3i}^T \bar{B}_i \end{bmatrix}^T$, which is an additional matrix as introduced by Lemma 1.1.

The following inequality is obtained from (1.110):

$$\begin{bmatrix} -Q_f - He(Y_{f1i}) & * & * \\ P_i^f + P_f + Y_{f1i}\bar{A}_i - Y_{f2i}^T & (2,2) & * \\ Y_{f1i}\bar{B}_i + Y_{f3i}^T \bar{B}_i & Y_{f2i}\bar{B}_i - \bar{A}_i^T Y_{f3i}^T \bar{B}_i & \beta^2 I - He(\bar{B}_i^T Y_{f3i} \bar{B}_i) \end{bmatrix} < 0, \tag{1.111}$$

where $(2,2) = \alpha P_i^f + \varpi_l^2 Q_f + He(Y_{f2i}\bar{A}_i) - \bar{C}_{fi}^T \bar{C}_{fi}$.

We can see that the block (2,2) in (1.111) is nonlinear for $\bar{C}_{fi}^T \bar{C}_{fi} = diag\{0, C_{fi}^T C_{fi}\}$, so we refer to [43] and introduce another constraint (1.104) which can guarantee

$$C_{fi}^T C_{fi} > \varsigma^2 I. \tag{1.112}$$

So, the block (2,2) in (1.111) can be replaced by $\alpha P_i^f + \varpi_l^2 Q_f + He(Y_{f2i}\bar{A}_i) - diag\{0, \varsigma^2 I\}$.

Next, we suppose Y_{1i}, Y_{2i}, Y_{3i} have the following forms:

$$Y_{f1i} = \begin{bmatrix} Z_{f1i} & N_i \\ M_{f1i} & N_i \end{bmatrix}, \quad Y_{f2i} = \begin{bmatrix} Z_{f2i} & N_i \\ M_{f2i} & N_i \end{bmatrix}, \quad Y_{f3i} = \begin{bmatrix} Z_{f3i} & N_i \\ M_{f3i} & N_i \end{bmatrix}. \tag{1.113}$$

Further,

$$Y_{fi}\bar{A}_i = \begin{bmatrix} Z_{fi}A_i + \hat{B}_{fi}C_i & \hat{A}_{fi} \\ M_{fi}A_i + \hat{B}_{fi}C_i & \hat{A}_{fi} \end{bmatrix}, \tag{1.114}$$

$$Y_{fi}\bar{B}_i = \begin{bmatrix} Z_{fi}B_i \\ M_{fi}B_i \end{bmatrix}, \tag{1.115}$$

$$\bar{B}_i^T Y_{f3i}\bar{B}_i = B_i^T Z_{f3i}B_i. \tag{1.116}$$

After that, we obtain

$$\begin{bmatrix} \Upsilon_{11i} & * & * \\ \Upsilon_{21i} & \Upsilon_{22i} & * \\ \Upsilon_{31i} & \Upsilon_{32i} & \Upsilon_{33i} \end{bmatrix} < 0, \tag{1.117}$$

where $T \in \mathbb{R}^n$, $\|T\|_2 = \varsigma$,

$$\Upsilon_{11i} = \begin{bmatrix} -Q_f^{11} - He(Z_{f1i}) & * \\ -Q_f^{21} - N_i - M_{f1i}^T & -Q_f^{22} - He(N_i) \end{bmatrix},$$

$$\Upsilon_{21i} = \begin{bmatrix} P_i^{f11} + P_f^{11} + Z_{f1i}A_i + \hat{B}_{fi}C_i - Z_{f2i}^T & \hat{A}_{fi} - M_{f2i}^T \\ M_{f1i}A_i + \hat{B}_{fi}C_i - N_i^T & P_i^{f22} + P_f^{22} + \hat{A}_{fi} - N_i^T \end{bmatrix},$$

$$\Upsilon_{31i} = \begin{bmatrix} Z_{f1i}B_i + Z_{f3i}^T B_i \\ M_{f1i}B_i + \hat{B}_i \end{bmatrix},$$

$$\Upsilon_{22i} = \begin{bmatrix} \varpi_l^2 Q_f^{11} + He(Z_{f2i}A_i + \hat{B}_{fi}C_i) & * \\ \varpi_l^2 Q_f^{12} + \hat{A}_{fi} + A_i^T M_{f2i}^T + C_i^T B_{fi}^T & \varpi_l^2 Q_f^{22} + He(\hat{A}_{fi}) - \varsigma^2 I \end{bmatrix},$$

$$\Upsilon_{32i} = \begin{bmatrix} Z_{f2i}B_i + A_i^T Z_{f3i}^T B_i - C_i^T \hat{B}_{fi}^T B_i \\ M_{f2i}B_i - \hat{A}_{fi}^T B_i \end{bmatrix},$$

$$\Upsilon_{33i} = \beta^2 I - He(B_i^T Z_{f3i}B_i).$$

Then $A_{fi} = N_i^{-1}\hat{A}_{fi}$, $B_{fi} = N_i^{-1}\hat{B}_{fi}$, and C_{fi} are the parameters of the filter (1.52). Then, applying (1.113)–(1.116) and (1.112) to (1.111), one can get (1.117). The proof is completed. □

3.2.2 FD filter design for high-frequency H_∞ performance

This subsection provides FD filter design conditions under which the finite-frequency performance (1.58) can be guaranteed.

Theorem 1.6 *Consider the continuous-time switched fuzzy system (1.48). For a given positive scalar α, γ, there exists an FD filter in the form of (1.52), such that the filter error system (1.54) with $f(t) = 0$ satisfies the finite-frequency performance (1.58) when ω belongs to (1.69) if there exist matrices*

$$P_d = \begin{bmatrix} P_d^{11} & 0 \\ 0 & P_d^{22} \end{bmatrix} > 0, \qquad Q_d = \begin{bmatrix} Q_d^{11} & Q_d^{12} \\ * & Q_d^{22} \end{bmatrix} > 0, \tag{1.118}$$

$$P_i^d = \begin{bmatrix} P_i^{d11} & 0 \\ 0 & P_i^{d22} \end{bmatrix} > 0, \quad i = 1, \ldots, r. \tag{1.119}$$

The matrices $\hat{A}_{di}, \hat{B}_{di}, C_{di}, M_{d1i}, M_{d2i}, M_{d3i}, Z_{d1i}, Z_{d2i}, Z_{d3i}$, *and* N_i, $i = 1, \ldots, r$ *and scalar* $\mu > 1$ *such that the following inequalities hold:*

$$P_i^d \leq \mu P_j^d, \quad i \neq j, \tag{1.120}$$

$$\begin{bmatrix} \Upsilon_{11i} & * & * & * \\ \Upsilon_{21i} & \Upsilon_{22i} & * & * \\ \Upsilon_{31i} & \Upsilon_{32i} & \Upsilon_{33i} & * \\ 0 & \Upsilon_{42i} & 0 & \Upsilon_{44i} \end{bmatrix} < 0, \tag{1.121}$$

where

$$\Upsilon_{11i} = \begin{bmatrix} Q_d^{11} - He(Z_{d1i}) & * \\ Q_d^{21} - N_i - M_{d1i}^T & Q_d^{22} - He(N_i) \end{bmatrix}, \Upsilon_{31i} = \begin{bmatrix} Z_{d1i} B_{di} + \hat{B}_{fi} D_{di} \\ M_{d1i} B_{di} + \hat{B}_{fi} D_{di} \end{bmatrix},$$

$$\Upsilon_{21i} = \begin{bmatrix} P_i^{d11} + P_d^{11} + Z_{d1i} A_i + \hat{B}_{fi} C_i - Z_{d2i}^T & \hat{A}_{fi} - M_{d2i}^T \\ M_{d1i} A_i + \hat{B}_{fi} C_i - N_i^T & P_i^{d22} + P_d^{22} + \hat{A}_{fi} - N_i^T \end{bmatrix},$$

$$\Upsilon_{22i} = \begin{bmatrix} \alpha P_i^{d11} - \varpi_h^2 Q_d^{11} + He(Z_{d2i} A_i + \hat{B}_{di} C_i) & * \\ -\varpi_h^2 Q_d^{21} + \hat{A}_{fi} + A_i^T M_{d2i}^T + C_i^T B_{fi}^T & \alpha P_i^{d22} - \varpi_h^2 Q_d^{22} + He(\hat{A}_{fi}) \end{bmatrix},$$

$$\Upsilon_{32i} = \begin{bmatrix} Z_{d2i} B_{di} + \hat{B}_{fi}^T D_{di} \\ M_{d2i} B_{di} + \hat{B}_{fi}^T D_{di} \end{bmatrix}, \Upsilon_{42i} = \begin{bmatrix} 0 & C_{fi}^T \end{bmatrix}, \ \Upsilon_{33i} = -\gamma^2 I, \ \Upsilon_{44i} = -I.$$

Then $A_{fi} = N_i^{-1} \hat{A}_{fi}$, $B_{fi} = N_i^{-1} \hat{B}_{fi}$, *and* C_{fi} *are the parameters of the filter (1.52).*

Proof Theorem 1.6 is proved by using a similar method to Theorem 1.4. □

1.2.5 Stability Condition

Under Assumption 1, we only have to provide some conditions to guarantee the stability of FD filter (1.52).

Theorem 1.7 *For a given positive scalar* α, *the FD filter in the form of (1.52) is stable with average dwell time* $T > T^T = \frac{ln\mu}{\alpha}$, *if there exist matrices* $P_i = \begin{bmatrix} P_i^{11} & 0 \\ 0 & P_i^{22} \end{bmatrix} > 0$, $i = 1, \ldots, r$, *matrices* $\hat{A}_{fi}, \hat{B}_{fi}, C_{fi}, M_{1i}, M_{2i}, Z_{1i}, Z_{2i}$, *and* N_i, $i = 1, \ldots, r$ *and scalar* $\mu > 1$ *such that the following inequalities hold:*

$$P_i^d \le \mu P_j^d, \quad i \neq j, \tag{1.122}$$

$$\begin{bmatrix} \Upsilon_{11i} & * & * \\ \Upsilon_{21i} & \Upsilon_{22i} & * \\ \Upsilon_{31i} & \Upsilon_{32i} & \Upsilon_{33i} \end{bmatrix} < 0, \tag{1.123}$$

where

$$\begin{aligned}
\Upsilon_{11i} &= \begin{bmatrix} -He(Z_{f1i}) & * \\ -N_i - M_{f1i}^T & -He(N_i) \end{bmatrix}, \\
\Upsilon_{21i} &= \begin{bmatrix} P_i^{11} + Z_{1i}A_i + \hat{B}_{fi}C_i - Z_{2i}^T & \hat{A}_{fi} - M_{2i}^T \\ M_{1i}A_i + \hat{B}_{fi}C_i - N_i^T & P_i^{22} + \hat{A}_{fi} - N_i^T \end{bmatrix}, \\
\Upsilon_{31i} &= \begin{bmatrix} Z_{1i}B_{di} + \hat{B}_{fi}D_{di} \\ M_{1i}B_{di} + \hat{B}_{fi}D_{di} \end{bmatrix}, \\
\Upsilon_{22i} &= \begin{bmatrix} \alpha P_i^{11} + He(Z_{2i}A_i + \hat{B}_{fi}C_i) & * \\ \hat{A}_{fi} + A_i^T M_{2i}^T + C_i^T B_{fi}^T & \alpha P_i^{22} + He(\hat{A}_{fi}) \end{bmatrix}, \\
\Upsilon_{32i} &= \begin{bmatrix} Z_{2i}B_{di} + \hat{B}_{fi}D_{di} \\ M_{2i}B_{di} + \hat{B}_{fi}D_{di} \end{bmatrix}, \\
\Upsilon_{33i} &= -\gamma^2 I.
\end{aligned}$$

So $A_{fi} = N_i^{-1}\hat{A}_{fi}$, $B_{fi} = N_i^{-1}\hat{B}_{fi}$, *and* C_{fi} *are the parameters of the filter (1.52).*

Proof First, with a similar method to the one used in proving Theorem 1.6, we have

$$P_i^d \le \mu P_j^d, \quad i \neq j, \tag{1.124}$$

$$\begin{bmatrix} \Delta_{11i} & * & * \\ \Delta_{21i} & \Delta_{22i} & * \\ \Delta_{31i} & \Delta_{32i} & \Delta_{33i} \end{bmatrix} < 0, \tag{1.125}$$

where

$$\begin{aligned}
\Upsilon_{11i} &= \begin{bmatrix} Q_d^{11} - He(Z_{f1i}) & * \\ Q_d^{21} - N_i - M_{f1i}^T & Q_d^{22} - He(N_i) \end{bmatrix}, \Upsilon_{31i} = \begin{bmatrix} Z_{1i}B_{di} + \hat{B}_{fi}D_{di} \\ M_{1i}B_{di} + \hat{B}_{fi}D_{di} \end{bmatrix}, \\
\Upsilon_{21i} &= \begin{bmatrix} P_i^{11} + Z_{1i}A_i + \hat{B}_{fi}C_i - Z_{2i}^T & \hat{A}_{fi} - M_{2i}^T \\ M_{1i}A_i + \hat{B}_{fi}C_i - N_i^T & P_i^{22} + \hat{A}_{fi} - N_i^T \end{bmatrix}, \\
\Upsilon_{22i} &= \begin{bmatrix} -\varpi_h^2 Q_d^{11} + He(Z_{2i}A_i + \hat{B}_{fi}C_i) & * \\ \varpi_h^2 Q_d^{21} + \hat{A}_{fi} + A_i^T M_{2i}^T + C_i^T B_{fi}^T & -\varpi_h^2 Q_d^{22} + He(\hat{A}_{fi}) \end{bmatrix}, \\
\Upsilon_{32i} &= \begin{bmatrix} Z_{2i}B_{di} + \hat{B}_{fi}D_{di} \\ M_{2i}B_{di} + \hat{B}_{fi}D_{di} \end{bmatrix}, \Upsilon_{33i} = -\gamma^2 I.
\end{aligned}$$

Equation (1.124) is equivalent to

$$\begin{aligned}\dot{V}(t) &= \dot{\eta}^T(t)P_i\eta(t) + \eta^T(t)P_i\dot{\eta}(t)\\ &= \eta^T(t)(\bar{A}_i^T P_i + P_i\bar{A}_i)\eta(t)\\ &= \eta^T(t)\begin{bmatrix}\bar{A}_i\\ I\end{bmatrix}^T\begin{bmatrix}0 & P_i\\ P_i & 0\end{bmatrix}\begin{bmatrix}\bar{A}_i\\ I\end{bmatrix}\eta(t),\end{aligned} \tag{1.126}$$

$$V_i \leq \mu V_j, \qquad i \neq j, \tag{1.127}$$

$$\begin{bmatrix}\bar{A}_i\\ I\end{bmatrix}^T\begin{bmatrix}0 & P_i\\ * & \alpha P_i\end{bmatrix}\begin{bmatrix}\bar{A}_i\\ I\end{bmatrix} < 0. \tag{1.128}$$

Next, from (1.126) and (1.128) we can obtain

$$\dot{V}_i(t) + \alpha V_i(t) = \eta^T(t)\begin{bmatrix}\bar{A}_i\\ I\end{bmatrix}^T\begin{bmatrix}0 & P_i\\ * & \alpha P_i\end{bmatrix}\begin{bmatrix}\bar{A}_i\\ I\end{bmatrix}\eta(t) \tag{1.129}$$

and

$$V_{i_k}(t) \leq \mu e^{-\alpha(t-t_k)}V_{i_{k-1}}(t_k^-) \leq e^{-(\alpha-\ln\mu/T)}V_{i_0}(t_0).$$

Obviously, we have

$$V(t) \geq a\|\tilde{\eta}_1(t)\|^2, \qquad V(0) \leq b\|\tilde{\eta}_1(0)\|^2. \tag{1.130}$$

So we obtain

$$\|\tilde{\eta}_1(t)\|^2 \leq 1/aV(t) \leq 1/ae^{-(\alpha-\ln\mu/T)}V(t_0) \leq b/ae^{-(\alpha-\ln\mu/T)}\|\tilde{\eta}_1(0)\|^2. \tag{1.131}$$

After that, $A_{fi} = N_i^{-1}\hat{A}_{fi}$, $B_{fi} = N_i^{-1}\hat{B}_{fi}$, and C_{fi} are the parameters of the filter (1.52).

Therefore, with Lemma 1.2 the filter (1.52) is stable. □

1.2.6 H_-/H_∞ FD Filter Design Algorithms

Now, based on Theorems 1.5, 1.6, and 1.7, when α, γ, and μ are given, we can derive the FD filter parameters by solving the optimization problem (1.62):

$$\begin{aligned}&max\ \beta\\ &s.t.(1.101), (1.102), (1.103), (1.104), (1.105), (1.117), (1.118),\\ &(1.119), (1.137), (1.121), (1.122).\end{aligned} \tag{1.132}$$

We have $A_{fi} = N_i^{-1}\hat{A}_{fi}$, $B_{fi} = N_i^{-1}\hat{B}_{fi}$. Then the filters which we need are defined by the gain matrices A_{fi}, B_{fi}, C_{fi}.

1.2.7 FD Threshold Design

The next step is to evaluate the residual signal and compare it with some threshold value to detect the fault in the system. In this section, the threshold for detecting faults is designed and the detection logic unit is based on the results proposed in [55].

The evaluation function based on the RMS energy of the residual signal is used in this paper. So we have

$$J_r(t) = \|r\|_{rms} := \sqrt{\frac{1}{t}\int_0^t r^T(\tau)r(\tau)d\tau}. \tag{1.133}$$

The threshold J_{th} is obtained by

$$J_{th} = \sup_{f(t)=0, d(t)\in L_2,} J_r(t). \tag{1.134}$$

Finally the occurrence of a fault can be detected by the following logic rule:

$$J_r > J_{th} \Rightarrow alarm, \tag{1.135}$$

$$J_r \leq J_{th} \Rightarrow no\ faults. \tag{1.136}$$

1.2.8 An Illustrative Example

A liquid-level control system, in this part, has been considered to test the effectiveness of the proposed approach on a simulation model built in the Simulink/MATLAB environment, which is described in Fig. 1.10. As in [56], the system is composed of two tanks: one flow source, two outlet pipes, and one connecting pipe. The pipes contain valves that can be opened or closed by an external controller. Based on the status of each valve (open or closed), there exist eight different system modes, however, just three valve configurations are considered as follows.

$$\begin{aligned} Model1 &: V\text{-}2\ ON,\ V\text{-}1\ and\ V\text{-}3\ OFF \\ Model2 &: V\text{-}1\ and\ V\text{-}2\ ON,\ V\text{-}3\ OFF \\ Model3 &: V\text{-}2\ and\ V\text{-}3\ ON,\ V\text{-}1\ OFF \end{aligned}$$

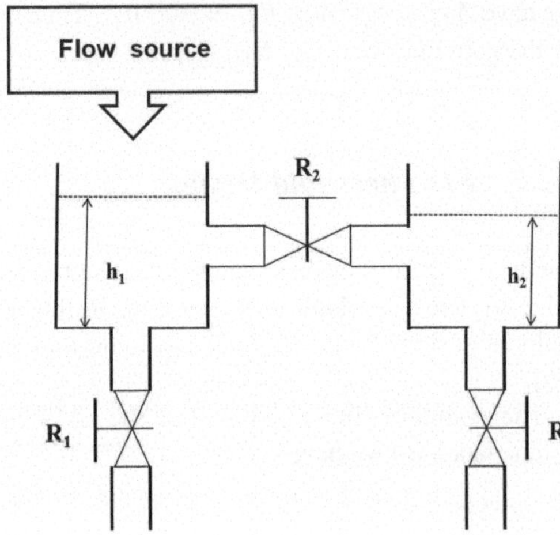

Fig. 1.10 A liquid-level control system

Consider the flow through the valves is laminar, which implies that the relation between the flow rate in the valves and the height of the liquid is linear. Depending on the value of the tank capacity C_T and the pipe resistance V in each mode, the behavior of the system is governed by a specific differential equation. The state-space representation of the system is given by

$$\dot{x}(t) = \sum_{i=1}^{3} \sigma_i(t)(A_i x(t) + B_i d(t) + F_i f(t))y(t) = \sum_{i=1}^{3} \sigma_i(t)(C_i x(t) + D_i d(t)), \tag{1.137}$$

where the state $x(t) = \left[h_1^T(t) \; h_2^T(t) \right]^T$ contains the heights of liquid in the tanks, and $d(t) = 2e^{-0.5t}(1 + sin3\pi t)$ is the input flow from the flow source to tank 1. The switching signal $\sigma_i(t)$ in this example is a piecewise constant function with the set of images equal to $\{1, 2, 3\}$ satisfies the constrained ADT condition, which is described in Fig. 1.11. The output $y(t)$ is the liquid in each tank while the valve configuration can jump between the three given modes. It is assumed that the system (1.116) is affected by disturbances and system faults. The following values for the system parameters is considered:

$$C_{T_1} = 5m, C_{T_2} = 5m, V\text{-}1 = V\text{-}2 = 300\frac{s}{m^2}, V\text{-}3 = 100\frac{s}{m^2}.$$

Consider the switched system (1.48) with the following parameters:

$$A_1 = \begin{bmatrix} -4.8 & 1 \\ 0.232 & -1.4 \end{bmatrix}, \; B_1 = \begin{bmatrix} 0.5 \\ -0.1 \end{bmatrix}, C_1 = \begin{bmatrix} -0.8 \\ 0.1 \end{bmatrix}, \; D_1 = 1, \; F_1 = \begin{bmatrix} -3.5 \\ -1.7 \end{bmatrix},$$

$$A_2 = \begin{bmatrix} -1.2 & 1 \\ -2 & -4.5 \end{bmatrix}, \; B_2 = \begin{bmatrix} 0.2 \\ 0.2 \end{bmatrix}, \; C_2 = \begin{bmatrix} 5 \\ -1 \end{bmatrix}, \; D_2 = 1, \; F_2 = \begin{bmatrix} -10.2 \\ -1.56 \end{bmatrix},$$

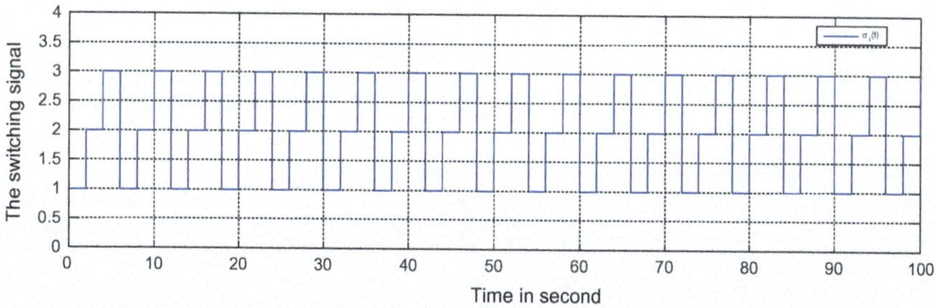

Fig. 1.11 The switching signal $\sigma_i(t)$

$$A_3 = \begin{bmatrix} -10 & 0.6 \\ -0.6 & -0.8 \end{bmatrix}, \quad B_3 = \begin{bmatrix} 0.2 \\ -0.2 \end{bmatrix}, \quad C_3 = \begin{bmatrix} 5 \\ -1 \end{bmatrix}, \quad D_3 = 1, \; F_3 = \begin{bmatrix} -3.4 \\ -1.7 \end{bmatrix}.$$

Choose $\mu = 1.2, \alpha = 0.2$. Then by (1.132), the gain matrices of the FD filter are as follows.

$$A_{f1} = \begin{bmatrix} -0.8642 & 0.7609 \\ 0.1323 & -1.1915 \end{bmatrix}, \quad B_{f1} = \begin{bmatrix} -5.1274 \\ -10.4909 \end{bmatrix}, \quad C_{f1} = \begin{bmatrix} -15.3563 \\ -15.7044 \end{bmatrix},$$
$$A_{f2} = \begin{bmatrix} -1.0693 & -0.8440 \\ -0.0042 & -1.9275 \end{bmatrix}, \quad B_{f2} = \begin{bmatrix} 0.3534 \\ 0.0454 \end{bmatrix}, \quad C_{f2} = \begin{bmatrix} -15.5732 \\ -15.6529 \end{bmatrix},$$
$$A_{f3} = \begin{bmatrix} -1.5063 & 0.7571 \\ 0.0175 & -1.3025 \end{bmatrix}, \quad B_{f3} = \begin{bmatrix} 1.7794 \\ 0.3428 \end{bmatrix}, \quad C_{f3} = \begin{bmatrix} -15.2551 \\ -15.7895 \end{bmatrix}.$$

In the simulation, we consider the following piecewise constant fault:

$$f_1(t) = \begin{cases} 0, & 0s \le t < 10s; \\ 5, & 10s \le t < 50s; \\ 0, & t > 50s; \end{cases}$$

which is shown in Fig. 1.12. The residual signal $r(t)$ are shown in Fig. 1.13 for the cases of with fault and without fault. Figure 1.14 describes the residual evaluation function $J_r(t)$ with the faulty and unfaulty cases. By (1.85), the threshold is set as $J_{th} = 0.8 * 10^6$, the residual evaluation function $J_r(t)$ in Fig. 1.14 indicates that the fault is detected at approximately $t = 11.97$s.

1.2.9 Summary

In this section, we have studied the actuator FD problem for switched systems in finite frequency. Based on Parseval's lemma and S-procedure, we have obtained

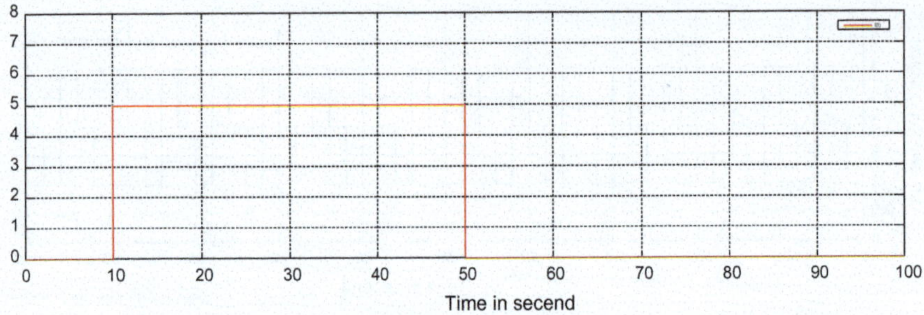

Fig. 1.12 The fault

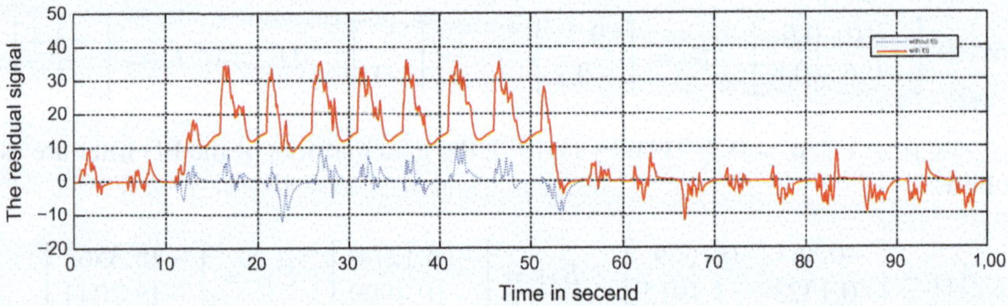

Fig. 1.13 The residual signal $r(t)$

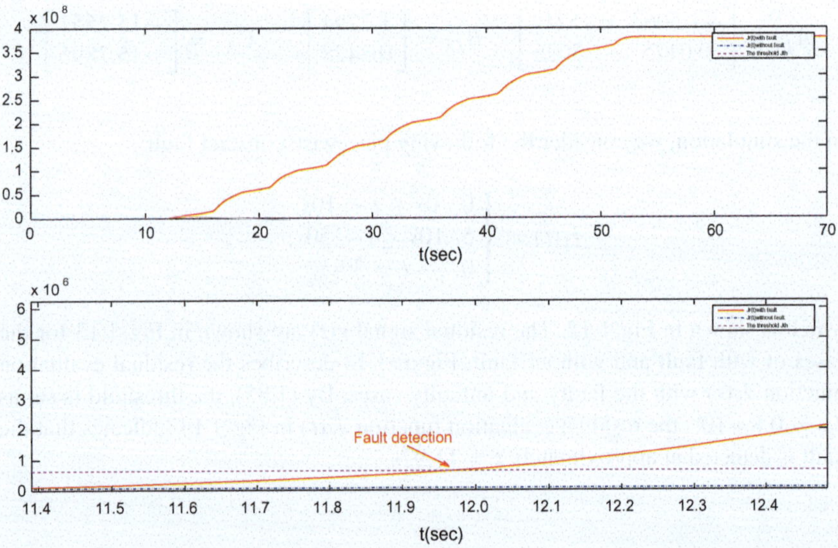

Fig. 1.14 The evaluation function $J_r(t)$

some sufficient conditions which ensure that the augmented filter system has the H_- fault affection level, the H_∞ disturbance attention level, and we have stability. The FD filter design conditions have been derived in terms of solving a set of LMIs. A practical example has been provided to demonstrate the effectiveness of the proposed method.

1.3 Finite-Frequency FD Filter Design

In this section, the finite-frequency FD filter design for discrete-time switched systems is investigated. The frequencies of the faults and the unknown disturbance input are assumed to be finite and in three known intervals, qualified as low-, middle-, and high-frequency intervals. Based on the switched Lyapunov function and the generalized Kalman–Yakubovic–Popov (KYP) lemma, efficient conditions are obtained to guarantee the existence of a finite-frequency FD filter and such that the error system is asymptotically stable with an H_∞/H_- performance index. Finally, a chemical reactor control system is employed to illustrate the obtained techniques.

1.3.1 Introduction

It is very important that the modern industrial systems demand high reliability and safety because they are subjected to potential abnormalities and faults, such as chemical processes, intelligent vehicle dynamics, power plants, etc. [57, 58] Therefore, it is a very necessary and significant researching task to find efficient methods to detect the fault before it can cause disastrous consequences. As a result, fruitful achievements of FD observer or filter design have been extensively exploited [59, 60]. Among them, the model-based FD approach is regarded as an efficient and popular method and has been extensively studied in recent 20 years [61–67]. Its principle is to design an observer or a filter as residual signal generator, such that the residual signal is not only sensitive to the faults but also robust to the unknown disturbance signals [68, 69].

A switched system, as a kind of hybrid system, is constituted by several continuous-time or discrete-time subsystems and a switching signal, which specifies which subsystem will be activated during the operation time [70, 71]. The applications of switched systems have been widely researched in many industrial fields, for example, traffic control systems, mechanical systems, chemical processes, electrical power systems, etc., in recent years, the FD techniques for such systems have been widely considered. For instance, [46, 72], respectively, studied the FD filter design problems for time-delayed continuous-time and discrete-time switched systems. Based on the descriptor observer method, [73] provided a sensor FD filter design method for discrete-time switched systems. Asynchronous FD filter design methods for switched systems were presented in [74, 75].

It can be seen that most of the obtained achievements are highlighted on the emphasis of the whole frequency domain. Nevertheless, in some real practical applications, the frequency ranges of the unknown disturbance signals and the faults are generally known beforehand. Therefore, the developed results for various dynamic systems in full frequency may be more conservative due to the overdesign, which motivates us to begin the meaningful work of developing a finite-frequency FD filter for switched systems. Actually, some finite-frequency FD filter or observer designs for various dynamic systems have been exploited in recent studies. In [76], a finite-frequency FD filter was designed for discrete-time switched systems with sensor faults, while the similar issue for T-S fuzzy systems was investigated in [77]. By using the generalized KYP Lemma, the multi-objective FD filter design in finite-frequency domain for discrete-time T-S fuzzy systems was considered in [78]. In [79], the finite-frequency FD problem for sampled-data systems was studied by using the lifting technique.

In this section, the problem we mainly consider is to design a finite-frequency FD filter for discrete-time switched systems. The unknown disturbance input and the fault are both supposed to be in finite-frequency domains. An H_∞/H_- finite-frequency FD filter is designed, which can ensure that the residual signal is simultaneously robust to the unknown disturbance signals and sensitive to the faults. Finally, a chemical reactor example is employed to verify the proposed techniques.

The contributions of this section consist in the following three aspects: firstly, the design procedures of a finite-frequency FD filter is given for discrete-time switched systems. Moreover, the proposed results can greatly reduce the computational burden. Secondly, the performance of the FD filter, in terms of robustness with regard to the unknown disturbance signals and sensitivity to the faults, is guaranteed by the design procedure which includes H_∞ and H_- criteria. Finally, a chemical reaction control system example is taken to illustrate the application of the obtained techniques and simulations are performed to show the efficiency of this method.

1.3.2 Problem Formulation and Preliminaries

The discrete-time switched system is described by

$$\begin{cases} x(t+1) = \sum\limits_{i=1}^{N} \alpha_i(t)(A_i x(t) + B_i u(t) + D_{1i} d(t) + F_{1i} f(t)) \\ y(t) = \sum\limits_{i=1}^{N} \alpha_i(t)(C_i x(t) + D_{2i} d(t) + F_{2i} f(t)), \end{cases} \tag{1.138}$$

where $x(t) \in \mathbb{R}^{n_x}$ is the state vector, $u(t) \in \mathbb{R}^{n_u}$ is the known input vector, $y(t) \in \mathbb{R}^{n_y}$ is the measured output vector; $d(t) \in \mathbb{R}^{n_d}$ and $f(t) \in \mathbb{R}^{n_f}$, respectively, represent the unknown disturbance input and the fault, which both belong to $l_2[0, \infty)$ and whose frequency ranges are assumed to be finite and known beforehand. $\vartheta_d \in \Theta_d$ and $\vartheta_f \in \Theta_f$, respectively, denote the frequencies of the disturbance and the fault,

Table 1.1 The description of the frequency ranges

Frequency	LF	MF	HF
Θ_d	$\lvert\vartheta_d\rvert \le \vartheta_{dl}$	$\vartheta_{dm1} \le \vartheta_d \le \vartheta_{dm2}$	$\lvert\vartheta_d\rvert \ge \vartheta_{dh}$
Θ_f	$\lvert\vartheta_f\rvert \le \vartheta_{fl}$	$\vartheta_{fm1} \le \vartheta_f \le \vartheta_{fm2}$	$\lvert\vartheta_f\rvert \ge \vartheta_{fh}$

which are divided into three ranges, namely low-frequency (LF), middle-frequency (MF), and high-frequency (HF) ranges as shown in the Table 1.1:
where ϑ_{dl}, ϑ_{dm1}, ϑ_{dm2}, ϑ_{dh}, ϑ_{fl}, ϑ_{fm1}, ϑ_{fm2}, and ϑ_{fh} are all known constants.

The piecewise function

$$\alpha_i(t) : [0, \infty) \to \{0, 1\}, \quad \sum_{i=1}^{N} \alpha_i(t) = 1 \tag{1.139}$$

is the switching signal. As in [80], we assume that the instantaneous value of the switching signal is available in real time. The matrices A_i, B_i, C_i, D_{1i}, D_{2i}, F_{1i}, and F_{2i} are known with appropriate dimensions.

The following filter is employed as a residual signal generator:

$$\begin{cases} \xi(t+1) = \sum_{i=1}^{N} \alpha_i(t)(N_i\xi(t) + G_i u(t) + L_i y(t)), \\ \hat{x}(t) = \sum_{i=1}^{N} \alpha_i(t)(\xi(t) - E_i y(t)), , \\ \hat{y}(t) = \sum_{i=1}^{N} \alpha_i(t) C_i \hat{x}(t). \end{cases} \tag{1.140}$$

where $\hat{x}(t) \in \mathbb{R}^{n_x}$ is the state estimation of systems (1.138), $\hat{y}(t) \in \mathbb{R}^{n_y}$ is the filter's output vector. N_i, G_i, L_i, and E_i are the filter's gain matrices to be determined later.

Define the state error signal $e(t)$ and the residual signal $r(t)$

$$e(t) = x(t) - \hat{x}(t), \quad r(t) = y(t) - \hat{y}(t). \tag{1.141}$$

Then, we have

$$\begin{cases} e(t) = \sum_{i=1}^{N} \alpha_i(t)(T_i x(t) - \xi(t) + E_i D_{2i} d(t) + E_i F_{2i} f(t)), \\ r(t) = \sum_{i=1}^{N} \alpha_i(t)(C_i e(t) + D_{2i} d(t) + F_{2i} f(t)), \end{cases} \tag{1.142}$$

where $T_i = I + E_i C_i$.

From (1.138), (1.140), and (1.142), the following error system can be derived:

$$\begin{cases} e(t+1) = \sum_{i=1}^{N} \alpha_i(t)(N_i e(t) + (T_i A_i - L_i C_i - N_i T_i)x(t) + (T_i B_i - G_i)u(t) \\ \qquad + E_i D_{2i} d(t+1) + (T_i D_{1i} - L_i D_{2i} - N_i E_i D_{2i})d(t) \\ \qquad + (T_i F_{1i} - L_i F_{2i} - N_i E_i F_{2i}) f(t) + E_i F_{2i} f(t+1)), \\ r(t) = \sum_{i=1}^{N} \alpha_i(t)(C_i x(t) + D_{2i} d(t) + F_{2i} f(t)). \end{cases} \tag{1.143}$$

If the following assumptions hold:

$$T_i A_i - L_i C_i - N_i T_i = 0; \tag{1.144}$$

$$T_i B_i - G_i = 0; \tag{1.145}$$

$$E_i \begin{bmatrix} D_{2i} & F_{2i} \end{bmatrix} = 0; \tag{1.146}$$

then the error dynamics (1.143) reduces to

$$\begin{cases} e(t+1) = \sum_{i=1}^{N} \alpha_i(t)(\bar{A}_i e(t) + \bar{D}_i d(t) + \bar{F}_i f(t)), \\ r(t) = \sum_{i=1}^{N} \alpha_i(t)(C_i e(t) + D_{2i} d(t) + F_{2i} f(t)), \end{cases} \tag{1.147}$$

where

$$\bar{A}_i = T_i A_i - K_i C_i; \tag{1.148}$$

$$\bar{D}_i = T_i D_{1i} - K_i D_{2i}; \tag{1.149}$$

$$\bar{F}_i = T_i F_{1i} - K_i F_{2i}; \tag{1.150}$$

$$K_i = L_i + N_i E_i. \tag{1.151}$$

Consequently, the following purposes of this section are formulated: to design a finite-frequency FD filter (1.140) ensuring the asymptotical stability of the error system (1.147) and such that the residual signal $r(t)$ is robust to the unknown disturbance $d(t)$ and sensible to the fault $f(t)$, where the unknown disturbance $d(t)$ and the fault $f(t)$ are both in finite-frequency domain.

Remark 1.8 By giving the assumption in (1.144)–(1.146), a standard error dynamic system (1.147) is obtained, which is only affected by the disturbances and the faults. Based on these assumptions, the main objectives of our section can be made.

Lemma 1.5 ([81]) *Consider the following discrete-time switched system Σ_d:*

$$\begin{cases} x(t+1) = \sum_{i=1}^{N} \alpha_i(t)(A_i(\lambda)x(t) + B_i(\lambda)d(t)), \\ r(t) = \sum_{i=1}^{N} \alpha_i(t)(C_i(\lambda)x(t) + D_i(\lambda)d(t)), \end{cases} \tag{1.152}$$

where $A_i(\lambda)$, $B_i(\lambda)$, $C_i(\lambda)$, and $D_i(\lambda)$ are matrices of appropriate dimensions with a set of constant parameters and defined in the following form:

$$
\begin{aligned}
&(A_i(\lambda), B_i(\lambda), C_i(\lambda), D_i(\lambda)) \\
&= \sum_{l=1}^{s} \lambda_l (A_{i,l}, B_{i,l}, C_{i,l}, D_{i,l}), \quad \sum_{l=1}^{s} \lambda_l = 1.
\end{aligned} \tag{1.153}
$$

For a given symmetric matrix

$$
\Pi = \begin{bmatrix} I & 0 \\ 0 & -\gamma^2 I, \end{bmatrix} \tag{1.154}
$$

the following statements are equivalent:
(i): The finite-frequency inequality

$$
\sigma_{max}(G^i_{dr}(e^{j\vartheta_d})) < \gamma, \tag{1.155}
$$

where $\sigma_{max}(G^i_{dr}(e^{j\vartheta_d}))$ represents the maximum singular value of the transfer function matrix $G^i_{dr}(e^{j\vartheta_d})$ for the $i$sth subsystem from the disturbance input $d(t)$ to error output $r(t)$;
(ii): There exist Hermitian matrix functions $P_i(\lambda)$, $Q_i(\lambda)$ satisfying $Q_i(\lambda) > 0$, and

$$
\begin{aligned}
&\begin{bmatrix} A_i(\lambda) & B_i(\lambda) \\ I & 0 \end{bmatrix}^T \Xi_i(\lambda) \begin{bmatrix} A_i(\lambda) & B_i(\lambda) \\ I & 0 \end{bmatrix} \\
&+ \begin{bmatrix} C_i(\lambda) & D_i(\lambda) \\ 0 & I \end{bmatrix}^T \Pi \begin{bmatrix} C_i(\lambda) & D_i(\lambda) \\ 0 & I \end{bmatrix} < 0,
\end{aligned} \tag{1.156}
$$

where

$$
\Xi_i(\lambda) = \begin{bmatrix} -P_i(\lambda) & Q_i(\lambda) \\ Q_i(\lambda) & P_i(\lambda) - 2cos(\vartheta_{dl})Q_i(\lambda) \end{bmatrix} \tag{1.157}
$$

for low-frequency range $|\vartheta_d| \le \vartheta_{dl}$,

$$
\begin{aligned}
&\Xi_i(\lambda) = \begin{bmatrix} -P_i(\lambda) & e^{j\vartheta_{dc}} Q_i(\lambda) \\ e^{-j\vartheta_{dc}} Q_i(\lambda) & P_i(\lambda) - 2cos(\vartheta_{dw})Q_i(\lambda) \end{bmatrix}, \\
&\vartheta_{dc} = \frac{\vartheta_{dm1} + \vartheta_{dm2}}{2}, \quad \vartheta_{dw} = \frac{\vartheta_{dm1} - \vartheta_{dm2}}{2},
\end{aligned} \tag{1.158}
$$

for middle-frequency range $\vartheta_{dm1} < \vartheta_d < \vartheta_{dm2}$, and

$$
\Xi_i(\lambda) = \begin{bmatrix} -P_i(\lambda) & -Q_i(\lambda) \\ -Q_i(\lambda) & P_i(\lambda) + 2cos(\vartheta_{dh})Q_i(\lambda) \end{bmatrix} \tag{1.159}
$$

for high-frequency range $|\vartheta_d| \geq \vartheta_{dh}$.

By a direct extension of Lemma 1.5, we can get the following lemma.

Lemma 1.6 *Consider the following discrete-time switched system* Σ_f*:*

$$\begin{cases} x(t+1) = \sum_{i=1}^{N} \alpha_i(t)(A_i(\lambda)x(t) + B_i(\lambda)f(t)) \\ r(t) = \sum_{i=1}^{N} \alpha_i(t)(C_i(\lambda)x(t) + D_i(\lambda)f(t)) \end{cases} \tag{1.160}$$

where $A_i(\lambda)$*,* $B_i(\lambda)$*,* $C_i(\lambda)$*, and* $D_i(\lambda)$ *are defined as in (1.153).*

For a given symmetric matrix

$$\Pi = \begin{bmatrix} -I & 0 \\ 0 & \beta^2 I \end{bmatrix} \tag{1.161}$$

the following statements are equivalent:
(i): The finite-frequency inequality

$$\sigma_{min}(G^i_{fr}(e^{j\vartheta_f})) > \beta, \tag{1.162}$$

where $\sigma_{min}(G^i_{fr}(e^{j\vartheta_f}))$ *is the minimum singular value of the transfer function matrix* $G^i_{fr}(e^{j\vartheta_f})$ *for the i-th subsystem from the fault* $f(t)$ *to error output* $r(t)$*;*
(ii): There exist Hermitian matrix functions $P_i(\lambda)$*,* $Q_i(\lambda)$ *satisfying* $Q_i(\lambda) > 0$*, and*

$$\begin{aligned} &\begin{bmatrix} A_i(\lambda) & B_i(\lambda) \\ I & 0 \end{bmatrix}^T \Xi_i(\lambda) \begin{bmatrix} A_i(\lambda) & B_i(\lambda) \\ I & 0 \end{bmatrix} \\ &+ \begin{bmatrix} C_i(\lambda) & D_i(\lambda) \\ 0 & I \end{bmatrix}^T \Pi \begin{bmatrix} C_i(\lambda) & D_i(\lambda) \\ 0 & I \end{bmatrix} < 0 \end{aligned} \tag{1.163}$$

where

$$\Xi_i(\lambda) = \begin{bmatrix} -P_i(\lambda) & Q_i(\lambda) \\ Q_i(\lambda) & P_i(\lambda) - 2cos(\vartheta_{fl})Q_i(\lambda) \end{bmatrix} \tag{1.164}$$

for low-frequency range $|\vartheta_f| \leq \vartheta_{fl}$*,*

$$\begin{aligned} \Xi_i(\lambda) &= \begin{bmatrix} -P_i(\lambda) & e^{j\vartheta_{fc}}Q_i(\lambda) \\ e^{-j\vartheta_{fc}}Q_i(\lambda) & P_i(\lambda) - 2cos(\vartheta_{fw})Q_i(\lambda) \end{bmatrix}, \\ \vartheta_{fc} &= \frac{\vartheta_{fm1} + \vartheta_{fm2}}{2}, \quad \vartheta_{fw} = \frac{\vartheta_{fm1} - \vartheta_{fm2}}{2}, \end{aligned} \tag{1.165}$$

for middle-frequency range $\vartheta_{fm1} < \vartheta_f < \vartheta_{fm2}$*, and*

$$\Xi_i(\lambda) = \begin{bmatrix} -P_i(\lambda) & -Q_i(\lambda) \\ -Q_i(\lambda) & P_i(\lambda) + 2cos(\vartheta_{fh})Q_i(\lambda) \end{bmatrix} \tag{1.166}$$

for high-frequency range $|\vartheta_f| \geq \vartheta_{fh}$.

Proof The condition (1.162) is equivalent to

$$G^i_{fr}(e^{j\vartheta_f})^T G^i_{fr}(e^{j\vartheta_f}) - \beta^2 I > 0, s \tag{1.167}$$

which can be written as

$$\begin{bmatrix} G^i_{fr}(e^{j\vartheta_f})^T \\ I \end{bmatrix}^T \begin{bmatrix} -I & 0 \\ 0 & \beta^2 I \end{bmatrix} \begin{bmatrix} G^i_{fr}(e^{j\vartheta_f})^T \\ I \end{bmatrix} > 0. \tag{1.168}$$

By applying the generalized KYP lemma in [82], the results are obtained directly. □

Remark 1.9 Because $\sigma_{max}(G^i_{dr}(e^{j\vartheta_d}))$ (or $\sigma_{min}(G^i_{fr}(e^{j\vartheta_f}))$) is the maximum (minimum) singular value of the transfer function matrix $G^i_{dr}(e^{j\vartheta_d})$ (or $G^i_{fr}(e^{j\vartheta_f})$) for the ith subsystem of the switched system Σ_d (or Σ_f) from the disturbance input $d(t)$ (or the fault $f(t)$) to the error output $r(t)$, which means that (1.155) (or (1.162)) is equivalent to minimize (or maximize) the error $r(t)$ over the energy bounded disturbance input $d(t)$ (or the fault $f(t)$), such that

$$\sup_{d(t)\neq 0} \frac{\|r(t)\|_2}{\|d(t)\|_2} < \gamma, (or \quad \inf_{f(t)\neq 0} \frac{\|r(t)\|_2}{\|f(t)\|_2} > \beta),$$

which is usually called H_∞ (or H_-) performance index.

Remark 1.10 In Lemma 1.5 (or Lemma 1.6), the matrices of system Σ_d (or Σ_f) are defined in (1.153). If we set $A_{i,1} = A_{i,2} = \cdots = A_{i,s} = A_i$, $B_{i,1} = B_{i,2} = \cdots = B_{i,s} = B_i$, $C_{i,1} = C_{i,2} = \cdots = C_{i,s} = C_i$, $D_{i,1} = D_{i,2} = \cdots = D_{i,s} = D_i$, then the system Σ_d (or Σ_f) becomes a common linear switched system. Accordingly, the parameters $P_i(\lambda)$ and $Q_i(\lambda)$ in Lemma 1.5 (or Lemma 1.7) become P_i and Q_i.

Lemma 1.7 (Projection Lemma [83]) *Given a symmetric matrix Σ and two matrices Γ, Λ of appropriate column dimensions, the problem*

$$\Sigma + \Gamma^T \Theta \Lambda + \Lambda^T \Theta^T \Gamma < 0 \tag{1.169}$$

is solvable with regard to the matrix Θ of compatible dimensions if and only if the following inequalities hold:

$$\mathcal{N}_\Gamma^T \Sigma \mathcal{N}_\Gamma < 0, \quad \mathcal{N}_\Lambda^T \Sigma \mathcal{N}_\Lambda < 0, \tag{1.170}$$

where $\mathcal{N}_\Gamma$ and $\mathcal{N}_\Lambda$ are, respectively, any basis of the nullspace of the matrices Γ and Λ.

Based on the conclusions in Lemmas 1.5 and 1.6, we give the following definition:

Definition 1.7 Consider the discrete-time switched system (1.138). Given two positive scalars γ and β, and the finite-frequency unknown disturbance input $d(t)$ and fault $f(t)$, the system (1.140) is called a finite-frequency H_∞/H_- FD filter for the system (1.138) if the following conditions are satisfied:

- the error system (1.147) is asymptotically stable with $d(t)=0$ and $f(t)=0$;
- $\sigma_{max}(G^i_{dr}(e^{j\vartheta_d})) < \gamma$;
- $\sigma_{min}(G^i_{fr}(e^{j\vartheta_f})) > \beta$.

1.3.3 Main Results

The main problem of this section is to design an H_∞/H_- finite-frequency FD filter for the discrete-time switched system (1.138).

1.3.3.1 Robust Performance Analysis

In this subsection, efficient conditions to reduce the effects of the unknown disturbance input $d(t)$ to the residual signal $r(t)$ will be established. If we set $f(t)=0$, then the error system (1.147) is transformed to

$$\begin{cases} e(t+1) = \sum\limits_{i=1}^{N} \alpha_i(t)(\bar{A}_i e(t) + \bar{D}_i d(t)) \\ r(t) = \sum\limits_{i=1}^{N} \alpha_i(t)(C_i e(t) + D_{2i} d(t)).s \end{cases} \tag{1.171}$$

Theorem 1.8 *For a given positive scalar γ, the system (1.171) is asymptotically stable with the H_∞ performance index γ if there exist positive definite matrices P_{1i}, P_{1j}, Hermitian matrices P_{2i}, $Q_{1i} > 0$, matrices S_i, W_i, and Ω_i, for $\forall i, j \in \{1, 2, \ldots, N\}$, which satisfy*

$$\begin{bmatrix} P_{1j} - \Omega_i - \Omega_i^T & (\Omega_i + S_i C_i) A_i - W_i C_i \\ * & -P_{1i} \end{bmatrix} < 0 \tag{1.172}$$

$$\begin{bmatrix} -P_{2i} & Q_{1i} - \Omega_i^T & 0 & 0 \\ * & \Psi_{1i} & \psi_i & C_i^T \\ * & * & -\gamma^2 I & D_{2i}^T \\ * & * & * & -I \end{bmatrix} < 0 \tag{1.173}$$

for the low-frequency disturbance $|\vartheta_d| \le \vartheta_{dl}$;

$$\begin{bmatrix} -P_{2i} & e^{j\vartheta_{dc}}Q_{1i} - \Omega_i^T & 0 & 0 \\ * & \Psi_{mi} & \psi_i & C_i^T \\ * & * & -\gamma^2 I & D_{2i}^T \\ * & * & * & -I \end{bmatrix} < 0 \tag{1.174}$$

for the middle-frequency disturbance $\vartheta_{dm1} \leq \vartheta_d \leq \vartheta_{dm2}$*;*

$$\begin{bmatrix} -P_{2i} & -Q_{1i} - \Omega_i^T & 0 & 0 \\ * & \Psi_{hi} & \psi_i & C_i^T \\ * & * & -\gamma^2 I & D_{2i}^T \\ * & * & * & -I \end{bmatrix} < 0 \tag{1.175}$$

for the high-frequency disturbance $|\vartheta_d| \geq \vartheta_{dh}$*;*
where

$$\begin{aligned} \Psi_{li} &= (\Omega_i + S_i C_i)A_i - W_i C_i + ((\Omega_i + S_i C_i)A_i - W_i C_i)^T c \\ &\quad + P_{2i} - 2cos(\vartheta_{dl})Q_{1i}; && (1.176) \\ \Psi_{mi} &= (\Omega_i + S_i C_i)A_i - W_i C_i + ((\Omega_i + S_i C_i)A_i - W_i C_i)^T \\ &\quad + P_{2i} - 2cos(\vartheta_{dw})Q_{1i}; && (1.177) \\ \Psi_{hi} &= (\Omega_i + S_i C_i)A_i - W_i C_i + ((\Omega_i + S_i C_i)A_i - W_i C_i)^T && (1.178) \\ &\quad + P_{2i} + 2cos(\vartheta_{dh})Q_{1i}; && (1.179) \\ \psi_i &= (\Omega_i + S_i C_i)D_{1i} - W_i D_{2i}; && (1.180) \\ \vartheta_{dc} &= \frac{\vartheta_{dm1} + \vartheta_{dm2}}{2}, \vartheta_{dw} = \frac{\vartheta_{dm1} - \vartheta_{m2}}{2}. && (1.181) \end{aligned}$$

Proof Firstly, we establish the stability of system (1.171).
Define

$$S_i = \Omega_i E_i, \quad W_i = \Omega_i K_i. \tag{1.182}$$

The condition (1.172) is equivalent to

$$\begin{bmatrix} P_{1j} & 0 \\ 0 & -P_{1i} \end{bmatrix} + \begin{bmatrix} \Omega_i \\ 0 \end{bmatrix} \begin{bmatrix} -I & T_i A_i - K_i C_i \end{bmatrix} + \begin{bmatrix} -I \\ (T_i A_i - K_i C_i)^T \end{bmatrix} \begin{bmatrix} \Omega_i^T & 0 \end{bmatrix} < 0. \tag{1.183}$$

By Lemma 1.7, if we set

$$\Sigma = \begin{bmatrix} P_{1j} & 0 \\ 0 & -P_{1i} \end{bmatrix}, \quad \Lambda = \begin{bmatrix} -I & T_i A_i - K_i C_i \end{bmatrix}, \Theta = I, \quad \Gamma = \begin{bmatrix} \Omega_i^T & 0 \end{bmatrix}, \tag{1.184}$$

and select

$$\mathcal{N}_\Lambda = \begin{bmatrix} T_i A_i - K_i C_i \\ I \end{bmatrix}, \tag{1.185}$$

then one can get from (1.183) that

$$\begin{aligned} \mathcal{N}_\Lambda^T \Sigma \mathcal{N}_\Lambda &= \begin{bmatrix} (T_i A_i - K_i C_i)^T & I \end{bmatrix} \begin{bmatrix} P_{1j} & 0 \\ 0 & -P_{1i} \end{bmatrix} \begin{bmatrix} T_i A_i - K_i C_i \\ I \end{bmatrix} \\ &= (T_i A_i - K_i C_i)^T P_{1j} (T_i A_i - K_i C_i) - P_{1i} < 0 \end{aligned} \tag{1.186}$$

Construct a switched Lyapunov function for the error dynamics (1.171) as

$$V(t, e(t)) = e^T(t) \left(\sum_{i=1}^{N} \alpha_i(t) P_{1i} \right) e(t). \tag{1.187}$$

From the Lyapunov stability theory, if the following inequality

$$\begin{aligned} &\Delta V(t) \\ &= V(t+1, e(t+1)) - V(t, e(t)) \\ &= e^T(t+1) \left(\sum_{i=1}^{N} \alpha_i(t+1) P_{1i} \right) e(t+1) \\ &\quad - e^T(t) \left(\sum_{i=1}^{N} \alpha_i(t) P_{1i} \right) e(t) < 0 \end{aligned} \tag{1.188}$$

holds, then the error dynamics (1.171) with $d(t) = 0$

$$e(t+1) = \sum_{i=1}^{N} \alpha_i(t) \bar{A}_i e(t) \tag{1.189}$$

is asymptotically stable. If we define the arbitrary switching signal as $\alpha_i(t) = i$, $\alpha_i(t+1) = j$, then the condition (1.186) can guarantee the asymptotically stability of the system (1.171).

The following proof of the finite-frequency H_∞ performance γ will be given. By Schur complement lemma, (1.173) can be shown as

$$\begin{bmatrix} -P_{2i} & Q_{1i} - \Omega_i^T & 0 \\ * & \Psi_{1i} + C_i^T C_i & (2,3) \\ * & * & -\gamma^2 I + D_{2i}^T D_{2i}, \end{bmatrix} < 0 \tag{1.190}$$

where $(2,3) = (\Omega_i + S_i C_i) D_{1i} - W_i D_{2i} + C_i^T D_{2i}$. (1.190) can be rewritten as

$$\begin{bmatrix} -P_{2i} & Q_{1i} & 0 \\ ** & P_{2i} - 2cos(\vartheta_{dl})Q_{1i} + C_i^T C_i & C_i^T D_{2i} \\ ** & * & -\gamma^2 I + D_{2i}^T D_{2i} \end{bmatrix} +$$
$$\begin{bmatrix} 0 \\ \Omega_i \\ 0 \end{bmatrix} \begin{bmatrix} -I \\ (T_i A_i - K_i C_i)^T \\ (T_i D_{1i} - K_i D_{2i})^T \end{bmatrix}^T + \begin{bmatrix} -I \\ (T_i A_i - K_i C_i)^T \\ (T_i D_{1i} - K_i D_{2i})^T \end{bmatrix} \begin{bmatrix} 0 \; \Omega_i \; 0 \end{bmatrix} < 0.s \tag{1.191}$$

By using Lemma 1.6 and defining

$$\Sigma = \begin{bmatrix} -P_{2i} & Q_{1i} & 0 \\ * & P_{2i} - 2cos(\vartheta_{dl})Q_{1i} + C_i^T C_i & C_i^T D_{2i} \\ * & * & -\gamma^2 I + D_{2i}^T D_{2i} \end{bmatrix},$$
$$\Lambda = \begin{bmatrix} -I \; T_i A_i - K_i C_i \; T_i D_{1i} - K_i D_{2i} \end{bmatrix}, \Theta = I, \quad \Gamma = \begin{bmatrix} 0 \; \Omega \; 0 \end{bmatrix}. \tag{1.192}$$

with N_Λ chosen as

$$\mathscr{N}_\Lambda = \begin{bmatrix} T_i A_i - K_i C_i & T_i D_{1i} - K_i D_{2i} \\ I & 0 \\ 0 & I \end{bmatrix}, \tag{1.193}$$

then from (1.191) one can get

$$\begin{aligned} &\mathscr{N}_\Lambda^T \Sigma \mathscr{N}_\Lambda \\ &= \begin{bmatrix} T_i A_i - K_i C_i & T_i D_{1i} - K_i D_{2i} \\ I & 0 \\ 0 & I \end{bmatrix}^T \\ &\times \begin{bmatrix} -P_{2i} & Q_{1i} & 0 \\ * & P_{2i} - 2cos(\vartheta_{dl})Q_{1i} + C_i^T C_i & C_i^T D_{2i} \\ * & * & -\gamma^2 I + D_{2i}^T D_{2i} \end{bmatrix} \\ &\times \begin{bmatrix} T_i A_i - K_i C_i & T_i D_{1i} - K_i D_{2i} \\ I & 0 \\ 0 & I \end{bmatrix} \\ &= \begin{bmatrix} \Pi_{11i} & \Pi_{12i} \\ \Pi_{12i}^T & \Pi_{22i} \end{bmatrix} < 0, \end{aligned} \tag{1.194}$$

where

$$\Pi_{11i} = -\bar{A}_i^T P_{2i} \bar{A}_i + Q_{1i} \bar{A}_i + \bar{A}_i^T Q_{1i} + P_{2i} - 2cos(\vartheta_{dl})Q_{1i} + C_i^T C_i; \tag{1.195}$$
$$\Pi_{12i} = -\bar{A}_i^T P_{2i} \bar{D}_i + Q_{1i} \bar{D}_i + C_i^T D_{2i}; \tag{1.196}$$
$$\Pi_{22i} = \bar{D}_i^T P_{2i} \bar{D}_i - \gamma^2 I + D_{2i}^T D_{2i}. \tag{1.197}$$

Therefore, if (1.173) holds, it means that (1.194) and (1.163) hold. According to Lemma 1.5, one can conclude that the system (1.171) has finite-frequency H_∞ performance γ. By following the same lines of Theorem 1.8, the proofs of conditions (1.174) and (1.175) can be deduced by setting

$$\Sigma = \begin{bmatrix} -P_{2i} & e^{j\vartheta_{dc}}Q_{1i} & 0 \\ e^{-j\vartheta_{dc}}Q_{1i} & P_{2i} - 2cos(\vartheta_{dw})Q_{1i} + C_i^T C_i & C_i^T D_{2i} \\ 0 & D_{2i}^T C_i & -\gamma^2 I + D_{2i}^T D_{2i} \end{bmatrix} \tag{1.198}$$

and

$$\Sigma = \begin{bmatrix} -P_{2i} & -Q_{1i} & 0 \\ -Q_{1i} & P_{2i} + 2cos(\vartheta_{dh})Q_{1i} + C_i^T C_i & C_i^T D_{2i} \\ 0 & D_{2i}^T C_i & -\gamma^2 I + D_{2i}^T D_{2i}. \end{bmatrix} \tag{1.199}$$

Then the proof is concluded. □

1.3.3.2 Sensitive Performance Analysis

In this subsection, the sensitivity conditions of the fault $f(t)$ to the residual signal $r(t)$ will be established. Then, if we set $d(t) = 0$, the error system (1.147) is transformed to the following form:

$$\begin{cases} e(t+1) = \sum_{i=1}^{N} \alpha_i(t)(\bar{A}_i e(t) + \bar{F}_i f(t)) \\ r(t) = \sum_{i=1}^{N} \alpha_i(t)(C_i e(t) + F_{2i} f(t)). \end{cases} \tag{1.200}$$

Theorem 1.9 *Given a scalar $\beta > 0$, the system (1.200) is asymptotically stable with the H_- performance index β if there exist positive definite matrices P_{1i}, P_{1j}, Hermitian matrices P_{3i}, $Q_{2i} > 0$, matrices S_i, W_i, and Ω_i, for $\forall i, j \in \{1, 2, \ldots, N\}$, which satisfy*

$$\begin{bmatrix} P_{1j} - \Omega_i - \Omega_i^T & (\Omega_i + S_i C_i)A_i - W_i C_i \\ * & -P_{1i} \end{bmatrix} < 0 \tag{1.201}$$

$$\begin{bmatrix} -P_{3i} & Q_{2i} - \Omega_i^T & 0 & 0 \\ * & \Upsilon_{fi} & \phi_i & 0 \\ * & * & -F_{2i}^T F_{2i} & \beta I \\ * & * & * & -I \end{bmatrix} < 0 \tag{1.202}$$

for the low-frequency fault $|\vartheta_f| \leq \vartheta_{fl}$;

$$\begin{bmatrix} -P_{3i} & e^{j\vartheta_{fc}}Q_{2i} - \Omega_i^T & 0 & 0 \\ * & \Upsilon_{mi} & \phi_i & 0 \\ * & * & -F_{2i}^T F_{2i} & \beta I \\ * & * & * & -I \end{bmatrix} < 0 \tag{1.203}$$

for the middle-frequency fault $\vartheta_{fm1} \leq \vartheta_f \leq \vartheta_{fm2}$*;*

$$\begin{bmatrix} -P_{3i} & -Q_{2i} - \Omega_i^T & 0 & 0 \\ * & \Upsilon_{hi} & \phi_i & 0 \\ * & * & -F_{2i}^T F_{2i} & \beta I \\ * & * & * & -I \end{bmatrix} < 0 \tag{1.204}$$

for the high-frequency fault $|\vartheta_f| \geq \vartheta_{fh}$*;*
where

$$\Upsilon_{li} = (\Omega_i + S_iC_i)A_i - W_iC_i + ((\Omega_i + S_iC_i)A_i - W_iC_i)^T - C_i^T C_i + P_{3i} - 2cos(\vartheta_{fl})Q_{2i}; \tag{1.205}$$

$$\Upsilon_{mi} = (\Omega_i + S_iC_i)A_i - W_iC_i + ((\Omega_i + S_iC_i)A_i - W_iC_i)^T - C_i^T C_i + P_{3i} - 2cos(\vartheta_{fw})Q_{2i}; \tag{1.206}$$

$$\Upsilon_{hi} = (\Omega_i + S_iC_i)A_i - W_iC_i + ((\Omega_i + S_iC_i)A_i - W_iC_i)^T - C_i^T C_i + P_{3i} + 2cos(\vartheta_{fh})Q_{2i}; \tag{1.207}$$

$$\phi_i = (\Omega_i + S_iC_i)F_{1i} - W_iF_{2i} + C_i^T F_{2i}; \tag{1.208}$$

$$\vartheta_{fc} = \frac{\vartheta_{fm1} + \vartheta_{fm2}}{2}, \quad \vartheta_{fw} = \frac{\vartheta_{fm1} - \vartheta_{fm2}}{2}. \tag{1.209}$$

Proof By using Lemma 1.6 and following the same process of proof as in Theorem 1.8, the conditions in (1.202)–(1.204) can be easily deduced. The detailed proof is thus omitted for the brevity. □

1.3.3.3 FD Filter Design

In this subsection, the finite-frequency FD filter in (1.140) is designed based on the prerequisites (1.144)–(1.146), (1.148)–(1.151), and (1.182).

Theorem 1.10 *Consider the discrete-time switched system (1.138). Given scalars* $\gamma > 0$ *and* $\beta > 0$*, there exists a finite-frequency FD filter (1.140) such that the error system (1.147) is asymptotically stable with* H_- *performance index* β *and* H_∞ *performance index* γ *if there exist positive definite matrices* P_{1i}*,* P_{1j}*, Hermitian matrices* P_{2i}*,* P_{3i}*,* $Q_{1i} > 0$*,* $Q_{2i} > 0$*, and matrices* S_i*,* W_i*,* Ω_i*, and* R_i*, for* $\forall i, j \in \{1, 2, \ldots, N\}$*, which satisfy the conditions (1.172), (1.173) (or (1.174), or (1.175)), (1.202) (or (1.203), or (1.204)), and*

$$(\Omega_i + S_i C_i)B_i - R_i = 0; \tag{1.210}$$
$$S_i \left[D_{2i} \; F_{2i} \right] = 0. \tag{1.211}$$

In this case, the parameters of the FD filter (1.140) are given by

$$E_i = \Omega_i^{-1} S_i; \tag{1.212}$$
$$G_i = \Omega_i^{-1} R_i; \tag{1.213}$$
$$N_i = (I + E_i C_i)A_i - \Omega_i^{-1} W_i C_i; \tag{1.214}$$
$$L_i = \Omega_i^{-1} W_i - N_i E_i. \tag{1.215}$$

Proof Define $R_i = \Omega_i G_i$, and pre-multiply Eq. (1.145) by Ω_i, then one can get (1.210). By the definition (1.182), one can get that the condition (1.211) is equivalent to (1.146). (1.212)–(1.215) can be easily deduced by Theorems 1.8 and 1.9. Then the proof is concluded. □

Remark 1.11 As a result, the FD filter design problem in Definition 1.7 can be summarized as follows: to minimize γ and maximize β such that the conditions (1.172), (1.210), (1.211), and

- (1.173) (or (1.174), or (1.175)), according to the frequency range of $d(t)$
- (1.202) (or (1.203), or (1.204)), according to the frequency range of $f(t)$

are satisfied.

Remark 1.12 One can note that complex parameters are involved in the H_∞/H_- conditions (1.174) and (1.203) for the middle-frequency range, which cannot be directly solved by using the LMI software. The following equivalent statements are employed to solve this problem:

- $\Xi < 0$
- $\begin{bmatrix} Re(\Xi) & Im(\Xi) \\ Im(\Xi) & Re(\Xi) \end{bmatrix} < 0,$

where $Re(\Xi)$ and $Im(\Xi)$, respectively, represent the real part and imaginary part of the complex matrix Ξ. Based on the equivalent substitution, the complex-valued matrix inequality can be transformed into strict LMI, which can be easily resolved.

1.3.4 FD Threshold Design

After completing the design of the FD filter, the last task is to construct the residual evaluation function $J(r(t))$ and a threshold J_{th} to evaluate whether a fault has happened or not. Firstly, the residual evaluation function is selected as

$$J(r(t)) = \|r(t)\|_2 = \sqrt{\frac{1}{t}\sum_{s=t_0}^{t_0+t} r^T(s)r(s)}, \tag{1.216}$$

where t_0 denotes the initial evaluation time instant, and t is the evaluation time step. Then the threshold is determined by

$$J_{th} = \sup_{d(t)\in l_2[0,\infty),\, f(t)=0} J(r(t)).s \tag{1.217}$$

Based on the above definitions, the occurrence of the fault can be detected by comparing $J(r(t))$ and J_{th} according to the following logical test:

$$J(r(t)) > J_{th} \Longrightarrow with\ \ faults \Longrightarrow alarm, \tag{1.218}$$

$$J(r(t)) \leq J_{th} \Longrightarrow no\ \ faults.s \tag{1.219}$$

Remark 1.13 Since the residual signal is generated by the filter, the threshold J_{th} is usually chosen a priori with a predesigned switching signal. Therefore, the threshold is a supremum of a chosen switching signal, $J(r(t))$ in fault-free conditions.

1.3.5 An Illustrative Example

In this section, a continuous stirred tank reactor (CSTR) control system is employed to verify the techniques represented in Fig. 1.15 [84]. A constant volume of the CSTR is supplied by a single inlet flow through a selector valve links up with two different source streams. At each time, the supervisor decides that the reactor will be supplied by one of the source streams. Suppose that the liquid volume is constant, the loss of heat can be negligible, perfectly mixing and a first-order reaction in reactant A, then the kinetic equations of the CSTR at each operating mode are described by

$$\dot{C}_A = \frac{q_i}{V}(C_{Afi} - C_A) - a_0 exp\left(-\frac{E}{RT}\right)C_A,$$

$$\dot{T} = \frac{q_i}{V}(T_{fi} - T) - a_1 exp\left(-\frac{E}{RT}\right)C_A + a_2(T_c - T),$$

where C_A is the reactant A concentration (mol/L), q_i, C_{Afi}, T_{fi} are, respectively, the ith mode's feed flow rate (L/min), concentration of the feed stream (mol/L) and reactor temperature of the feed stream (K). V, E, R, T, and T_c are the reactor' volume (L), the activation energy, the gas constant $(J/(mol\cdot K))$, the reactor temperature (K), and the coolant temperature (K), respectively. We select the nominal operating conditions corresponding to an unstable equilibrium point as $T_C^* = 300\,\text{K}$, $C_A^* = 0.5\,\text{mol/L}$, and $T^* = 350\,\text{K}$ for both modes. By setting a certain sampling time $T_s = 1\,\text{s}$ and considering the existence of a disturbance input and the fault, one can get

$$A_1 = \begin{bmatrix} 0.8021 & 0.2895 \\ -0.8685 & -0.2111 \end{bmatrix},\ A_2 = \begin{bmatrix} 1.2600 & 0.4445 \\ -3.5558 & -0.9624 \end{bmatrix},$$

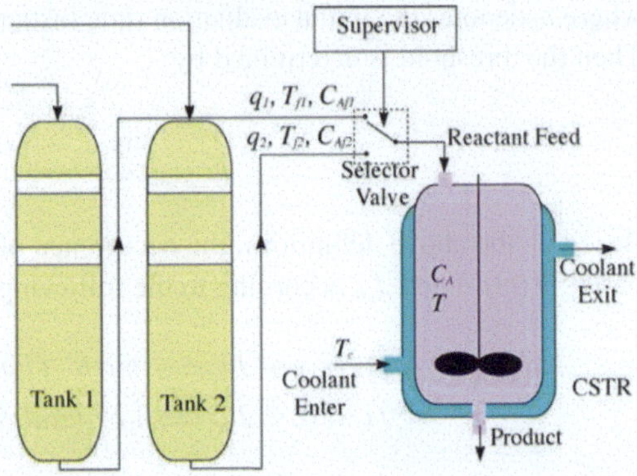

Fig. 1.15 Schematic diagram of the CSTR

and the other system matrices in (1.138) as

$$B_1 = \begin{bmatrix} 0.1 \\ 0.1 \end{bmatrix}, C_1 = \begin{bmatrix} 0.05 & 0.02 \\ 0 & 0.05 \end{bmatrix}, D_{11} = \begin{bmatrix} 0.2 \\ 0.1 \end{bmatrix},$$

$$D_{21} = \begin{bmatrix} 1 \\ 1 \end{bmatrix}, F_{11} = \begin{bmatrix} 1 \\ 1 \end{bmatrix}, F_{21} = \begin{bmatrix} 0.5 \\ 0.5 \end{bmatrix};$$

$$B_2 = \begin{bmatrix} 0.5 \\ 0.5 \end{bmatrix}, C_2 = \begin{bmatrix} 0.03 & 0.06 \\ 0 & 0.03 \end{bmatrix}, D_{12} = \begin{bmatrix} 0.1 \\ 0.2 \end{bmatrix},$$

$$D_{22} = \begin{bmatrix} 0.5 \\ 0.5 \end{bmatrix}, F_{12} = \begin{bmatrix} 0.5 \\ 0.5 \end{bmatrix}, F_{22} = \begin{bmatrix} 1 \\ 1 \end{bmatrix}.$$

The switching signal is plotted in Fig. 1.18. We assume that the disturbance input $d(t)$ is in high frequency, i.e., $|\vartheta_d| \geq \vartheta_{dh} = \frac{5\pi}{6}$, and the frequency of the fault $f(t)$ belongs to low-frequency domain, i.e., $|\vartheta_f| \leq \vartheta_{fl} = \frac{\pi}{6}$. Then by solving the conditions (1.175), (1.202), (1.210), and (1.211) in Theorem 1.10, one can get the minimum H_∞ performance $\gamma = 2.0176$ and maximum H_- performance $\beta = 0.8019$ with a set of feasible solutions as follows:

$$P_{11} = \begin{bmatrix} 55.6319 & 19.7398 \\ 19.7398 & 7.0535 \end{bmatrix}, \quad P_{12} = \begin{bmatrix} 0.8383 & 0.1480 \\ 0.1480 & 1.7907 \end{bmatrix},$$

$$P_{21} = \begin{bmatrix} 0.6901 & -0.7625 \\ -0.7625 & 3.3265 \end{bmatrix}, \quad P_{22} = \begin{bmatrix} 130.0480 & 38.9379 \\ 38.9379 & 12.0052 \end{bmatrix},$$

$$P_{31} = \begin{bmatrix} 0.9740 & 0.2101 \\ 0.2101 & 6.7280 \end{bmatrix}, \quad P_{32} = \begin{bmatrix} 202.3401 & 62.6920 \\ 62.6920 & 20.0366 \end{bmatrix},$$

$$Q_{11} = \begin{bmatrix} -1.4366 & 0.9463 \\ 0.9463 & -4.8304 \end{bmatrix}, \quad Q_{12} = \begin{bmatrix} -164.0461 & -52.2480 \\ -52.2480 & -17.2203 \end{bmatrix},$$

$$Q_{21} = \begin{bmatrix} -1.4366 & 0.9463 \\ 0.9463 & -4.8304 \end{bmatrix}, \quad Q_{22} = \begin{bmatrix} -164.0461 & -52.2480 \\ -52.2480 & -17.2203 \end{bmatrix},$$

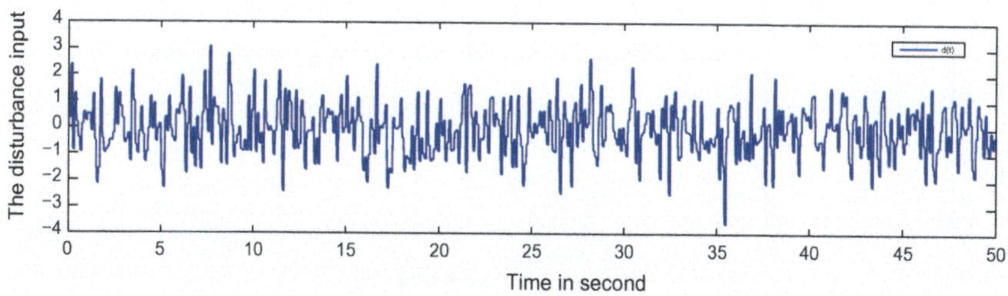

Fig. 1.16 The disturbance input $d(t)$

$$W_1 = \begin{bmatrix} 7.7003 & -6.8097 \\ -6.8097 & 9.7627 \end{bmatrix}, \quad W_2 = \begin{bmatrix} 26.8121 & 12.8175 \\ 12.8175 & 1.2735 \end{bmatrix},$$
$$\Omega_1 = \begin{bmatrix} 0.7555 & -0.2870 \\ -0.2870 & 1.8703 \end{bmatrix}, \quad \Omega_2 = \begin{bmatrix} 60.4719 & 21.2686 \\ 21.2686 & 7.5628 \end{bmatrix},$$
$$S_1 = \begin{bmatrix} -0.5 & 0.5 \\ 1 & -1 \end{bmatrix}, \quad S_2 = \begin{bmatrix} 1 & -1 \\ 0.5 & -0.5 \end{bmatrix}.$$

Then the filter's parameters in (1.140) can be calculated as

$$N_1 = \begin{bmatrix} 0.3022 & 0.4656 \\ -0.7278 & -0.3611 \end{bmatrix}, \quad N_2 = \begin{bmatrix} 1.7229 & 0.8744 \\ -4.9130 & -2.2793 \end{bmatrix},$$
$$G_1 = \begin{bmatrix} 0.2483 \\ 0.0785 \end{bmatrix}, \quad G_2 = \begin{bmatrix} -18.2384 \\ 51.7939 \end{bmatrix},$$
$$L_1 = \begin{bmatrix} 9.2875 & -7.3987 \\ -2.3938 & 4.2625 \end{bmatrix}, \quad L_2 = \begin{bmatrix} -14.5266 & 14.5295 \\ 42.1822 & -40.3270 \end{bmatrix},$$
$$E_1 = \begin{bmatrix} -0.4871 & 0.4871 \\ 0.4599 & -0.4599 \end{bmatrix}, \quad E_2 = \begin{bmatrix} -0.6164 & 0.6164 \\ 1.7996 & -1.7996 \end{bmatrix}.$$

For the simulation, the unknown disturbance input $d(t)$ is shown in Fig. 1.16, and the fault is set as

$$f(t) = \begin{cases} 0; & 0 \le t < 10 \\ 10; & 10 \le t, \end{cases} \tag{1.220}$$

which is described in Fig. 1.17. Figure 1.18 is the switching signal which specifies the switching between the two subsystems. Once a fault has happened as in (1.220), Fig. 1.19 describes the state vector $x(t)$ and its estimation $\hat{x}(t)$, it is clear that the observer gives a good estimation performance. The residual signal $r(t)$ is shown in Fig. 1.20 for the faulty and unfaulty cases. Figure 1.21 describes the residual evaluation function $J(r(t))$ with the faulty and unfaulty cases. By (1.217), one can choose the thresholds $J_{th}(r_1(t)) = 2.0615$ and $J_{th}(r_2(t)) = 2.6354$. According to the logical rules in (1.218) and (1.219) and selecting the threshold $J_{th}(r(t)) = \min\{J_{th}(r_1(t)), J_{th}(r_2(t))\}$, then the fault can be successfully detected.

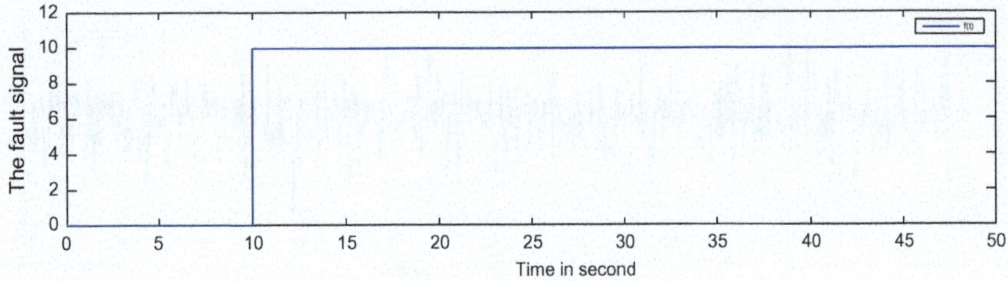

Fig. 1.17 The fault $f(t)$

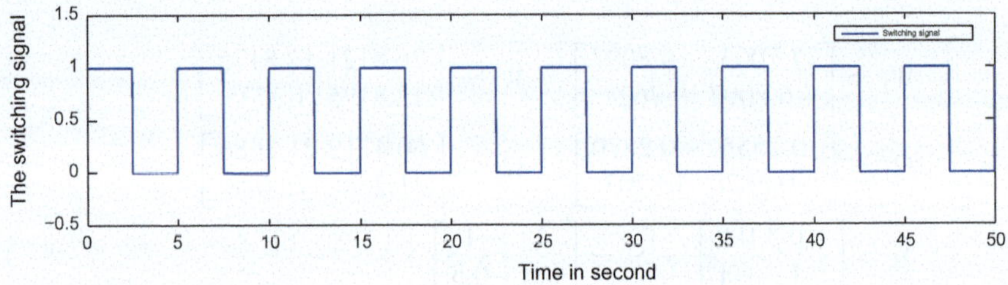

Fig. 1.18 The switching signal

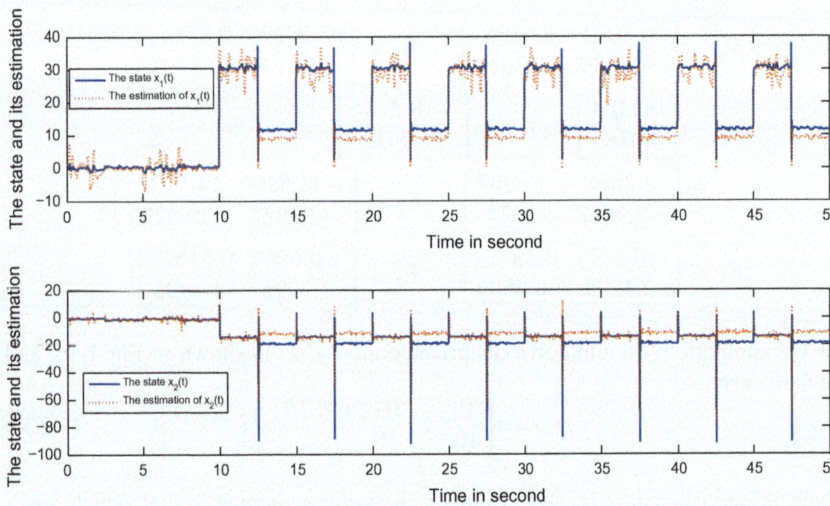

Fig. 1.19 The state $x(t)$ and its estimation $\hat{x}(t)$

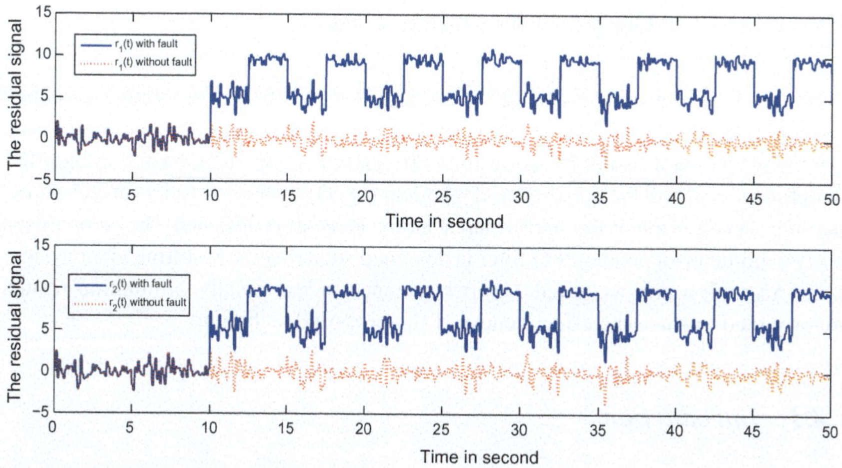

Fig. 1.20 The residual signal $r(t)$ with faulty and unfaulty cases

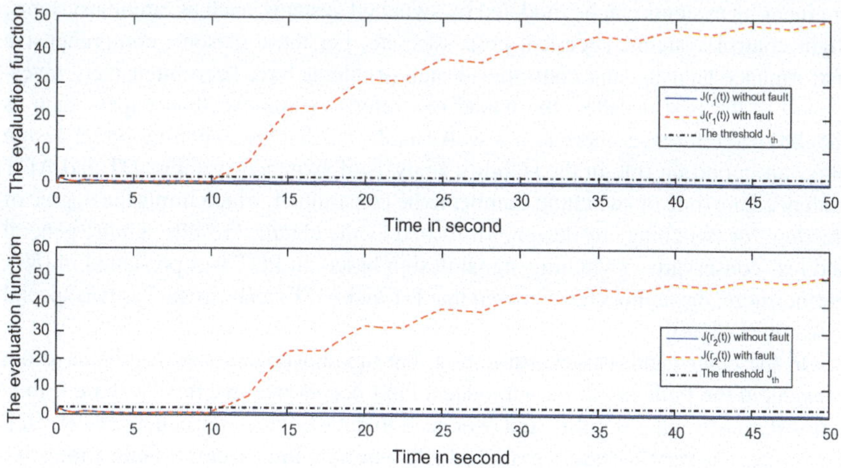

Fig. 1.21 The evaluation function $J(r(t))$

1.3.6 Summary

In this section, the issue of finite-frequency FD filter design for discrete-time switched systems is addressed. Based on the KYP lemma for switched systems, a novel finite-frequency FD filter design technique is derived, which can guarantee as robust as possible to the disturbance signal while being as sensitive as possible to the fault signal. Finally, the efficiency of the proposed techniques is shown in simulation results.

1.4 FD Filter Design in Nonlinear Case

In this section, the problem of FD filter design for a class of nonlinear switched systems is investigated. Firstly, the nonlinear switched system is transferred into a T-S fuzzy switched model by using fuzzy IF-THEN rules. Then, based on the PDT switching signal and the quasi-time-dependent (QTD) Lyapunov function technique, an efficient condition of the performance analysis result is obtained. Based on this, a fuzzy-parameter-dependent FD filter is designed such that the resulting error system is GUAS with a non-weighted H_∞ performance index. Finally, simulation results are provided to show the effectiveness of the proposed technique.

1.4.1 Introduction

Switched systems are a class of hybrid systems, which consists a collection of subsystems and a switching signal deciding the switching among them [2]. Many practical engineering systems can be modeled by switched systems, such as circuits systems, flight control systems, chemical processes, etc. For these reasons, comprehensive performance analysis and controller design problems have been intensively investigated in the past decades; the reader can refer to some excellent papers such as [85–88] and references therein. It is well established that the switching signal design plays an important role in the stability analysis of switched systems. DT and ADT both need the time or switching number to be constrained, which limits the degree of freedom for switching law design. Motivated by the obtained results, a more general and less conservative switching signal design based on PDT was presented in [89]. Furthermore, the authors have proved that DT and ADT can be treated as two special cases of PDT [90].

On the other hand, it is worth noting that the abovementioned results have not considered the fault cases. In fact, once a fault occurs in a practical system, it may seriously destroy the stability and even lead to catastrophic consequences [91, 92]. Therefore, it is very necessary to find an efficient technique to detect faults timely for the increasing demands of high safety and reliability in modern industry. In recent years, model-based FD techniques have been widely studied, and efficient approaches have been developed in the past few years [93–95]. Besides, many researchers have devoted themselves to investigate FD problems for switched system. In [72], an efficient condition of FD filter design for discrete-time linear switched systems with state delay under arbitrary switching signal was considered. By an ADT method, the results in [72] were extended to the time-varying delay case [96]. Meanwhile, similar results were provided for continuous-time switched systems in [97]. The classical technique to deal with FD problems consists in constructing an FD filter as residual generator. The residual signal is then compared to a predefined threshold. If it exceeds the threshold, then it is decided that a fault has occurred. Compared with fruitful research results for linear switched systems, there are few results devoted to nonlinear switched systems. By using the T-S fuzzy model and ADT techniques, a fuzzy FD filter design was proposed for continuous-time nonlinear switched stochastic systems in [60]. However, to our knowledge, the problem of FD for nonlinear discrete-time switched system has not been considered yet, which motivates us to investigate this challenging issue.

In this section, our attention is focused on designing an FD filter for nonlinear discrete-time switched systems under PDT switching. In order to design the FD filter, we firstly transfer the nonlinear switched system into a T-S fuzzy switched model through T-S fuzzy rules [98]. Then an observer-based fuzzy FD filter is proposed as a residual generator. Based on PDT switching and QTD Lyapunov function method, a model-dependent FD filter is designed such that the error-switched system is GUAS and with non-weighted H_∞ performance. Finally, to demonstrate the feasibility and effectiveness of the proposed method, a simulation example is included. The main contributions of this section can be summarized as follows:

- In comparison with ADT switching adopted in [60], a more general and less conservative PDT switching signal is utilized such that the error dynamic system is GUAS and fulfills a prescribed performance index;
- An FD filter design technique for continuous-time nonlinear switched systems is developed in [60]. In this section, we extend the results to the discrete-time case.

1.4.2 Model Description and Preliminaries

Consider the following class of nonlinear discrete-time switched systems:

$$\begin{cases} x(t+1) = \varphi_{\sigma(t)}(x(t), u(t), d(t), f(t)), \\ y(t) = \psi_{\sigma(t)}(x(t), u(t), d(t), f(t)), \end{cases} \tag{1.221}$$

where $x(t) \in \mathbb{R}^{n_x}$ is the state vector, $u(t) \in \mathbb{R}^{n_u}$, $d(t) \in \mathbb{R}^{n_d}$, and $f(t) \in \mathbb{R}^{n_f}$ are, respectively, the control input, the disturbance signal and the fault signal, which belong to $l_2[0, \infty)$, $y(t) \in \mathbb{R}^{n_y}$ is the measurement output vector. The piecewise constant function $\sigma(t) : [0, \infty) \rightarrow \phi = \{1, 2, \ldots, N\}$ is the switching signal, which is dependent on time t, state $x(t)$, or other logical signals, N is the number of the subsystems. In this section, we assume that the switching instants $t_0, t_1, t_2, \ldots, t_k, \ldots$ are known a priori. When $t \in [t_k, t_{k+1})$, it is said that the $\sigma(t_k)$th subsystem is activated in the time interval $[t_k, t_{k+1})$. φ_i and ψ_i $(\sigma(t) = i \in \phi)$ are a set of nonlinear functions.

In order to solve the FD problem, a T-S fuzzy model is employed to deal with the nonlinear switched systems (1.221). The T-S fuzzy model is described by fuzzy IF-THEN rules, whose collection represents the approximation of the nonlinear switched systems (1.221). The jth rule of the fuzzy model is

Rule j : IF $\theta_1^{[i]}(t)$ is $\mu_{j1}^{[i]}$ and $\theta_2^{[i]}(t)$ is $\mu_{j2}^{[i]}$ and $\cdots$ and $\theta_p^{[i]}(t)$ is $\mu_{jp}^{[i]}$, THEN

$$\begin{cases} x(t+1) = \sum_{j=1}^{q} h_j^{[i]}\left(\theta^{[i]}(t)\right)\Big[A_j^{[i]}x(t) + B_{uj}^{[i]}u(t) \\ \qquad\qquad + B_{dj}^{[i]}d(t) + E_{fj}^{[i]}f(t)\Big],\ i \in \phi \\ y(t) = \sum_{j=1}^{q} h_j^{[i]}\left(\theta^{[i]}(t)\right)\Big[C_j^{[i]}x(t) + D_{uj}^{[i]}u(t) \\ \qquad\qquad + D_{dj}^{[i]}d(t) + F_{fj}^{[i]}f(t)\Big],\ i \in \phi \end{cases} \tag{1.222}$$

where $A_j^{[i]}$, $B_{uj}^{[i]}$, $B_{dj}^{[i]}$, $E_{fj}^{[i]}$, $C_j^{[i]}$, $D_{uj}^{[i]}$, $D_{dj}^{[i]}$, $F_{fj}^{[i]}$ are constant real matrices with appropriate dimensions. $\theta^{[i]}(t) = \left[\theta_1^{[i]}(t), \theta_2^{[i]}(t), \ldots, \theta_p^{[i]}(t)\right]$, $\theta_k^{[i]}(t)$ $(k = 1, 2, \ldots, p)$ are the premise variables, $\mu_{jk}^{[i]}$ $(j = 1, 2, \ldots, q; k = 1, 2, \ldots, p)$ are the fuzzy sets, which are characterized by membership functions, q is the number of IF-THEN rules, p is the number of premise variables.

$$h_j^{[i]}\left(\theta^{[i]}(t)\right) = \frac{\upsilon_j^{[i]}\left(\theta^{[i]}(t)\right)}{\sum_{j=1}^{q} \upsilon_j^{[i]}\left(\theta^{[i]}(t)\right)}, \tag{1.223}$$

$$\upsilon_j^{[i]}\left(\theta^{[i]}(t)\right) = \prod_{k=1}^{p} \mu_{jk}^{[i]}\left(\theta_k^{[i]}(t)\right), \tag{1.224}$$

and $\mu_{jk}^{[i]}\left(\theta_k^{[i]}(t)\right)$ is the grade of membership of $\theta_k^{[i]}(t)$ and $\mu_{jk}^{[i]}$. We assume

$$\upsilon_j^{[i]}\left(\theta^{[i]}(t)\right) \geq 0, \quad \sum_{j=1}^{q} \upsilon_j^{[i]}\left(\theta^{[i]}(t)\right) > 0 \tag{1.225}$$

for any t. Hence, $h_j^{[i]}\left(\theta^{[i]}(t)\right)$ satisfies

$$h_j^{[i]}\left(\theta^{[i]}(t)\right) \geq 0, \quad \sum_{j=1}^{q} h_j^{[i]}\left(\theta^{[i]}(t)\right) = 1, \quad j = 1, 2, \ldots, q. \tag{1.226}$$

In this section, we mainly focus on the issue of FD for nonlinear switched systems (1.221) under PDT switching. In the following, the definition of PDT is recalled:

Definition 1.8 ([89]) Consider switching instants $t_0, t_1, \ldots$ with $t_0 = 0$. A positive constant τ is said to be a persistent dwell time if there exists an infinite number of disjoint intervals of length not smaller than τ on which $\sigma(t)$ is a constant, and consecutive intervals with this property are separated by no more than T, where T is called the period of persistence.

As illustrated in Fig. 1.22, the interval υ-stage includes two parts: τ-portion and T-portion in the PDT switching. During the τ-portion, a certain subsystem is activated and the T-portion can be regarded as a stage of arbitrary switchings. The running

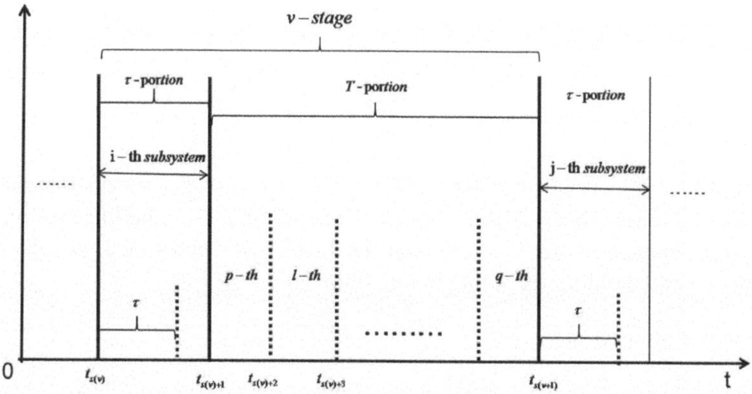

Fig. 1.22 PDT switching ($T^{(v)} \leq T$)

time of the subsystem in τ-portion is at least τ. In the T-portion, if we denote the actual running time at the v-stage as $T^{(v)}$, then the following equation holds:

$$T^{(v)} = \sum_{r=1}^{\mathcal{Q}(t_{s(v)+1}, t_{s(v+1)})} , T_{\sigma(t_{s(v)+r})} \leq T \tag{1.227}$$

where $t_{s(v+1)} = t_{s(v)+1} + T^{(v)}$, $T_{\sigma(t_{s(v)+r})}$ denotes the running time in T-portion of the subsystem activated at the switching instant $t_{s(v)+r} \in [t_{s(v)+1}, t_{s(v+1)})$; $\mathcal{Q}(t_{s(v)+1}, t_{s(v+1)})$ stands for the total number of switching within $[t_{s(v)+1}, t_{s(v+1)})$; $t_{s(v)+1}$ denotes the next switching instant after $t_{s(v)}$ at the v-stage and $t_{s(v+1)}$ stands the instant switching into the $(v+1)$-stage.

Remark 1.14 From Definition 1.8, one can obtain that the PDT switching limits the length of switching instants, which permits that the switching signal is arbitrary in the T-portion. Therefore, the PDT switching is easier to design and less conservative than ADT switching and DT switching. Besides, the possible negative impacts caused by the arbitrary switching in the T-portion can be tolerated by the τ-portion. Therefore, PDT switching is more general to guarantee the desired control objective of switched dynamic systems.

In order to detect the fault, we propose the following FD filter as the residual signal generator. The FD filter's premise variable is assumed to be the same as the plant's premise variable. Based on the same IF-THEN rules, the fuzzy parameter dependent filter is proposed:

Rule j : IF $\theta_1^{[i]}$ is $\mu_{j1}^{[i]}$ and $\theta_2^{[i]}$ is $\mu_{j2}^{[i]}$ and $\cdots$ and $\theta_p^{[i]}$ is $\mu_{jp}^{[i]}$, THEN

$$\begin{cases} x_F(t+1) = \sum_{j=1}^{q} h_j^{[i]}\left(\theta^{[i]}(t)\right)\left[A_{Fj}^{[i]}(\rho(t))x_F(t) + B_{Fj}^{[i]}(\rho(t))y(t)\right],\ i \in \phi \\ \tilde{r}(t) = \sum_{j=1}^{q} h_j^{[i]}\left(\theta^{[i]}(t)\right)\left[C_{Fj}^{[i]}(\rho(t))x_F(t) + D_{Fj}^{[i]}(\rho(t))y(t)\right],\ i \in \phi, \end{cases} \tag{1.228}$$

where $x_F(t) \in \mathbb{R}^{n_{x_F}}$ is the filter's state, $\tilde{r}(t) \in \mathbb{R}^{n_r}$ is the so-called residual signal. The matrices $A_{Fj}^{[i]}(\rho(t))$, $B_{Fj}^{[i]}(\rho(t))$, $C_{Fj}^{[i]}(\rho(t))$, and $D_{Fj}^{[i]}(\rho(t))$ are the filter parameters to be determined, and $\rho(t)$ is a scheduler for the activated subsystem, which can be simply computed online according to the following rules:

(i:) in the τ-portion,

$$\rho(t) = \begin{cases} t - t_{s(v)}, & t \in [t_{s(v)}, t_{s(v)} + \tau) \\ \tau, & t \in [t_{s(v)} + \tau, t_{s(v)+1}) \end{cases} \tag{1.229}$$

(ii:) in the T-portion,

$$\rho(t) = t - H_r, t \in [t_{s(v)+1}, t_{s(v+1)}), \tag{1.230}$$

where

$$\begin{aligned} H_r = \arg\{&\max(t_{s(v)+r} \mid t \geq t_{s(v)+r}, t_{s(v)+r} \in \\ &[t_{s(v)+1}, t_{s(v+1)}), 1 \leq r \leq \mathcal{Q}(t_{s(v)+1}, t_{s(v+1)}))\}. \end{aligned} \tag{1.231}$$

By the definition in (1.231), the actual running time of the $\sigma(t)$th subsystem in the T-portion, $T_{\sigma(t)} \in [1, \min(\tau - 1, T^{(v)})]$, satisfies that

$$T_{\sigma(t)} = \begin{cases} t_{s(v)+r+1} - H_r, & t \in [t_{s(v)+1}, t_{s(v+1)-1}), \\ t_{s(v+1)} - H_r, & t \in [t_{s(v+1)-1}, t_{s(v+1)}). \end{cases} \tag{1.232}$$

In order to detect the fault $f(t)$ more conveniently and fastly, sometimes one is more interested in the fault signal of a certain frequency interval, which can be formulated as $\hat{f}(z) = W_f(z)f(z)$ with $W_f(z)$ being a given stable weighting matrix. One minimal realization of $\hat{f}(z) = W_f(z)f(z)$ is

$$\begin{cases} \bar{x}(t+1) = A_w\bar{x}(t) + B_w f(t) \\ \hat{f}(t) = C_w\bar{x}(t) + D_w f(t) \end{cases} \tag{1.233}$$

where $\bar{x}(t) \in \mathbb{R}^{n_{\bar{x}}}$ is the state of weighted fault, $\hat{f}(t) \in \mathbb{R}^{n_f}$ is the weighted fault. A_w, B_w, C_w, and D_w are known constant matrices.

Remark 1.15 The introduction of $W_f(z)$ can limit the frequency ranges of interest, but the system performance would be improved, and the frequency characteristics required to reflect the emphases of different frequency ranges could be captured. Therefore, we are interested in minimizing the error between the residual signal and the weighted fault signal, rather than the error between the residual and the original fault signal.

Denoting $e(t) = \tilde{r}(t) - \hat{f}(t)$ and augmenting the model of system (1.221) to include the states of (1.228) and (1.233), the following augmented system is obtained:

$$\begin{cases} \xi(t+1) = \sum_{j=1}^{q} h_j^{[i]}\left(\theta^{[i]}(t)\right) \sum_{l=1}^{q} h_l^{[i]}\left(\theta^{[i]}(t)\right) \\ \qquad\qquad \left[A_{jl}^{[i]}(\rho(t))\xi(t) + B_{jl}^{[i]}(\rho(t))\omega(t)\right], i \in \phi \\ e(t) = \sum_{j=1}^{q} h_j^{[i]}\left(\theta^{[i]}(t)\right) \sum_{l=1}^{q} h_l^{[i]}\left(\theta^{[i]}(t)\right) \\ \qquad\qquad \left[C_{jl}^{[i]}(\rho(t))\xi(t) + D_{jl}^{[i]}(\rho(t))\omega(t)\right], i \in \phi \end{cases} \tag{1.234}$$

where

$$\xi(t) = \left[\bar{x}^T(t)\ x^T(t)\ x_F^T(t)\right]^T, \omega(t) = \left[u^T(t)\ d^T(t)\ f^T(t)\right]^T,$$

$$A_{jl}^{[i]} = \begin{bmatrix} A_w & 0 & 0 \\ 0 & A_j^{[i]} & 0 \\ 0 & B_{Fl}^{[i]}(\rho(t))A_j^{[i]} & A_{Fl}^{[i]}(\rho(t)) \end{bmatrix},$$

$$B_{jl}^{[i]} = \begin{bmatrix} 0 & 0 & B_w \\ B_{uj}^{[i]} & B_{dj}^{[i]} & E_{fj}^{[i]} \\ B_{Fl}^{[i]}(\rho(t))D_{uj}^{[i]} & B_{Fl}^{[i]}(\rho(t))D_{dj}^{[i]} & B_{Fl}^{[i]}(\rho(t))F_{fj}^{[i]} \end{bmatrix},$$

$$C_{jl}^{[i]} = \left[-C_w\ D_{Fl}^{[i]}(\rho(t))C_j^{[i]}\ C_{Fl}^{[i]}(\rho(t))\right],$$

$$D_{jl}^{[i]} = \left[D_{Fl}^{[i]}(\rho(t))D_{uj}^{[i]}\ D_{Fl}^{[i]}(\rho(t))D_{dj}^{[i]}\ D_{Fl}^{[i]}(\rho(t))F_{fj}^{[i]} - D_w\right].$$

Definition 1.9 ([99]) The switched system (1.234) with $\omega(t) = 0$ is GUAS under a certain switching signal $\sigma(t)$ if for initial condition $\xi(t_0)$, there exists a class of K_∞ function κ such that the solution of the system (1.221) satisfies the inequality $\|\xi(t)\| \le \kappa(\|\xi(t_0)\|)$, $\forall t \in \mathbb{Z}$ and $\|\xi(t)\| \to 0$ as $t \to \infty$.

Now, we formulate the problem of FD filter design as an H_∞ filter design problem: design filter (1.228) for system (1.221) such that the augmented system (1.234) is GUAS when $\omega(t) = 0$ and, under zero-initial condition, the minimum of $\bar{\gamma}$ is made in the feasibility of

$$\sup_{\omega(t)\neq 0} \frac{\|e(t)\|_2}{\|\omega(t)\|_2} < \bar{\gamma}.\ \ \bar{\gamma} > 0 \tag{1.235}$$

After designing the residual generator, the last step for a successful FD is the residual evaluation stage which includes an evaluation function and a threshold. In this section, the threshold J_{th} and the residual evaluation function $J_L(\tilde{r}(t))$ are selected as

$$J_L(\tilde{r}(t)) = \|\tilde{r}(t)\|_2 = \left(\sum_{t=t_0}^{t_0+L} \tilde{r}^T(t)\tilde{r}(t)\right)^{\frac{1}{2}}, \tag{1.236}$$

$$J_{th} = \sup_{d(t)\in l_2, f(t)=0} \|\tilde{r}(t)\|_2, \tag{1.237}$$

where t_0 denotes the initial evaluation time instant, and L is the evaluation time step. Based on this, the occurrence of faults can be detected by comparing $J_L(\tilde{r}(t))$ and J_{th} according to the following logical test:

$$J_L(\tilde{r}(t)) > J_{th} \Longrightarrow with \ \ faults \Longrightarrow alarm, \tag{1.238}$$

$$J_L(\tilde{r}(t)) \leq J_{th} \Longrightarrow no \ \ faults. \tag{1.239}$$

1.4.3 Main Results

1.4.3.1 System Performance Analysis

In this subsection, we will investigate the sufficient conditions of GUAS with non-weighted H_∞ performance for the system (1.234) under PDT switching.

Theorem 1.11 *Consider the augmented system (1.234) with* $0 < \alpha < 1$, $\mu \geq 1$, *and* $\gamma > 0$ *are given constant scalars, where* $(1-\alpha)\mu \geq 1$. *For a prescribed period of persistence* $T \geq 0$, *if there exist matrices* $P^{[i]}(\varsigma) > 0$ *and* Ω, $\varsigma = 0, 1, \ldots, T_j, \ldots, \tau - 1$, *such that the following inequalities hold*

$$P^{[i]}(0) \leq \mu P^{[\iota]}(T_j), \ P^{[i]}(0) \leq \mu P^{[\iota]}(\tau), \ \forall i, \iota \in \phi \tag{1.240}$$

$$\Phi_{jj}^{[i]}(\varsigma) < 0, \quad j = 1, \ldots, q, \ i \in \phi \tag{1.241}$$

$$\Phi_{jj}^{[i]}(\tau) < 0, \quad j = 1, \ldots, q, \ i \in \phi \tag{1.242}$$

$$\frac{1}{q-1}\Phi_{jj}^{[i]}(\varsigma) + \frac{1}{2}(\Phi_{jl}^{[i]}(\varsigma) + \Phi_{lj}^{[i]})(\varsigma) < 0, 1 \leq j < l \leq q, \ i \in \phi \tag{1.243}$$

$$\frac{1}{q-1}\Phi_{jj}^{[i]}(\tau) + \frac{1}{2}(\Phi_{jl}^{[i]}(\tau) + \Phi_{lj}^{[i]})(\tau) < 0, 1 \leq j < l \leq q, \ i \in \phi \tag{1.244}$$

where

$$\Phi_{jl}^{[i]}(\zeta) = \begin{bmatrix} P^{[i]}(\zeta+1) - He(\Omega) & 0 & \Omega^T A_{jl}^{[i]}(\zeta) & \Omega^T A_{jl}^{[i]}(\zeta) \\ * & -I & C_{jl}^{[i]}(\zeta) & D_{jl}^{[i]}(\zeta) \\ * & * & -(1-\alpha)P^{[i]}(\zeta+1) & 0 \\ * & * & * & -\gamma^2 I \end{bmatrix},$$

$$\Phi_{jl}^{[i]}(\tau) = \begin{bmatrix} P^{[i]}(\tau) - He(\Omega) & 0 & \Omega^T A_{jl}^{[i]}(\tau) & \Omega^T A_{jl}^{[i]}(\tau) \\ * & -I & C_{jl}^{[i]}(\tau) & D_{jl}^{[i]}((\tau) \\ * & * & -(1-\alpha)P^{[i]}(\tau) & 0 \\ * & * & * & -\gamma^2 I \end{bmatrix},$$

then the system (1.234) is GUAS and guarantees the performance index condition in (1.235) with

$$\bar{\gamma} = \gamma \sqrt{\mu^{[(T+1)+\frac{T+1}{T+\tau}]} \frac{\alpha}{1-(1-\alpha)\mu^{\frac{T+1}{T+\tau}}}} \tag{1.245}$$

for the PDT switching signals satisfying

$$(T+1)ln\mu + (T+\tau)\ln(1-\alpha) < 0 \tag{1.246}$$

Proof Based on the switching signal and the fuzzy basis function, we can get from the conditions in (1.241)–(1.244) that

$$\sum_{j=1}^{q} h_j^{[i]} \sum_{l=1}^{q} h_l^{[i]} \begin{bmatrix} P^{[i]}(\zeta+1) - He(\Omega) & 0 & \Omega^T A_{jl}^{[i]}(\zeta) & \Omega^T A_{jl}^{[i]}(\zeta) \\ * & -I & C_{jl}^{[i]}(\zeta) & D_{jl}^{[i]}(\zeta) \\ * & * & -(1-\alpha)P^{[i]}(\zeta+1) & 0 \\ * & * & * & -\gamma^2 I \end{bmatrix} < 0, \tag{1.247}$$

$$\sum_{j=1}^{q} h_j^{[i]} \sum_{l=1}^{q} h_l^{[i]} \begin{bmatrix} P^{[i]}(\tau) - He(\Omega) & 0 & \Omega^T A_{jl}^{[i]}(\tau) & \Omega^T A_{jl}^{[i]}(\tau) \\ * & -I & C_{jl}^{[i]}(\tau) & D_{jl}^{[i]}((\tau) \\ * & * & -(1-\alpha)P^{[i]}(\tau) & 0 \\ * & * & * & -\gamma^2 I \end{bmatrix} < 0. \tag{1.248}$$

We first establish the stability of system (1.234). To this end, we assume that $\omega(t) = 0$, and choose the following quasi-time-dependent Lyapunov function:

$$V_{\sigma(t)}(\rho(t), t) = \xi^T(t) P^{[\sigma(t)]}(\rho(t)) \xi(t) \tag{1.249}$$

we have

$$
\begin{aligned}
&V_{\sigma(t_{s(v+1)})}(\rho(t_{s(v+1)}), t_{s(v+1)}) \\
&\leq \mu V_{\sigma(t_{s(v)+r})}(\rho(t_{s(v)+1} + T^{(v)}), t_{s(v)+1} + T^{(v)}) \\
&\leq \mu(1-\alpha)^{T_{\sigma(t_{s(v)+r})}} V_{\sigma(t_{s(v)+r})}(\rho(t_{s(v)+r}), t_{s(v)+r}) \\
&\leq \mu^2(1-\alpha)^{T_{\sigma(t_{s(v)+r})}} V_{\sigma(t_{s(v)+r-1})}(\rho(t_{s(v)+r}), t_{s(v)+r}) \\
&\leq \mu^2(1-\alpha)^{T_{\sigma(t_{s(v)+r})}} (1-\alpha)^{T_{\sigma(t_{s(v)+r-1})}} \\
&\quad \times V_{\sigma(t_{s(v)+r-1})}(\rho(t_{s(v)+r-1}), t_{s(v)+r-1}) \\
&\leq \cdots\cdots\cdots \\
&\leq \mu^{\mathcal{Q}(t_{s(v)}, t_{s(v+1)})}(1-\alpha)^{t_{s(v+1)} - t_{s(v)+1}} \\
&\quad \times (1-\alpha)^{t_{s(v)+1} - t_{s(v)}} V_{\sigma(t_{s(v)})}(\rho(t_{s(v)}), t_{s(v)}).
\end{aligned} \tag{1.250}
$$

By Definition 1.8, we have

$$
\begin{aligned}
&V_{\sigma(t_{s(v+1)})}(\rho(t_{s(v+1)}), t_{s(v+1)}) \\
&\leq \mu^{\mathcal{Q}(t_{s(v)}, t_{s(v+1)})}(1-\alpha)^{T^{(v)}}(1-\alpha)^{\tau} V_{\sigma(t_{s(v)})}(\rho(t_{s(v)}), t_{s(v)}) \\
&\leq \mu^{T^{(v)}}(1-\alpha)^{T^{(v)}}(1-\alpha)^{\tau} V_{\sigma(t_{s(v)})}(\rho(t_{s(v)}), t_{s(v)}).
\end{aligned} \tag{1.251}
$$

According to $(1-\alpha)\mu \geq 1$ and $\mu \geq 1$, we have

$$
1 \leq \mu^{T^{(v)}}(1-\alpha)^{T^{(v)}} \leq \mu^{T}(1-\alpha)^{T}, \tag{1.252}
$$

which gives

$$
\begin{aligned}
&V_{\sigma(t_{s(v+1)})}(\rho(t_{s(v+1)}), t_{s(v+1)}) \\
&\leq \mu^{T}(1-\alpha)^{T}(1-\alpha)^{\tau} V_{\sigma(t_{s(v)})}(\rho(t_{s(v)}), t_{s(v)}) \\
&\leq \mu^{T+1}(1-\alpha)^{T+\tau} V_{\sigma(t_{s(v)})}(\rho(t_{s(v)}), t_{s(v)})
\end{aligned} \tag{1.253}
$$

Let $\mu^{T+1}(1-\alpha)^{T+\tau} = \lambda$, we have

$$
\begin{aligned}
&V_{\sigma(t_{s(v+1)})}(\rho(t_{s(v+1)}), t_{s(v+1)}) \\
&\leq \lambda V_{\sigma(t_{s(v)})}(\rho(t_{s(v)}), t_{s(v)}) \\
&\leq \cdots\cdots\cdots \\
&\leq \lambda^{v-1} V_{\sigma(t_{s(0)})}(\rho(t_{s(v)}), t_{s(0)}).
\end{aligned} \tag{1.254}
$$

It can be concluded that if (1.246) holds, we get $\lambda < 1$. Then based on Definitions 1.8 and 1.9, the system (1.234) is GUAS when $v \to \infty$.

For any nonzero $\omega(t) \in l_2[0, \infty)$ and zero-initial condition $\xi(0) = 0$, one has

$$\begin{aligned}&\Delta V_{\sigma(t)}(\rho(t),t)+\alpha V_{\sigma(t)}(\rho(t),t)\\&=V_{\sigma(t+1)}(\rho(t+1),t+1)-(1-\alpha)V_{\sigma(t)}(\rho(t),t).\end{aligned} \tag{1.255}$$

For $\sigma(t_{s(v)})=i$ and $\sigma(t_{s(v+1)})=\iota$, $i\neq\iota\in\phi$, we have

$$V_{\sigma(t_{s(v+1)})}(\rho(t_{s(v+1)}),t_{s(v+1)})=V_{\sigma(t_{s(v+1)})}(\rho(t_{s(v)}+T^{(v)},t_{s(v)}+T^{(v)})). \tag{1.256}$$

Then, for $\forall t\in[t_{s(v)},t_{s(v+1)})$ and $\rho(t)$, define

$$\Delta V_{\sigma(t)}(t-t_{s(v)},t)+\alpha V_{\sigma(t)}(t-t_{s(v)},t)+\Gamma(t)<0,\quad t\in[t_{s(v)},t_{s(v)}+\tau) \tag{1.257}$$

$$\Delta V_{\sigma(t)}(\tau,t)+\alpha V_{\sigma(t)}(\tau,t)+\Gamma(t)<0,\qquad t\in[t_{s(v)}+\tau,t_{s(v)+1}) \tag{1.258}$$

$$\begin{aligned}\Delta V_{\sigma(t)}(t-t_{s(v)+r},t)+\alpha V_{\sigma(t)}(t-t_{s(v)+r},t)+\Gamma(t)<0,\\ t\in[t_{s(v)+1},t_{s(v+1)}),\end{aligned} \tag{1.259}$$

where

$$\Gamma(t)=\tilde{r}^T(t)\tilde{r}(t)-\gamma^2\omega^T(t)\omega(t). \tag{1.260}$$

When $t\in[t_{s(v)+1},t_{s(v+1)})$, we have

$$\rho(t)=t-t_{s(v)+r}\leq T_{\sigma(t)}\leq\min(\tau-1,T^{(v)})\leq\tau-1. \tag{1.261}$$

From the conditions in (1.247) and (1.248), the conditions in (1.257)–(1.259) can be guaranteed. Then, based on (1.240) and (1.257)–(1.259), we have

$$\begin{aligned}&V_{\sigma(t_{s(v+1)})}(\rho(t_{s(v+1)}),t_{s(v+1)})\\&=V_{\sigma(t_{s(v+1)})}(\rho(t_{s(v)+1}+T^{(v)}),t_{s(v)+1}+T^{(v)})\\&\leq\mu V_{\sigma(t_{s(v)+r})}(\rho(t_{s(v)+1}+T^{(v)}),t_{s(v)+1}+T^{(v)})\\&\leq\mu(1-\alpha)^{T_{\sigma(t_{s(v)+r})}}V_{\sigma(t_{s(v)+r})}(\rho(t_{s(v)+1}+T^{(v)}\\&\quad-T_{\sigma(t_{s(v)+r})}),t_{s(v)+1}+T^{(v)}-T_{\sigma(t_{s(v)+r})})\\&\quad+\mu\sum_{s=t_{s(v)+r}}^{t_{s(v+1)}-1}(1-\alpha)^{t_{s(v+1)}-s-1}\Gamma(s)\\&\leq\mu^2(1-\alpha)^{T_{\sigma(t_{s(v)+r})}}V_{\sigma(t_{s(v)+r-1})}(\rho(t_{s(v)+1}+T^{(v)}\end{aligned}$$

$$
\begin{aligned}
& -T_{\sigma(t_{s(v)+r})}), t_{s(v)+1} + T^{(v)} - T_{\sigma(t_{s(v)+r})}) \\
& +\mu \sum_{s=t_{s(v)+r}}^{t_{s(v+1)}-1} (1-\alpha)^{t_{s(v+1)}-s-1} \Gamma(s) \\
\leq\ & \mu^2 (1-\alpha)^{T_{\sigma(t_{s(v)+r})}+T_{\sigma(t_{s(v)+r-1})}} V_{\sigma(t_{s(v)+r-1})} \\
& (\rho(t_{s(v)+1} + T^{(v)} - T_{\sigma(t_{s(v)+r})}) - T_{\sigma(t_{s(v)+r-1})}), \\
& t_{s(v)+1} + T^{(v)} - T_{\sigma(t_{s(v)+r})} - T_{\sigma(t_{s(v)+r-1})})) \\
& +\mu^2 (1-\alpha)^{T_{\sigma(t_{s(v)+r})}} \\
& \times \sum_{s=t_{s(v)+r-1}}^{t_{s(v)+r}-1} (1-\alpha)^{t_{s(v)+r-1}-s-1} \Gamma(s) \\
& +\mu \sum_{s=t_{s(v)+r}}^{t_{s(v+1)}-1} (1-\alpha)^{t_{s(v+1)}-s-1} \Gamma(s).
\end{aligned} \tag{1.262}
$$

Noting that $T_{\sigma(t_{s(v)+r})} = t_{s(v+1)} - t_{s(v)+r}$, one can get

$$
\begin{aligned}
& V_{\sigma(t_{s(v+1)})}(\rho(t_{s(v+1)}), t_{s(v+1)}) \\
\leq\ & \mu^{\mathcal{Q}(t_{s(v)+1}, t_{s(v+1)})} (1-\alpha)^{\sum_{m=1}^{\mathcal{Q}(t_{s(v)+1}, t_{s(v+1)})} T_{\sigma(t_{s(v)+m})}} \\
& \times V_{\sigma(t_{s(v)+1})}(\rho(t_{s(v+1)}), t_{s(v+1)}) \\
& + \mu^r \sum_{s=t_{s(v)+1}}^{t_{s(v)+2}-1} (1-\alpha)^{t_{s(v+1)}-t_{s(v)+2}+t_{(s(v)+2)}-s-1} \Gamma(s) \\
& + \cdots\cdots\cdots\cdots \\
& + \mu \sum_{s=t_{s(v)+r}}^{t_{s(v+1)}-1} (1-\alpha)^{t_{s(v+1)}-s-1} \Gamma(s).
\end{aligned} \tag{1.263}
$$

It follows

$$
\begin{aligned}
& V_{\sigma(t_{s(v+1)})}(\rho(t_{s(v+1)}), t_{s(v+1)}) \\
\leq & \sum_{s=t_{s(v)}}^{t_{s(v+1)}-1} \mu^{\mathcal{Q}(s, t_{s(v+1)})} (1-\alpha)^{t_{s(v+1)}-s-1} \Gamma(s) \\
& +\mu^{\mathcal{Q}(t_{s(v)}, t_{s(v+1)})} (1-\alpha)^{t_{s(v+1)}-t_{s(v)}} V_{\sigma(t_{s(v)})}(\rho(t_{s(v)}), t_{s(v)}).
\end{aligned} \tag{1.264}
$$

Let $t_{s(1)} = t_0$, we have

$$
\begin{aligned}
&V_{\sigma(t_{s(v)})}(\rho(t_{s(v)}), t_{s(v)}) \\
&\leq \sum_{s=t_{s(1)}}^{t_{s(v)}-1} \mu^{\mathscr{Q}(s,t_{s(v)})}(1-\alpha)^{t_{s(v)}-s-1}\Gamma(s) \\
&+\mu^{\mathscr{Q}(t_{s(1)},t_{s(v)})}(1-\alpha)^{t_{s(v)}-t_{s(1)}} V_{\sigma(t_{s(1)})}(\rho(t_{s(1)}), t_{s(1)}).
\end{aligned}
\tag{1.265}
$$

Under $\xi(0) = 0$ and $V_{\sigma(t_{s(v)})}(\rho(t_{s(v)}), t_{s(v)}) \geq 0$, one can get

$$
\sum_{s=t_{s(1)}}^{t_{s(v)}-1} \mu^{\mathscr{Q}(s,t_{s(v)})}(1-\alpha)^{t_{s(v)}-s-1}\Gamma(s) \leq 0 \tag{1.266}
$$

then it follows that

$$
\begin{aligned}
&\sum_{s=t_0}^{t-1} \mu^{\mathscr{Q}(s,t)}(1-\alpha)^{t-s-1}\tilde{r}^T(s)\tilde{r}(s) \\
&\leq \gamma^2 \sum_{s=t_0}^{t-1} \mu^{\mathscr{Q}(s,t)}(1-\alpha)^{t-s-1}\omega^T(s)\omega(s).
\end{aligned}
\tag{1.267}
$$

Noting that $\mu > 1$ and $\mathscr{Q}(s,t) \geq 0$, we have $\mu^{\mathscr{Q}(s,t)} \geq 1$. It means

$$
\begin{aligned}
&\sum_{s=t_0}^{t-1} (1-\alpha)^{t-s-1}\tilde{r}^T(s)\tilde{r}(s) \\
&\leq \gamma^2 \sum_{s=t_0}^{t-1} \mu^{\mathscr{Q}(s,t)}(1-\alpha)^{t-s-1}\omega^T(s)\omega(s)
\end{aligned}
\tag{1.268}
$$

Based on Definition 1.8, we have

$$
0 \leq \mathscr{Q}(s,t) \leq \frac{t-s}{\tau+T}(T+1) \leq \left[\frac{t-s}{\tau+T}+1\right](T+1). \tag{1.269}
$$

Then, it follows that

$$
\begin{aligned}
&\sum_{s=t_0}^{t-1} (1-\alpha)^{t-s-1}\tilde{r}^T(s)\tilde{r}(s) \\
&\leq \gamma^2 \sum_{s=t_0}^{t-1} \mu^{[\frac{t-s}{\tau+T}+1](T+1)}(1-\alpha)^{t-s-1}\omega^T(s)\omega(s) \\
&= \gamma^2 \sum_{s=t_0}^{t-1} \mu^{[\frac{t-s-1}{\tau+T}+\frac{1}{\tau+T}+1](T+1)}(1-\alpha)^{t-s-1}\omega^T(s)\omega(s)
\end{aligned}
$$

$$= \gamma^2 \sum_{s=t_0}^{t-1} \mu^{\left[\frac{t-s-1}{\tau+T}(T+1)\right]} \mu^{\frac{(T+1)+(T+1)(\tau+T)}{\tau+T}} (1-\alpha)^{t-s-1} \omega^T(s)\omega(s)$$

$$= \gamma^2 \mu^{\frac{(T+1)+(T+1)(\tau+T)}{\tau+T}} \sum_{s=t_0}^{t-1} \left[\mu^{\frac{T+1}{\tau+T}}(1-\alpha)\right]^{t-s-1} \omega^T(s)\omega(s). \tag{1.270}$$

Thus, we have

$$\sum_{t=t_0+1}^{\infty} \sum_{s=t_0}^{t-1} (1-\alpha)^{t-s-1} \tilde{r}^T(s)\tilde{r}(s) \tag{1.271}$$

$$\leq \gamma^2 \mu^{\frac{(T+1)+(T+1)(\tau+T)}{\tau+T}} \sum_{t=t_0+1}^{\infty} \sum_{s=t_0}^{t-1} \left[\mu^{\frac{T+1}{\tau+T}}(1-\alpha)\right]^{t-s-1} \omega^T(s)\omega(s). \tag{1.272}$$

Noting that $0 < 1-\alpha < 1$, (1.246) guarantees that

$$\mu^{\frac{T+1}{\tau+T}}(1-\alpha) < 1. \tag{1.273}$$

Then, (1.272) yields that

$$\sum_{t=t_0}^{\infty} \tilde{r}^T(t)\tilde{r}(t) \leq \gamma^2 \frac{\alpha}{1-\mu^{\frac{T+1}{\tau+T}}(1-\alpha)} \mu^{\frac{(T+1)+(T+1)(\tau+T)}{\tau+T}} \sum_{t=t_0}^{\infty} \omega^T(t)\omega(t), \tag{1.274}$$

which gives

$$\sum_{t=t_0}^{\infty} \tilde{r}^T(t)\tilde{r}(t) \leq \bar{\gamma}^2 \sum_{t=t_0}^{\infty} \omega^T(t)\omega(t), \tag{1.275}$$

which $\bar{\gamma}$ is defined in (1.245). (1.275) means that the performance defined in (1.235) is satisfied. □

1.4.3.2 FD Filter Design

Based on performance analysis in Theorem 1.11, in the following we propose a sufficient condition for the existence an FD filter in the form of (1.228).

Theorem 1.12 *Consider scalars* $0 < \alpha < 1$, $\mu \geq 1$, $\gamma > 0$, *and* $T > 0$ *with* $(1-\alpha)\mu \geq 1$. *Under PDT switching signal satisfying (1.246), the system (1.234) is GUAS and satisfies the condition in (1.235) with* $\bar{\gamma}$ *defined in (1.245), if there exist positive definite matrices* $\mathscr{P}^{[i]}(\delta) = \begin{bmatrix} \mathscr{P}^{[1i]}(\delta) & \mathscr{P}^{[2i]}(\delta) \\ \mathscr{P}^{[2i]T}(\delta) & \mathscr{P}^{[3i]}(\delta) \end{bmatrix}$, *matrices* $\mathscr{A}_{Fj}^{[i]}(\delta)$, $\mathscr{B}_{Fj}^{[i]}(\delta)$, $\mathscr{C}_{Fj}^{[i]}(\delta)$, $\mathscr{D}_{Fj}^{[i]}(\delta)$, R, S, *and* W ($\delta = 0, 1, \ldots, \tau$, $\zeta = 0, 1, \ldots, \tau-1$), *which satisfy*

$$\mathscr{P}^{[i]}(0) \leq \mu \mathscr{P}^{[\iota]}(T_j), \quad \mathscr{P}^{[i]}(0) \leq \mu \mathscr{P}^{[\iota]}(\tau), \quad \forall i, \iota \in \phi \tag{1.276}$$

$$\Pi_{jj}^{[i]}(\zeta) < 0, \qquad j = 1, \ldots, q, \ i \in \phi \tag{1.277}$$

$$\Pi_{jj}^{[i]}(\tau) < 0, \qquad j = 1, \ldots, q, \ i \in \phi \tag{1.278}$$

$$\frac{1}{q-1}\Pi_{jj}^{[i]}(\zeta) + \frac{1}{2}(\Pi_{jl}^{[i]}(\zeta) + \Pi_{lj}^{[i]})(\zeta) < 0, 1 \leq j < l \leq q, \ i \in \phi \tag{1.279}$$

$$\frac{1}{q-1}\Pi_{jj}^{[i]}(\tau) + \frac{1}{2}(\Pi_{jl}^{[i]}(\tau) + \Pi_{lj}^{[i]})(\tau) < 0, 1 \leq j < l \leq q, \ i \in \phi \tag{1.280}$$

where

$$\Pi_{jl}^{[i]}(\zeta) = \begin{bmatrix} \mathscr{P}^{[i]}(\zeta+1) - \Psi & 0 & \Upsilon_{13jl}^{[i]}(\zeta) & \Upsilon_{14jl}^{[i]}(\zeta) \\ * & -I & \Upsilon_{23jl}^{[i]}(\zeta) & \Upsilon_{24jl}^{[i]}(\zeta) \\ * & * & -(1-\alpha)\mathscr{P}^{[i]}(\zeta+1) & 0 \\ * & * & * & -\gamma^2 I \end{bmatrix},$$

$$\Pi_{jl}^{[i]}(\tau) = \begin{bmatrix} \mathscr{P}^{[i]}(\tau) - \Psi & 0 & \Upsilon_{13jl}^{[i]}(\tau) & \Upsilon_{14jl}^{[i]}(\tau) \\ * & -I & \Upsilon_{23jl}^{[i]}(\tau) & \Upsilon_{24jl}^{[i]}(\tau) \\ * & * & -(1-\alpha)\mathscr{P}^{[i]}(\tau) & 0 \\ * & * & * & -\gamma^2 I \end{bmatrix},$$

$$\mathscr{P}^{[i]}(\delta) = \begin{bmatrix} \mathscr{P}^{[1i]}(\delta) & \mathscr{P}^{[2i]}(\delta) \\ \mathscr{P}^{[2i]T}(\delta) & \mathscr{P}^{[3i]}(\delta) \end{bmatrix},$$

$$\Psi = \begin{bmatrix} R + R^T & S + W^T \\ W + S^T & W + W^T \end{bmatrix},$$

$$\Upsilon_{13jl}^{[i]}(\delta) = \begin{bmatrix} R^T \bar{A}_j^{[i]}(\delta)^T + \mathscr{B}_{Fl}^{[i]}(\delta)\bar{C}_j^{[i]}(\delta) & \mathscr{A}_{Fl}^{[i]}(\delta) \\ S^T \bar{A}_j^{[i]}(\delta)^T + \mathscr{B}_{Fl}^{[i]}(\delta)\bar{C}_j^{[i]}(\delta) & \mathscr{A}_{Fl}^{[i]}(\delta) \end{bmatrix},$$

$$\Upsilon_{14jl}^{[i]} = \begin{bmatrix} R^T \bar{B}_j^{[i]}(\delta) + \mathscr{B}_{Fl}^{[i]}(\delta)\bar{D}_j^{[i]}(\delta) \\ S^T \bar{B}_j^{[i]}(\delta) + \mathscr{B}_{Fl}^{[i]}(\delta)\bar{D}_j^{[i]}(\delta) \end{bmatrix},$$

$$\Upsilon_{23jl}^{[i]}(\delta) = \begin{bmatrix} \mathscr{D}_{Fl}^{[i]}(\delta)\bar{C}_j^{[i]}(\delta) - \begin{bmatrix} C_w & 0 \end{bmatrix} \mathscr{C}_{Fl}^{[i]}(\zeta) \end{bmatrix},$$

$$\Upsilon_{24jl}^{[i]}(\delta) = \mathscr{D}_{Fl}^{[i]}(\delta)\bar{D}_j^{[i]}(\delta) - \begin{bmatrix} 0 & 0 & D_w \end{bmatrix},$$

$$\bar{A}_j^{[i]}(\delta) = \begin{bmatrix} A_w & 0 \\ 0 & A_j^{[i]}(\delta) \end{bmatrix}, \bar{C}_j^{[i]}(\delta) = \begin{bmatrix} 0 & C_j^{[i]}(\delta) \end{bmatrix}$$

$$\bar{B}_j^{[i]}(\delta) = \begin{bmatrix} 0 & 0 & B_w \\ \bar{B}_{uj}^{[i]}(\delta) & \bar{B}_{dj}^{[i]}(\delta) & \bar{E}_{fj}^{[i]}(\delta) \end{bmatrix},$$

$$\bar{D}_j^{[i]}(\delta) = \begin{bmatrix} \bar{D}_{uj}^{[i]}(\delta) & \bar{D}_{dj}^{[i]}(\delta) & \bar{F}_{fj}^{[i]}(\delta) \end{bmatrix}.$$

In this case, a suitable FD filter in the form of (1.228) has the following parameters:

$$\begin{bmatrix} A_{Fj}^{[i]} & B_{Fj}^{[i]} \\ C_{Fj}^{[i]} & D_{Fj}^{[i]} \end{bmatrix} = \begin{bmatrix} \Omega_4^{-T} & 0 \\ 0 & I \end{bmatrix} \begin{bmatrix} \mathscr{A}_{Fj}^{[i]} & \mathscr{B}_{Fj}^{[i]} \\ \mathscr{C}_{Fj}^{[i]} & \mathscr{D}_{Fj}^{[i]} \end{bmatrix} \begin{bmatrix} \Omega_4^{-1}\Omega_3 & 0 \\ 0 & I, \end{bmatrix} \tag{1.281}$$

where Ω_3 and Ω_4 are any nonsingular matrices satisfying $\Omega_4^T\Omega_3^{-1}\Omega_4 = W$.

Proof Suppose that the LMIs (1.277) and (1.280) hold, we can get that $0 < \mathscr{P}^{[3i]}(\zeta+1) < W + W^T$, which means that matrix W is nonsingular. Therefore, one can find square and nonsingular matrices Ω_3 and Ω_4 satisfying $\Omega_4^T\Omega_3^{-1}\Omega_4 = W$. Define the following matrices

$$\begin{aligned} \Omega &= \begin{bmatrix} \Omega_1 & \Omega_2 \\ \Omega_4 & \Omega_3 \end{bmatrix}, \quad \Theta = \begin{bmatrix} I & 0 \\ 0 & \Omega_3^{-1}\Omega_4 \end{bmatrix}, \\ R &= \Omega_1, \quad S = \Omega_2\Omega_3^{-1}\Omega_4. \end{aligned} \tag{1.282}$$

From (1.281), we obtain that

$$\begin{bmatrix} \mathscr{A}_{Fj}^{[i]} & \mathscr{B}_{Fj}^{[i]} \\ \mathscr{C}_{Fj}^{[i]} & \mathscr{D}_{Fj}^{[i]} \end{bmatrix} = \begin{bmatrix} \Omega_4^{T} & 0 \\ 0 & I \end{bmatrix} \begin{bmatrix} A_{Fj}^{[i]} & B_{Fj}^{[i]} \\ C_{Fj}^{[i]} & D_{Fj}^{[i]} \end{bmatrix} \begin{bmatrix} \Omega_3^{-1}\Omega_4 & 0 \\ 0 & I. \end{bmatrix}$$

Without loss of generality, we assume

$$\mathscr{P}^{[i]}(\zeta) = \Theta^T P^{[i]}(\zeta)\Theta. \tag{1.283}$$

Substituting (1.282)–(1.283) yields

$$\begin{aligned} \Upsilon_{13jl}^{[i]}(\zeta) &= \left[\begin{matrix} \Omega_1^T\bar{A}_j^{[i]}(\zeta)^T + \Omega_4^T B_{Fl}^{[i]}(\zeta)\bar{C}_j^{[i]}(\zeta) \\ \Omega_4^T\Omega_3^{-T}\Omega_2^T\bar{A}_j^{[i]}(\zeta)^T + \Omega_4^T B_{Fl}^{[i]}(\zeta)\bar{C}_j^{[i]}(\zeta) \end{matrix}, \right. \\ &\qquad \left. \begin{matrix} \Omega_4^T A_{Fl}^{[i]}(\zeta)\Omega_3^{-1}\Omega_4 \\ \Omega_4^T\Omega_3^{-T}\Omega_2^T\bar{A}_j^{[i]}(\zeta)^T + \Omega_4^T B_{Fl}^{[i]}(\zeta)\bar{C}_j^{[i]}(\zeta) \end{matrix} \right], \\ \Upsilon_{14jl}^{[i]}(\zeta) &= \begin{bmatrix} \Omega_1^T\bar{B}_j^{[i]}(\zeta) + \Omega_4^T B_{Fl}^{[i]}(\zeta)\bar{D}_j^{[i]}(\zeta) \\ \Omega_4^T\Omega_3^{-T}\Omega_2^T\bar{B}_j^{[i]}(\zeta) + \Omega_4^T B_{Fl}^{[i]}(\zeta)\bar{D}_j^{[i]}(\zeta) \end{bmatrix}, \\ \Upsilon_{23jl}^{[i]}(\zeta) &= \left[D_{Fl}^{[i]}(\zeta)\bar{C}_j^{[i]}(\zeta) - \left[C_w \; 0\right] C_{Fl}^{[i]}(\zeta)\Omega_3^{-1}\Omega_4 \right], \\ \Upsilon_{24jl}^{[i]}(\zeta) &= \left[D_{Fl}^{[i]}(\zeta)\bar{D}_j^{[i]}(\zeta) - \left[0 \; 0 \; D_w\right] \right]. \end{aligned}$$

Then, one can conclude that the inequality in (1.277) is equivalent to

$$\bar{\Theta}^T \begin{bmatrix} P^{[i]}(\zeta+1) - (\Omega+\Omega^T) & 0 & \Omega^T A_{jl}^{[i]}(\zeta) & \Omega^T A_{jl}^{[i]}(\zeta) \\ * & -I & C_{jl}^{[i]}(\zeta) & D_{jl}^{[i]}(\zeta) \\ * & * & -(1-\alpha)P^{[i]}(\zeta+1) & 0 \\ * & * & * & -\gamma^2 I \end{bmatrix} \bar{\Theta} < 0 \tag{1.284}$$

where $\bar{\Theta} = diag\{\Theta, I, \Theta, I\}$,, which clearly guarantees the inequality in (1.247). By similar line, one can get that (1.278) is equivalent to (1.248). Thus, the proof is completed. □

1.4.4 An Illustrative Example

In this section, we introduce a single link robot arm example to illustrate the proposed technique [100]. The dynamic equation is given as follows:

$$\ddot{x}(t) = -\frac{m^k gl}{J^k} sin(x(t)) - \frac{D^k}{J^k} + \frac{1}{J^k} u(t),$$

where $x(t)$ is the angle of the arm, $u(t)$ is the control input. m^k is the mass, which makes the robot arm a switched system. J^k is the moment and D^k is the damping coefficient. The discrete state parameters $q^k = (m^k, J^k, D^k)$ are changing depending on the angle parameter $x(t)$. q^k takes the three values

$$q^1 = (1, 1, 2),$$
$$q^2 = (5, 5, 2),$$
$$q^3 = (10, 10, 2).$$

Let us transform the nonlinear switched robot arm system into a discrete-time switching fuzzy system with the local fuzzy weighting functions described in Fig. 1.23.

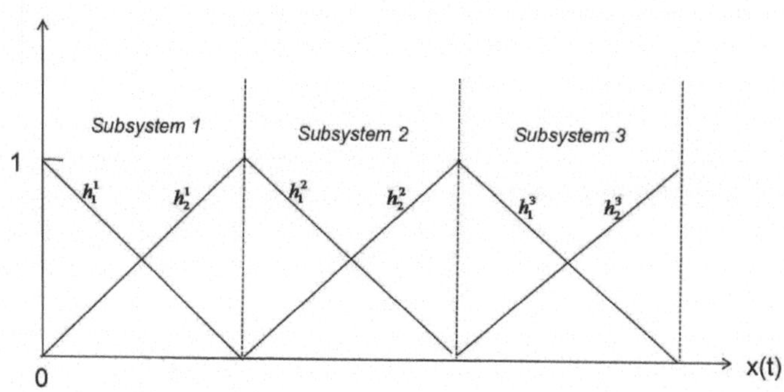

Fig. 1.23 The local fuzzy functions of robot arm

By linearization and the discretization with sampling time $T = 0.1$ s, one can get

$$\begin{cases} x(t+1) = \sum_{j=1}^{2} h_j^{[i]}\left(\theta^{[i]}(t)\right)\left[A_j^{[i]}x(t) + B_{uj}^{[i]}u(t) + B_{dj}^{[i]}d(t) + E_{fj}^{[i]}f(t)\right], \quad i = 1, 2, 3, \\ y(t) = \sum_{j=1}^{2} h_j^{[i]}\left(\theta^{[i]}(t)\right)\left[C_j^{[i]}x(t) + D_{uj}^{[i]}u(t) + D_{dj}^{[i]}d(t) + F_{fj}^{[i]}f(t)\right], \quad i = 1, 2, 3, \end{cases}$$

where *Subsystem 1 parameters are*

$$A_1^{[1]} = \begin{bmatrix} 1 & 0.1 \\ -0.49 & 0.8 \end{bmatrix}, B_{u1}^{[1]} = \begin{bmatrix} 0 \\ 0.1 \end{bmatrix}, B_{d1}^{[1]} = \begin{bmatrix} 0.1 \\ 0.1 \end{bmatrix}, E_{f1}^{[1]} = \begin{bmatrix} 0.1 \\ 0.1 \end{bmatrix},$$
$$C_1^{[1]} = \begin{bmatrix} 1 & 2 \end{bmatrix}, D_{u1}^{[1]} = 0.1, D_{d1}^{[1]} = 0.1, F_{f1}^{[1]} = 0.1;$$
$$A_2^{[1]} = \begin{bmatrix} 1 & 0.1 \\ 0.42 & 0.8 \end{bmatrix}, B_{u2}^{[1]} = \begin{bmatrix} 0 \\ 0.1 \end{bmatrix}, B_{d2}^{[1]} = \begin{bmatrix} 0.1 \\ 0.1 \end{bmatrix}, E_{f2}^{[1]} = \begin{bmatrix} 0.1 \\ 0.1 \end{bmatrix},$$
$$C_2^{[1]} = \begin{bmatrix} 1 & 2 \end{bmatrix}, D_{u2}^{[1]} = 0.1, D_{d2}^{[1]} = 0.1, F_{f2}^{[1]} = 0.1.$$

Subsystem 2 parameters are

$$A_1^{[2]} = \begin{bmatrix} 1 & 0.1 \\ -0.43 & 0.96 \end{bmatrix}, B_{u1}^{[2]} = \begin{bmatrix} 0 \\ 0.02 \end{bmatrix}, B_{d1}^{[2]} = \begin{bmatrix} 0.1 \\ 0.1 \end{bmatrix}, E_{f1}^{[2]} = \begin{bmatrix} 0.1 \\ 0.1 \end{bmatrix},$$
$$C_1^{[2]} = \begin{bmatrix} 1 & 2 \end{bmatrix}, D_{u1}^{[2]} = 0.1, D_{d1}^{[2]} = 0.1, F_{f1}^{[2]} = 0.1;$$
$$A_2^{[2]} = \begin{bmatrix} 1 & 0.1 \\ -0.25 & 0.96 \end{bmatrix}, B_{u2}^{[2]} = \begin{bmatrix} 0 \\ 0.02 \end{bmatrix}, B_{d2}^{[2]} = \begin{bmatrix} 0.1 \\ 0.1 \end{bmatrix}, E_{f2}^{[2]} = \begin{bmatrix} 0.1 \\ 0.1 \end{bmatrix},$$
$$C_2^{[2]} = \begin{bmatrix} 1 & 2 \end{bmatrix}, D_{u2}^{[2]} = 0.1, D_{d2}^{[2]} = 0.1, F_{f2}^{[2]} = 0.1.$$

Subsystem 3 parameters are:

$$A_1^{[3]} = \begin{bmatrix} 1 & 0.1 \\ -0.25 & 0.98 \end{bmatrix}, B_{u1}^{[3]} = \begin{bmatrix} 0 \\ 0.01 \end{bmatrix}, B_{d1}^{[3]} = \begin{bmatrix} 0.1 \\ 0.1 \end{bmatrix}, E_{f1}^{[3]} = \begin{bmatrix} 0.1 \\ 0.1 \end{bmatrix},$$
$$C_1^{[3]} = \begin{bmatrix} 1 & 2 \end{bmatrix}, D_{u1}^{[3]} = 0.1, D_{d1}^{[3]} = 0.1, F_{f1}^{[3]} = 0.1;$$
$$A_2^{[3]} = \begin{bmatrix} 1 & 0.1 \\ 0.01 & 0.98 \end{bmatrix}, B_{u2}^{[3]} = \begin{bmatrix} 0 \\ 0.1 \end{bmatrix}, B_{d2}^{[3]} = \begin{bmatrix} 0.1 \\ 0.1 \end{bmatrix}, E_{f2}^{[3]} = \begin{bmatrix} 0.1 \\ 0.1 \end{bmatrix},$$
$$C_2^{[3]} = \begin{bmatrix} 1 & 2 \end{bmatrix}, D_{u2}^{[3]} = 0.1, D_{d2}^{[3]} = 0.1, F_{f2}^{[3]} = 0.1.$$

Choose the fault weighting matrix

$$W(s) = \frac{0.5s}{s - 0.5}$$

which leads to $A_w = 0.5$, $B_w = 0.25$, $C_w = 1.0$, and $D_w = 0.5$.

To guarantee $\mu(1-\alpha) \geq 1$, we choose $\mu = 1.3$ and $\alpha = 0.2$. If we set the running time in τ-portion is $\tau = 4$ s, then according to the constrained switching condition

Table 1.2 The running time of τ-portion and T-portion

τ-portion	0s-4s	9s-13s	18s-22s	27s-30s
T-portion	4s-9s	13s-18s	22s-27s	—

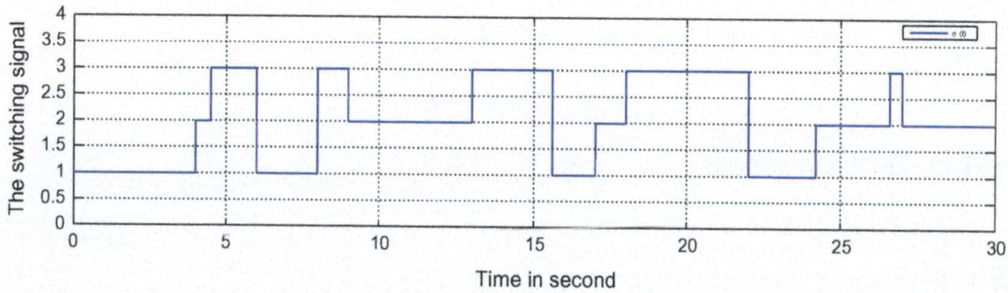

Fig. 1.24 The PDT switched signal $\sigma(t)$

(1.246), one can get $T = 5$ s satisfies the condition in (1.246). The actual running time of the PDT switching is described in the Table 1.2, and the switching signal is shown by Fig. 1.24.

By using MATLAB LMI Control Toolbox, we can get when the minimum performance index $\gamma = 0.96$, we obtain the parametric matrices of the desired FD filter as follows:

$$A_{F1}^{[1]} = \begin{bmatrix} 0.7643 & 5.9357 \\ 0.3479 & 0.3690 \end{bmatrix}, B_{F1}^{[1]} = \begin{bmatrix} -1.9654 \\ -0.2467 \end{bmatrix},$$
$$C_{F1}^{[1]} = \begin{bmatrix} -0.6321 & -0.4699 \end{bmatrix}, D_{F1}^{[1]} = -0.1533;$$
$$A_{F2}^{[1]} = \begin{bmatrix} 0.2589 & 3.8964 \\ 0.9752 & 0.3695 \end{bmatrix}, B_{F2}^{[1]} = \begin{bmatrix} -1.3679 \\ -0.9654 \end{bmatrix},$$
$$C_{F2}^{[1]} = \begin{bmatrix} -0.2467 & -0.0754 \end{bmatrix}, D_{F2}^{[1]} = -0.0221.$$
$$A_{F1}^{[2]} = \begin{bmatrix} 0.3679 & 2.0456 \\ 0.5683 & 0.3573 \end{bmatrix}, B_{F1}^{[2]} = \begin{bmatrix} -1.3687 \\ -0.1102 \end{bmatrix},$$
$$C_{F1}^{[2]} = \begin{bmatrix} -0.2467 & -0.6832 \end{bmatrix}, D_{F1}^{[2]} = -0.3578;$$
$$A_{F2}^{[2]} = \begin{bmatrix} 0.8965 & 3.0875 \\ 0.3578 & 0.8642 \end{bmatrix}, B_{F2}^{[2]} = \begin{bmatrix} -0.9326 \\ -0.1102 \end{bmatrix},$$
$$C_{F2}^{[2]} = \begin{bmatrix} -0.6421 & -0.8436 \end{bmatrix}, D_{F2}^{[2]} = -0.1469.$$
$$A_{F1}^{[3]} = \begin{bmatrix} 0.6739 & 2.8421 \\ 0.8313 & 0.2579 \end{bmatrix}, B_{F1}^{[3]} = \begin{bmatrix} -1.3687 \\ -0.1102 \end{bmatrix},$$
$$C_{F1}^{[3]} = \begin{bmatrix} -0.0024 & -0.0006 \end{bmatrix}, D_{F1}^{[3]} = -0.1421;$$
$$A_{F2}^{[3]} = \begin{bmatrix} 0.5739 & 1.9753 \\ 0.2589 & 0.2471 \end{bmatrix}, B_{F2}^{[3]} = \begin{bmatrix} -1.3687 \\ -0.1102 \end{bmatrix},$$

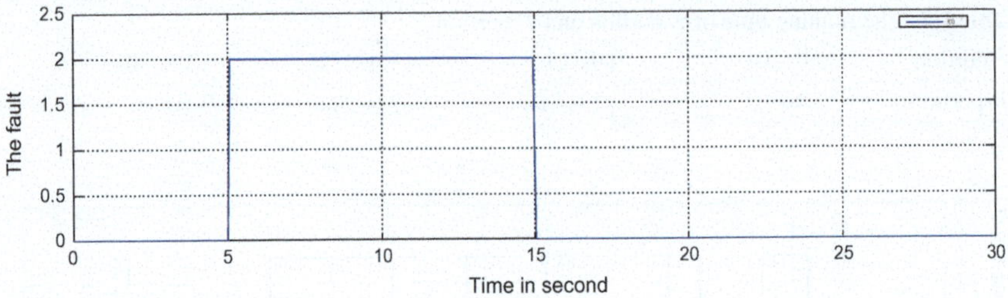

Fig. 1.25 The fault signal $f(t)$

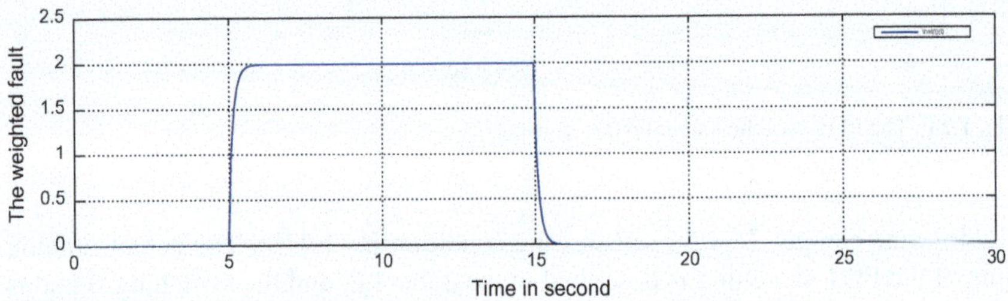

Fig. 1.26 The weighted fault signal $\hat{f}(t)$

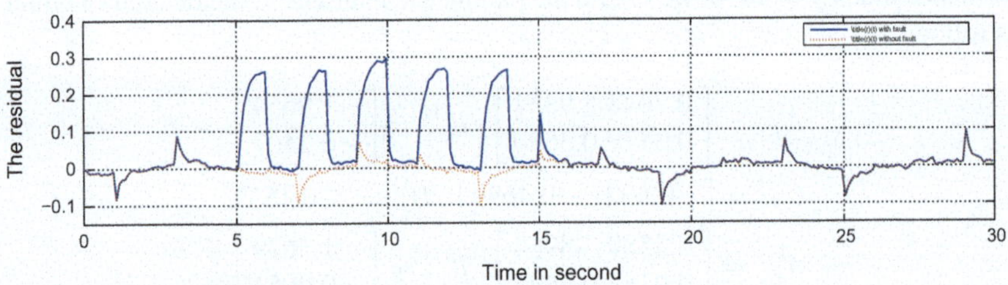

Fig. 1.27 The residual signal $\tilde{r}(t)$

$$C_{F2}^{[3]} = \begin{bmatrix} -0.3124 & -0.2156 \end{bmatrix}, \; D_{F2}^{[3]} = -0.0784.$$

We set the control input $u(t) = 0.5cos(t)$, the output disturbance signal $d(t)$ is considered as $d(t) = rand[0, 1]$, and the fault is set up as

$$f(t) = \begin{cases} 2, & 5 \le t \le 15 \\ 0, & others \end{cases}$$

then the fault signal $f(t)$ and weighted fault signal $\hat{f}(t)$ are, respectively, described in Figs. 1.25 and 1.26. Figure 1.27 shows the response of residual signal $\tilde{r}(t)$ with fault and without fault $f(t)$. Figure 1.28 presents the residual evaluation function $J_L(\tilde{r}(t))$

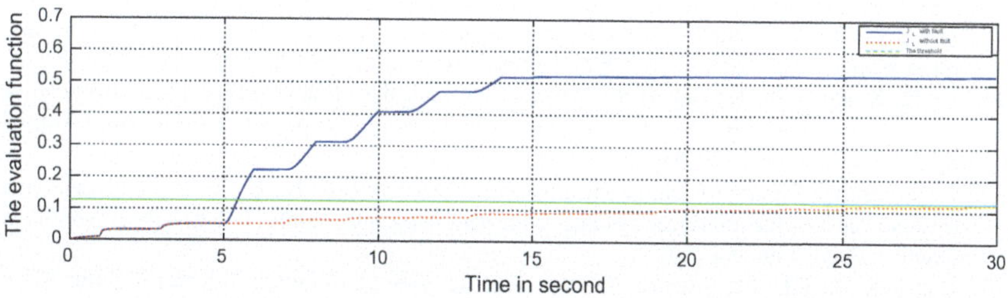

Fig. 1.28 The evaluation function $J_L(\tilde{r}(t))$

for both the faulty case (solid line) and fault-free case (dashed line). With a selected threshold $J_{th} = 0.125$, the simulation results show that $J_L(\tilde{r}(15.6)) = 0.1294 > J_{th}$, which means that the fault $f(t)$ can be successfully detected one time step after its occurrence.

1.4.5 Summary

In this section, we propose an FD filter design method for nonlinear switched discrete-time systems. Based on PDT and QTD Lyapunov function techniques, sufficient conditions are obtained to make the error system GUAS with a prescribed performance. Finally, a simulation example is provided to illustrate the proposed technique.

References

1. Branicky, M.S.: Multiple Lyapunov functions and other analysis tools for switched and hybrid systems. IEEE Trans. Autom. Control **43**(4), 475–482 (1998)
2. Liberzon, D., Morse, A.S.: Basic problems in stability and design of switched systems. IEEE Control Syst. Mag. **19**(5), 59–70 (1999)
3. Morse, A.S.: Supervisory control of families of linear set-point controllers part I: exact matching. IEEE Trans. Autom. Control **41**(10), 1413–1431 (1996)
4. Yang, H., Jiang, B.: On stability of non-linear and switched parabolic systems. IET Control Theory Appl. **7**(5), 749–756 (2013)
5. Zhang, B., Zheng, W., Xu, S.: Filtering of Markovian jump delay systems based on a new performance index. IEEE Trans. Circuits Syst.-I: Regul. Pap. **60**(5), 1250–1263 (2013)
6. Qin, C., Xiang, Z.R., Karimi, H.R.: Finite-time H_∞ control for switched systems with time-varying delay using delta operator approach. Int. J. Control Autom. Syst. **12**(6), 1150–1159 (2014)
7. Hou, L.L., Zong, G.D., Wu, Y.Q.: Finite-time control for discrete-time switched systems with time delay. Int. J. Control Autom. Syst. **10**(4), 855–860 (2012)
8. Zong, G.D., Ren, H., Hou, L.L.: Finite-time stability of interconnected impulsive switched systems. IET Control Theory Appl. **10**(6), 648–654 (2016)
9. Aouaouda, S., Chadli, M., Karimi, H.R.: Robust static output-feedback controller design against sensor failure for vehicle dynamics. IET Control Theory Appl. **8**(9), 728–737 (2014)

10. Jiang, J., Yu, X.: Fault-tolerant control systems: s comparative study between active and passive approaches. IFAC Annu. Rev. Control **36**(1), 60–72 (2012)
11. Li, S., Xiang, Z.R., Karimi, H.R.: Mixed $l - l_1$ fault detection observer design for positive switched systems with time-varying delay via delta operator approach. Int. J. Control Autom. Syst. **12**(4), 709–721 (2014)
12. Aouaouda, S., Chadli, M., Shi, P., Karimi, H.R.: Discrete-time $H_{-}H_{\infty}$ sensor fault detection observer design for nonlinear systems with parameter uncertainty. Int. J. Robust Nonlinear Control **25**(3), 339–361 (2015)
13. Busch, R., Peddle, I.K.: Active fault detection for open loop stable LTI SISO systems. Int. J. Control Autom. Syst. **12**(2), 324–332 (2014)
14. Yin, Y., Shi, P., Liu, F., Teo, K.L.: Robust fault detection for discrete-time stochastic systems with non-homogeneous jump processes. IET Control Theory Appl. **8**(1), 1–10 (2014)
15. Chen, J., Patton, R.J.: Standard H_{∞} filtering formulation of robust fault detection. In: Proc. Safe Process, Budapest, Hungary, pp. 256–261 (2000)
16. Takahashi, M., Takagi, T., Miyawaki, Y.: Fault diagnosis of joint-sensors in parallel manipulators. ICIC Express Lett. **3**(4), 1309–1314 (2009)
17. Jin, X., Yang, G.: Robust adaptive fault-tolerant compensation control with actuator failures and bounded disturbances. Acta Automatica Sinica **35**(3), 305–309 (2009)
18. Steven, X.D., Zhong, M., Tang, B., Zhang, P.: An LMI approach to the design of fault detection filter for time-delay LTI systems with unknown inputs. In: Proceedings of the American Control Conference Arlington, pp. 25–27 (2001)
19. Zhong, M., Ye, H., Shi, P., Wang, G.: Fault detection for Markovian jump systems. IEE Proc. Control Theory Appl. **152**(4), 397–402 (2005)
20. Zhao, Y., James, L., Gao, H.: Fault detection for fuzzy systems with intermittent measurements. IEEE Trans. Fuzzy Syst. **17**(2), 398–410 (2009)
21. Nguang, S.K., Shi, P.: Delay dependent fault estimation for uncertian time delay nonlinear systems: an LMI approach. Int. J. Robust Nonlinear Control **16**(18), 913–933 (2006)
22. Casavola, A., Famularo, D., Franzè, G.: A robust deconvolution scheme for fault detection and isolation of uncertain linear systems: an LMI approach. Automatica **41**(8), 1463–1472 (2005)
23. Zhong, M., Ding, S.X., Lam, J., Wang, H.: An LMI approach to design robust fault detection filter for uncertain LTI systems. Automatica **39**(3), 543–550 (2003)
24. Dai, L.: Singular Control Systems. Springer, Berlin (1989)
25. Ma, S., Zhang, C., Wu, Z.: Delay-dependent stability and H_{∞} control for uncertain discrete switched singular systems with time-delay. Appl. Math. Comput. **206**(1), 413–424 (2008)
26. Shi, S., Zhang, Q., Yuan, Z., Liu, W.: Hybrid impulsive control for switched singular systems. IET Control Theory Appl. **5**(1), 103–111 (2011)
27. Hwang, I., Kim, S., Seah, Y.: A survey of fault detection, isolation, and reconfiguration methods. IEEE Trans. Control Syst. Technol. **18**, 636–653 (2010)
28. Wang, H., Yang, G.: A finite frequency domain approach to fault detection for linear discrete-time systems. Asian J. Control **81**(7), 1162–1171 (2008)
29. Li, X., Yang, J.: Fault detection for T-S fuzzy systems with unknown membership functions. IEEE Trans. Fuzzy Syst. **22**, 139–152 (2014)
30. Model-based fault detection and diagnosis status and applications: Lsennann. Annu. Rev. Control **29**, 71–85 (2005)
31. Yeu, I., Kim, T., Seong, H., Shigeyasu, K.: Fault detection, isolation and reconstruction for descriptor systems. Asian J. Control **7**(4), 356–C367 (2005)
32. Nguang, S.K., Shi, P.: Fault detection for uncertain fuzzy systems: an LMI approach. IEEE Trans. Fuzzy Syst. **15**, 1251–1262 (2007)
33. Khosrowjerdi, M.J., Nikoukhah, R., Safari-Shad, N.: A mixed H_2/H_{∞} approach to simultaneous fault detection and control. Automatica **40**, 261–267 (2004)
34. Gertler: A survey of model-based failure detection and isolation in complex plants. IEEE Control Syst. Mag. **8**, 3–11 (1988)

35. Gao, N.: Robust fault detection for descriptor Markovian jump systems. In: Proceedings of IEEE International Conference on Control and Automation, pp. 835–840 (2007)
36. Wu, L.: Weighted H_∞ filtering of switched systems with time-varying delay: average dwell time approach. Circuits Syst. Signal Process. **28**, 1017–1036 (2009)
37. Zhang, K., Jiang, B., Cocquempot, V.: Distributed fault estimation observer design for multi-agent systems with switching topologies. IET Control Theory Appl. **11**(16), 2801–2807 (2017)
38. Liu, L., Guo, R., Ma, S.: Input/output to state stability of switched nonlinear systems with an improved average dwell time approach. Int. J. Control Autom. Syst. **14**(2), 461–468 (2016)
39. Farhat, A., Koenig, D.: Robust fault detection for uncertain switched systems. IFAC-PapersOnLine **50**(1), 15265–15270 (2017)
40. Hespanha, J.P., Morse, A.S.: Stability of switched systems with average dwell time. IEEE Conf. Decis. Control **3**(3), 2655–2660 (2002)
41. Gritli, W., Gharsallaoui, H., Benrejeb, M.: Fault detection based on unknown input observers for switched discrete-time systems. In: 2017 International Conference on Advanced Systems and Electric Technologies, Hammamet, pp. 436–441 (2017)
42. Du, D., Xu, S.: Actuator fault detection for discrete-time switched linear systems with output disturbance. Int. J. Control Autom. Syst. **15**(6), 2590–2598 (2017)
43. Jiang, B., Du, D., Cocquempot, V.: Fault detection for discrete-time switched systems with interval time-varying delays. Int. J. Control Autom. Syst. **9**(2), 396–401 (2011)
44. Benzaouia, A., Eddoukali, Y.: Robust fault detection and control for continuous-time switched systems with average dwell time. Circuits Syst. Signal Process. **37**(6), 2357–2373 (2018)
45. Wang, J., Shen, Y., Zhang, M.: Fault detection for linear switched systems using unknown input observers. In: Proceedings of the 33rd Chinese Control Conference, Nanjing, pp. 3219–3223 (2014)
46. Wang, D., Shi, P., Wang, W.: Robust fault detection for continuous-time switched delay systems: an linear matrix inequality approach. IET Control Theory Appl. **4**(1), 100–108 (2010)
47. Zhong, G., Yang, G.: Robust control and fault detection for continuous-time switched systems subject to a dwell time constraint. Int. J. Robust Nonlinear Control **25**(18), 3799–3817 (2016)
48. Zhai, D., Li, L., Zhang, J.: Fault detection for stochastic parameter-varying Markovian jump systems with application to networked control systems. Appl. Math. Model. **40**, 2368–2383 (2016)
49. Hespanha, J.P., Morse, A.S.: Stability of switched systems with average dwell time. In: Proc. 38th Conf. decision control, Phoenix, AZ, pp. 2655–2660 (1999)
50. Iwasaki, T., Hara, S., Fradkov, A., LoConte, L.: Restricted frequency inequality is equivalent to restricted dissipativity. In: 43rd IEEE Conference on Decision and Control, pp. 426–431 (2004)
51. Gahinet, P., Apkarian, P.: An linear matrix inequality approach to H_∞ control. Int. J. Robust Nonlinear Control **4**, 422–448 (1994)
52. Zhou, L., Ho, D.W.C., Zhai, G.: Stability analysis of switched linear singular systems. Automatica **49**, 1481–1487 (2013)
53. Tseng, C.S., Chen, B.S.: L_∞Z gain fuzzy control for nonlinear dynamic systems with persistent bounded disturbance. In: Proceedings of the 13th IEEE International Conference on Fuzzy Systems. Budapest, Hungary, Luly, pp. 783–788 (2004)
54. He, D., Xi, C., Lu, A.: Fault detection for switched T-S fuzzy systems in finite frequency domain. In: Advances in Difference Equations (2016)
55. Fran, P.M., Ding, X.: Survey of robust residual generation and evaluation methods in observer-based fault detection systems. J. Process Control **7**, 402–424 (1997)
56. Mahmoudi, A., Momeni, A., Aghdam, A.G., Gohari, P.: Switching between finite-time observers. Eur. J. Control **14**(4), 297–307 (2008)
57. Jin, X., Wang, S., Qin, J., Zheng, W., Kang, Y.: Adaptive fault-tolerant consensus for a class of uncertain nonlinear second-order multi-agent systems with circuit implementation. IEEE Trans. Circuits Syst. I: Regul. Pap. **65**(12), 2243–2255 (2018)

58. Qin, J., Zhang, G., Zheng, W., Kang, Y.: Adaptive sliding mode consensus tracking for second-order nonlinear multiagent systems with actuator faults. IEEE Trans. Cybern. **49**(5), 1605–1615 (2019)
59. Wang, Y., Shi, P., Lim, C.C., Liu, Y.: Event-triggered fault detection filter design for a continuous-time networked control system. IEEE Trans. Cybern. **46**(12), 3414–3426 (2016)
60. Su, X., Shi, P., Wu, L., Song, Y.: Fault detection filtering for nonlinear switched stochastic systems. IEEE Trans. Autom. Control **61**(5), 1310–1315 (2016)
61. Li, L., Ding, S.X., Qiu, J., Yang, Y.: Real-time fault detection approach for nonlinear systems and its asynchronous T-S fuzzy observer-based implementation. IEEE Trans. Cybern. **47**(2), 283–294 (2017)
62. Poon, J., Jain, P., Konstantakopoulos, I.C., Spanos, C., Panda, S.K., Sanders, S.R.: Model-based fault detection and identification for switching power converters. IEEE Trans. Power Electron. **32**(2), 1419–1430 (2017)
63. Luo, Y., Wang, Z., Wei, G., Alsaadi, F.E.: H_∞ fuzzy fault detection for uncertain 2-D systems under round-robin scheduling protocol. IEEE Trans. Syst. Man Cybern. Syst. **99**, 1–13 (2016)
64. Zhong, M., Ding, S.X., Zhou, D.: A new scheme of fault detection for linear discrete time-varying systems. IEEE Trans. Autom. Control **61**(9), 2597–2602 (2016)
65. Mao, Z., Jiang, B., Shi, P.: Protocol and fault detection design for nonlinear networked control systems. IEEE Trans. Circuits Syst. II: Express Briefs **56**(3), 255–259 (2009)
66. Wang, H., Liu, X., Li, S.: Robust adaptive fuzzy fault-tolerant control for a class of non-lower-triangular nonlinear systems with actuator failures. Inf. Sci. **336**(1), 60–74 (2016)
67. Zhang, L., Lin, J., Karim, R.: Sliding window-based fault detection from high-dimensional data streams. IEEE Trans. Syst. Man Cybern. Syst. **47**(2), 289–303 (2017)
68. Li, J., Park, J.H., Ye, D.: Fault detection filter design for switched systems with quantisation effects and packet dropout. IET Control Theory Appl. **11**(2), 182–193 (2017)
69. Li, L., Ding, S.X., Qiu, J., Yang, Y., Zhang, Y.: Weighted fuzzy observer-based fault detection approach for discrete-time nonlinear systems via piecewise-fuzzy Lyapunov functions. IEEE Trans. Fuzzy Syst. **24**(6), 1320–1333 (2016)
70. Zhao, X., Shi, P., Yin, Y., Nguang, S.K.: New results on stability of slowly switched systems: a multiple discontinuous Lyapunov function approach. IEEE Trans. Autom. Control **62**(7), 3502–3509 (2017)
71. Zhao, X., Yin, Y., Niu, B., Zheng, X.: Stabilization for a class of switched nonlinear systems with novel average dwell time switching by T-S fuzzy modeling. IEEE Trans. Cybern. **46**(8), 1952–1957 (2016)
72. Wang, D., Wang, W., Shi, P.: Robust fault detection for switched linear systems with state delays. IEEE Trans. Syst. Man Cybern. B Cybern. **39**(3), 800–805 (2009)
73. Du, D.: Fault detection for discrete-time linear systems based on descriptor observer approach. Appl. Math. Comput. **293**(15), 575–585 (2017)
74. Du, D., Jiang, B., Shi, P., Karimi, H.R.: Fault detection for continues-time switched systems under asynchronous switching. Int. J. Robust Nonlinear Control **24**(11), 1694–1706 (2014)
75. Li, J., Yang, G.: Asynchronous fault detection filter design approach for discrete-time switched linear systems. Int. J. Robust Nonlinear Control **24**(1), 70–96 (2014)
76. Zhao, X., Li, J., Ye, D.: Fault detection for switched systems with finite-frequency specifications. Nonlinear Dyn. **70**(1), 409–420 (2012)
77. Li, X., Yang, G.: Fault detection in finite frequency domain for Takagi-Sugeno fuzzy systems with sensor faults. IEEE Trans. Cybern. **44**(8), 1446–1458 (2014)
78. Yang, H., Xia, Y., Liu, B.: Fault detection for T-S fuzzy discrete systems in finite-frequency. IEEE Trans. Syst. Man Cybern. Part B-Cybern. **41**(4), 911–920 (2011)
79. Xue, S., Yang, X., Li, Z., Gao, H.: An approach to fault detection for multirate sampled-data systems with frequency specifications. IEEE Trans. Syst. Man Cybern. Syst. **99**, 1–11 (2017)
80. Daafouz, J., Riedinger, P., Iung, C.: Stability analysis and control synthesis for switched systems: a switched Lyapunov function approach. IEEE Trans. Autom. Control **47**(11), 1883–1887 (2002)

81. Ding, D., Yang, G.: Finite frequency H_∞ filtering for uncertain discrete-time switched linear systems. Progress Nat. Sci. **19**, 1625–1633 (2009)
82. Iwasaki, T., Hara, S.: Generalized KYP lemma: unified frequency domain inequalities with design applications. IEEE Trans. Autom. Control **50**(1), 41–59 (2005)
83. Gahinet, P., Apkarian, P.: A linear matrix inequality approach to H_∞ control. Int. J. Robust Nonlinear Control **4**(4), 421–448 (1994)
84. Yazdi, M.B., Jahed-Motlagh, M.R.: Stabilization of a CSTR with two arbitrarily switching modes using modal state feedback linearization. Chem. Eng. J. **155**(3), 838–843 (2009)
85. Shi, P., Su, X., Li, F.: Dissipativity-based filtering for fuzzy switched systems with stochastic perturbation. IEEE Trans. Autom. Control **61**(6), 1694–1699 (2016)
86. Xiang, W., Xiao, J.: Stabilization of switched continuous-time systems with all modes unstable via dwell time switching. Automatica **50**(3), 940–945 (2014)
87. Geromel, J.C., Colaneri, P., Bolzern, P.: Dynamic output feedback control of switched linear systems. IEEE Trans. Autom. Control **53**(3), 720–733 (2008)
88. Sun, X., Zhao, J., Hill, D.J.: Stability and L2-gain analysis for switched delay systems: a delay-dependent method. Automatica **42**(10), 1769–1774 (2006)
89. Hespanha, J.P.: Uniform stability of switched linear systems extensions of Lasalles invariance principle. IEEE Trans. Autom. Control **49**(4), 470–482 (2004)
90. Zhang, L., Zhuang, S., Shi, P.: Non-weighted quasi-time-dependent H_∞ filtering for switched linear systems with persistent dwell-time. Automatica **54**, 201–209 (2015)
91. Shen, Q., Jiang, B., Shi, P.: Adaptive fault diagnosis for TCS fuzzy systems with sensor faults and system performance aAnalysis. IEEE Trans. Fuzzy Syst. **22**(2), 274–285 (2014)
92. Jiang, B., Zhang, K., Shi, P.: Integrated fault estimation and accommodation design for discrete-time Takagi-Sugeno fuzzy systems with actuator faults. IEEE Trans. Fuzzy Syst. **19**(2), 291–304 (2011)
93. Li, H., Chen, Z., Wu, L., Lam, H., Du, H.: Event-triggered fault detection of nonlinear networked systems. IEEE Trans. Cybern. **47**(4), 1041–1052 (2017)
94. Zhao, Y., Lam, J., Gao, H.: Fault detection for fuzzy systems with intermittent measurements. IEEE Trans. Fuzzy Syst. **17**(2) (2009)
95. Longhi, S., Monteriu, A.: Fault detection and isolation of linear discrete-time periodic systems using the geometric approach. IEEE Trans. Autom. Control **62**(3), 1518–1523 (2017)
96. Zhang, D., Yu, L., Zhang, W.: Delay-dependent fault detection for switched linear systems with time-varying delaysthe average dwell time approach. Signal Process. **91**(4), 832–840 (2011)
97. Wang, D., Wang, W., Shi, P.: Exponential H_∞ filtering for switched linear systems with interval time-varying delay. Int. J. Robust Nonlinear Control **19**(5), 532–551 (2009)
98. Takagi, T., Sugeno, M.:'uzzy identification of systems and its application to modeling and control. IEEE Trans. Syst., Man, Cybern. SMC **SMC-15**(1), pp. 116–132 (1985)
99. Liberzon, D.: Switching in Systems and Control. Birkhauser, Berlin (2003)
100. Palm, R., Driankov, D.: Fuzzy switched hybrid systemsmodeling and identification. In: Proc. IEEE ISIC/CIRA/ISAS Joint Conf., pp. 130–135 (1998)

Chapter 2
PIO-Based Non-fragile Fault Diagnosis for Continuous-Time Switched Systems

2.1 Non-fragile Actuator FEA

This chapter deals with the problems of non-fragile FEA for continuous-time switched systems and they are investigated. Firstly, a non-fragile PIO is designed to achieve actuator FE for continuous-time switched systems. Based on the ADT method and the switched Lyapunov function technique, a new actuator fault estimator is obtained. Then, based on the online FE information, a PIO-based non-fragile state feedback controller is designed to compensate for the effect of the actuator fault by stabilizing the closed-loop switched system. Finally, an F-18 aircraft example is borrowed to illustrate the obtained contributions.

2.1.1 Introduction

With the rapid development of modern industry, the equipment of various systems are becoming more and more complex and expensive. If a sensor fault or an actuator fault occurs in a practical engineering system, it may lead to system performance degradation or even to system instability. To overcome this concern, the development of FTC strategies provides an efficient approach to achieve the system objectives in spite of faults and to avoid catastrophic behavior. In the past few decades, model-based FTC techniques have drawn much attention and fruitful methods have been proposed, for detailed information, the reader can refer to [1–6] and the references therein. Generally speaking, FTC techniques can be divided into two categories: passive and active. Passive FTC is to design a controller such that the closed-loop system is robust for all possible faults, while active FTC needs three steps: FD, identification, and accommodation.

In another research direction, the study of switched system has caused increasing interests due to the significance in modeling many man-made or physical systems as

D. Du et al., *Observer-Based Fault Diagnosis and Fault-Tolerant Control for Switched Systems*, Studies in Systems, Decision and Control 323,
https://doi.org/10.1007/978-981-15-9073-3_2

switched systems, such as chemical processes, traffic control systems, mechanical systems, etc. Great research efforts have been made on the issues of stability analysis and control synthesis under arbitrary switching signal or constrained switching signal [7–11]. Currently, the proposed constrained switching signal mainly includes: dwell-time switching [12], ADT switching [13], and PDT switching [14]. Among them, ADT switching is the most commonly used to deal with control synthesis analysis for switched systems, and has been proved in [13] to be more powerful and less conservative than dwell-time switching and arbitrary switching. Therefore, we focus on the FTC design problem for switched systems under ADT in this paper.

However, in many industrial applications implementations, it is inevitable that inaccuracies or uncertainties occur in the observer or controller, which may result in degraded system performance even in system instability [15]. Accordingly, it is very important to devise non-fragile observers or controllers, ensuring the desired performance with powerful capabilities to tolerate the admissible uncertainties. In [16], a non-fragile fault-tolerant controller was designed for switched systems with actuator faults such that the closed-loop system had an H_2 performance and circular disk pole constraints. Under the condition of unavailable states, a non-fragile sliding mode controller design for switched systems was considered in [17]. By using ADT method, non-fragile observer-based H_∞ control for discrete-time switched systems with data missing and time delay was studied in [18]. To the best of our knowledge, the issues of non-fragile observer-based FE and non-fragile controller-based FA for switched systems have not yet been considered.

Following the above discussions, this section investigates the problems of non-fragile FEA for continuous-time switched systems with actuator faults. The contributions of this paper can be summarized as the following aspects: (i:) By using ADT method, a non-fragile PIO is proposed such that the states and actuator faults can be simultaneously estimated; (ii:) Based on the states and actuator faults estimation information, a PIO-based non-fragile state feedback controller is designed to tolerate the faults, which can guarantee that the closed-loop system is asymptotically stable; (iii:) The obtained results are successfully illustrated to a flight control switched system of the F-18 aircraft.

2.1.2 Model Description and Preliminaries

Consider the following continuous-time switched system:

$$\begin{cases} \dot{x}(t) = \sum\limits_{i=1}^{N} \sigma_i(t)(A_i x(t) + B_i u(t) + G_i f(t)) \\ y(t) = \sum\limits_{i=1}^{N} \sigma_i(t) C_i x(t), \end{cases} \tag{2.1}$$

where $x(t) \in \mathbb{R}^{n_x}$, $u(t) \in \mathbb{R}^{n_p}$, $f(t) \in \mathbb{R}^{n_f}$, and $y(t) \in \mathbb{R}^{n_y}$ are, respectively, the state, the control input, the actuator fault, and the measured output. The piecewise constant function

$$\sigma_i(t) : [0, \infty) \to \{0, 1\}, \quad \sum_{i=1}^{N} \sigma_i(t) = 1 \tag{2.2}$$

is the switching signal, which specifies which subsystem is activated at the switching instant. A_i, B_i, C_i, and G_i are constant real matrices with appropriate dimensions.

The following assumptions are given:

Assumption 2 The actuator faults are piecewise constant faults, i.e., $\dot{f}(t) = 0$.

Assumption 3 The matrices B_i, $i = 1, \ldots, N$ are of full column rank, and

$$rank(\begin{bmatrix} B_i & G_i \end{bmatrix}) = rank(B_i) \tag{2.3}$$

Remark 2.1 From the perspective of practical engineering, this type of fault is very common. From this opinion, the consideration of this section is general and practical. Moreover, FE for this kind of fault has been widely investigated and can be extended other fault cases [19, 20].

For system (2.1), we employ the following non-fragile PIO to realize FE:

$$\begin{cases} \dot{\hat{x}}(t) = \sum_{i=1}^{N} \sigma_i(t)(A_i\hat{x}(t) + B_i u(t) + L_{\Delta i}(y(t) - \hat{y}(t)) + G_i \hat{f}(t)) \\ \hat{y}(t) = \sum_{i=1}^{N} \sigma_i(t) C_i \hat{x}(t) \\ \dot{\hat{f}}(t) = \sum_{i=1}^{N} \sigma_i(t) H_i (y(t) - \hat{y}(t)), \end{cases} \tag{2.4}$$

where $\hat{x}(t) \in \mathbb{R}^{n_x}$ and $\hat{f}(t) \in \mathbb{R}^{n_f}$ are, respectively, the estimation of the state and the actuator fault of system (2.1). $L_{\Delta i}$, $i = 1, \ldots, N$ are the proportional observer gains with uncertainties due to implementation reasons and satisfy

$$L_{\Delta i} = L_i + \Delta L_i, \tag{2.5}$$

$$\Delta L_i = M_i F_i(t) N_i, \tag{2.6}$$

$$F_i^T(t) F_i(t) \leq I, \tag{2.7}$$

where the matrices M_i and N_i, $i = 1, \ldots, N$ specify the uncertainties on the observer gains, and H_i, $i = 1, \ldots, N$ are the integral observer gains.

Remark 2.2 In practical applications, the occurrence of parameter uncertainties are unavoidable, which will affect the stability and performances of the dynamics if they are not treated properly. In this paper, the uncertainty term is assumed to be norm-bounded. This form was first proposed in [21] and has been widely used to deal with non-fragile control issues [22, 23].

Define

$$e(t) = x(t) - \hat{x}(t), \quad e_f(t) = f(t) - \hat{f}(t). \tag{2.8}$$

From (2.1) and (2.4), the derivative of the state error and of the actuator fault error can be obtained:

$$\dot{e}(t) = \dot{x}(t) - \dot{\hat{x}}(t) = \sum_{i=1}^{N} \sigma_i(t)((A_i - L_iC_i - \Delta L_iC_i)e(t) + G_ie_f(t)), \tag{2.9}$$

$$\dot{e}_f(t) = \dot{f}(t) - \dot{\hat{f}}(t) = \sum_{i=1}^{N} \sigma_i(t)H_iC_ie(t). \tag{2.10}$$

Then, one can get the following augmented system:

$$\dot{\bar{e}}(t) = \sum_{i=1}^{N} \sigma_i(t)(\bar{A}_i - \bar{L}_i\bar{C}_i - \bar{M}_iF_i(t)\bar{N}_i)\bar{e}(t), \tag{2.11}$$

where

$$\bar{e}(t) = \begin{bmatrix} e(t) \\ e_f(t), \end{bmatrix}, \ \bar{A}_i = \begin{bmatrix} A_i & G_i \\ 0 & 0, \end{bmatrix}, \ \bar{L}_i = \begin{bmatrix} L_i \\ H_i, \end{bmatrix}, \ \bar{C}_i = \begin{bmatrix} C_i & 0, \end{bmatrix},$$
$$\bar{M}_i = \begin{bmatrix} M_i \\ 0, \end{bmatrix}, \ \bar{N}_i = \begin{bmatrix} N_iC_i & 0. \end{bmatrix}. \tag{2.12}$$

The following definition and lemmas are recalled:

Definition 2.1 ([24]) For any switching signal $\sigma_i(t)$ and any $t_2 > t_1 > 0$, let $N_{\sigma_i(t)}$ (t_1, t_2) denote the number of switchings $\sigma_i(t)$ on an interval (t_1, t_2). If

$$N_{\sigma_i(t)}(t_1, t_2) \leq N_0 + \frac{t_2 - t_1}{\tau_a}$$

holds for a given $N_0 \geq 0$ and $\tau_a > 0$, then the constant τ_a is called the ADT and N_0 is the chattering bound.

Lemma 2.1 ([25]) *Consider the continuous-time switched system* $\dot{x}(t) = f_{\sigma_i(t)}(x(t))$, *and let* $\alpha > 0$, $\mu > 1$ *be given constants. Suppose that there exist* $\mathscr{C}^1$ *functions* $V_{\sigma_i(t)} : \mathbb{R}^n \to \mathbb{R}$ *and two class* $\mathscr{K}_\infty$ *functions* κ_1 *and* κ_2 *such that*

$$\kappa_1(|x(t)|) \leq V_{\sigma_i(t)}(x(t)) \leq \kappa_2(|x(t)|)$$

$$\dot{V}_{\sigma_i(t)}((t)) \leq -\alpha V_{\sigma_i(t)}(x(t))$$

and $\forall(\sigma_i(t_l) = i, \sigma_i(t_l^-) = j) \in N \times N, i \neq j$,

$$V_i(x(t)) \leq \mu V_j(x(t))$$

then the switched system $\dot{x}(t) = f_{\sigma_i(t)}(x(t))$ *is GUAS for any switching signal with average dwell time*

$$\tau_a > \tau_a^* = \frac{\ln \mu}{\alpha}. \tag{2.13}$$

Lemma 2.2 ([26]) *For any matrices* M, N, *a constant* $\varepsilon > 0$ *and a time-varying matrix* $F(t)$ *satisfying* $F^T(t)F(t) \leq I$ *for any* $t \geq 0$, *we have*

$$M^T F(t)N + N^T F^T(t)M \leq \varepsilon^{-1} M^T M + \varepsilon N^T N. \tag{2.14}$$

2.1.3 Non-fragile PIO Design

In this section, the design of the non-fragile PIO (2.4) is investigated.

Theorem 2.1 *For given scalars* $\alpha > 0$, $\mu > 1$, *and* $\varepsilon > 0$, *if there exist positive definite matrices* P_{1i}, P_{1j}, *matrices* Q_i, $\forall i, j \in \{1, 2, \ldots, N\}$, *which satisfy*

$$P_{1i} \leq \mu P_{1j}, \tag{2.15}$$
$$\Pi_i < 0, \tag{2.16}$$

where

$$\Pi_i = \begin{bmatrix} \Phi_i & P_{1i}\bar{M}_i & \bar{N}_i^T \\ * & -\varepsilon I & 0 \\ * & * & -\varepsilon^{-1} I \end{bmatrix}, \tag{2.17}$$
$$\Phi_i = \bar{A}_i^T P_{1i} + P_{1i}\bar{A}_i - Q_i\bar{C}_i - \bar{C}_i^T Q_i^T + \alpha P_{1i}, \tag{2.18}$$

then, for the switching signal $\sigma_i(t)$ *with ADT constraint (2.13), there exists a non-fragile PIO with gains* $\bar{L}_i = P_{1i}^{-1} Q_i$ *such that the augmented system (2.11) is asymptotically stable.*

Proof Firstly, we construct a switched Lyapunov function for the augmented dynamics (2.11) as

$$V_{1i}(t, \bar{e}(t)) = \bar{e}^T(t)\left(\sum_{i=1}^{N}\sigma_i(t)P_{1i}\right)\bar{e}(t). \tag{2.19}$$

Then along the trajectory of system (2.11), one has

$$\begin{aligned}
&\dot{V}_{1i}(t, \bar{e}(t)) + \alpha V_{1i}(t, \bar{e}(t)) \\
&= \dot{\bar{e}}^T(t)\left(\sum_{i=1}^{N}\sigma_i(t)P_{1i}\right)\bar{e}(t) + \bar{e}^T(t)\left(\sum_{i=1}^{N}\sigma_i(t)P_{1i}\right)\dot{\bar{e}}(t) + \bar{e}^T(t)\left(\sum_{i=1}^{N}\sigma_i(t)\alpha P_{1i}\right)\bar{e}(t) \\
&= \sum_{i=1}^{N}\sigma_i(t)\bar{e}^T(t)((\bar{A}_i - \bar{L}_i\bar{C}_i)^T P_{1i} + P_{1i}(\bar{A}_i - \bar{L}_i\bar{C}_i) + \alpha P_{1i} \\
&\quad -(\bar{M}_i F_i(t)\bar{N}_i)^T P_{1i} - P_{1i}\bar{M}_i F_i(t)\bar{N}_i)\bar{e}(t).
\end{aligned} \tag{2.20}$$

By Lemma 2.2, one can get that

$$\begin{aligned}
&\dot{V}_{1i}(t, \bar{e}(t)) + \alpha V_{1i}(t, \bar{e}(t)) \\
&\leq \sum_{i=1}^{N}\sigma_i(t)\bar{e}^T(t)((\bar{A}_i - \bar{L}_i\bar{C}_i)^T P_{1i} + P_{1i}(\bar{A}_i - \bar{L}_i\bar{C}_i) + \alpha P_{1i} \\
&\quad +\varepsilon^{-1}P_{1i}\bar{M}_i\bar{M}_i^T P_{1i} + \varepsilon\bar{N}_i^T\bar{N}_i)\bar{e}(t).
\end{aligned} \tag{2.21}$$

From the condition (2.16) and Schur complement lemma, one can conclude that $\dot{V}_{1i}(t, \bar{e}(t)) < -\alpha V_{1i}(t, \bar{e}(t))$. According to Lemma 2.1 and condition (2.15), one can get that the system (2.11) is asymptotically stable. (□)

Remark 2.3 By resolving the conditions in Theorem 2.1, one can get the non-fragile PIO gains L_i and H_i by calculating

$$L_i = \left[\, I_{n_x}\ 0 \,\right]\bar{L}_i, \quad H_i = \left[\, 0\ I_{n_f} \,\right]\bar{L}_i, \tag{2.22}$$

where $\bar{L}_i = P_{1i}^{-1}Q_i$.

2.1.4 Non-fragile Controller Design

In this section, PIO-based non-fragile state feedback controller is designed. We assume that the non-fragile state feedback controller has the following form:

$$u(t) = \sum_{i=1}^{N}\sigma_i(t)(K_{\Delta i}\hat{x}(t) - B_i^{\dagger}G_i\hat{f}(t)), \tag{2.23}$$

where $\hat{x}(t)$ and $\hat{f}(t)$ are the estimation from the non-fragile PIO (2.4). $B_i^{\dagger}$, $i = 1, \ldots, N$ denote the left inverses of the matrices B_i, $K_{\Delta i}$, $i = 1, \ldots, N$ are the controller gains with uncertainties and satisfy

$$K_{\Delta i} = K_i + \Delta K_i, \tag{2.24}$$

$$\Delta K_i = \tilde{M}_i \tilde{F}_i(t) \tilde{N}_i, \tag{2.25}$$

$$\tilde{F}_i^T(t) \tilde{F}_i(t) \leq I, \tag{2.26}$$

where the matrices $\tilde{M}_i$ and $\tilde{N}_i$, $i = 1, \ldots, N$ specify the uncertainties of the controller gains.

Lemma 2.3 *If there exist matrices B_i and G_i satisfying the condition in Assumption 2, then we have $B_i B_i^{\dagger} G_i = G_i$.*

Proof By

$$rank([\, B_i \;\; G_i \,]) = rank(B_i) \tag{2.27}$$

one can get that there exist some matrices Ω_i such that

$$B_i \Omega_i = G_i. \tag{2.28}$$

Then one has

$$B_i B_i^{\dagger} G_i = B_i B_i^{\dagger} B_i \Omega_i = B_i \Omega_i = G_i. \tag{2.29}$$

The proof is concluded.

By Lemma 2.3 and substituting (2.23) into the switched systems (2.1), one can get

$$\begin{aligned}
\dot{x}(t) &= \sum_{i=1}^{N} \sigma_i(t)(A_i x(t) + B_i(K_{\Delta i}\hat{x}(t) - B_i^{\dagger} G_i \hat{f}(t)) + G_i f(t)) \\
&= \sum_{i=1}^{N} \sigma_i(t)((A_i + B_i K_{\Delta i})x(t) - B_i K_{\Delta i} e(t) + G_i e_f(t)), \\
&= \sum_{i=1}^{N} \sigma_i(t)((A_i + B_i K_{\Delta i})x(t) + (\bar{K}_i - B_i \Delta K_i \tilde{I})\bar{e}(t)),
\end{aligned} \tag{2.30}$$

where $\bar{K}_i = [\, -B_i K_i \;\; G_i \,]$, $\tilde{I} = [\, I \;\; 0 \,]$.

The following theorem expresses how to design the non-fragile state feedback controller (2.23).

Theorem 2.2 *For given scalars $\alpha > 0$, $\mu > 1$, $\varepsilon_1 > 0$, and $\varepsilon_2 > 0$, if there exist positive definite matrices X_{1i}, X_{1j}, X_{2i}, X_{2j}, matrices W_i, $\forall i, j \in \{1, 2, \ldots, N\}$, which satisfy*

$$X_{1j} \leq \mu X_{1i}, \quad X_{2j} \leq \mu X_{2i}, \tag{2.31}$$

$$\begin{bmatrix} \tilde{\Phi}_i & B_i \tilde{M}_i & X_{2i} \tilde{N}_i^T \\ * & -(\frac{1}{\varepsilon_1} + \frac{1}{\varepsilon_2})^{-1} I & 0 \\ * & * & -\varepsilon_2^{-1} I \end{bmatrix} < 0 \tag{2.32}$$

where $\tilde{\Phi}_i = A_i^T X_{2i} + X_{2i} A_i + B_i W_i + W_i^T B_i^T + \alpha X_{2i}$, then, for the switching signal $\sigma_i(t)$ with ADT constraint (2.13), there exists a non-fragile state feedback controller (2.23) with gains $K_i = W_i X_{2i}^{-1}$ such that the closed-loop system (2.30) is asymptotically stable.

Proof For system (2.30), we consider the following switched Lyapunov function:

$$V_i(t, x(t), \bar{e}(t)) = \rho V_{1i}(t, \bar{e}(t)) + V_{2i}(t, x(t)), \tag{2.33}$$

where $V_{1i}(t, \bar{e}(t))$ is defined in (2.19), $\rho > 0$ is a constant scalar and will be determined later, and

$$V_{2i}(t, x(t)) = x^T(t) \left(\sum_{i=1}^{N} \sigma_i(t) P_{2i} \right) x(t), \tag{2.34}$$

where $X_{1i} = P_{1i}^{-1}$, $X_{2i} = P_{2i}^{-1}$.

Then, one can get

$$\begin{aligned} &\dot{V}_i(t, x(t), \bar{e}(t)) + \alpha V_i(t, x(t), \bar{e}(t)) \\ &= \rho \dot{V}_{1i}(t, \bar{e}(t)) + \dot{V}_{2i}(t, x(t)) + \beta V_{1i}(t, \bar{e}(t)) + \alpha V_{2i}(t, x(t)), \end{aligned} \tag{2.35}$$

where $\beta = \alpha\rho$.

Then along the trajectory of the closed-loop system (2.30), one has

$$\begin{aligned} &\dot{V}_{2i}(t, x(t)) + \beta V_{1i}(t, \bar{e}(t)) + \alpha V_{2i}(t, x(t)) \\ &= \sum_{i=1}^{N} \sigma_i(t) \Big[2x^T(t) P_{2i}((A_i + B_i K_{\Delta i}) x(t) + (\bar{K}_i - B_i \Delta K_i \tilde{I}) \bar{e}(t)) \\ &\quad + \beta V_{1i}(t, \bar{e}(t)) + \alpha V_{2i}(t, x(t)) \Big]. \end{aligned} \tag{2.36}$$

By Lemma 2.2, one can get

$$\begin{aligned} &-2x^T(t) P_{2i} B_i \Delta K_i \tilde{I} \bar{e}(t) \\ &= -2x^T(t) P_{2i} B_i \tilde{M}_i \tilde{F}_i(t) \tilde{N}_i \tilde{I} \bar{e}(t) \\ &\leq \frac{1}{\varepsilon_1} x^T(t) P_{2i} B_i \tilde{M}_i \tilde{M}_i^T B_i^T P_{2i} x(t) + \varepsilon_1 \bar{e}(t) \tilde{I}^T \tilde{N}_i^T \tilde{N}_i \tilde{I} \bar{e}(t). \end{aligned} \tag{2.37}$$

$$
\begin{aligned}
& 2x^T(t)P_{2i}B_i\Delta K_i x(t) \\
&= 2x^T(t)P_{2i}B_i\tilde{M}_i\tilde{F}_i(t)\tilde{N}_i x(t) \\
&\le \frac{1}{\varepsilon_2}x^T(t)P_{2i}B_i\tilde{M}_i\tilde{M}_i^T B_i^T P_{2i}x(t) + \varepsilon_2 x(t)\tilde{N}_i^T\tilde{N}_i x(t).
\end{aligned} \tag{2.38}
$$

Therefore, we have

$$
\begin{aligned}
& \dot{V}_{2i}(t, x(t)) + \beta V_{1i}(t, \bar{e}(t)) + \alpha V_{2i}(t, x(t)) \\
&\le \sum_{i=1}^{N}\sigma_i(t)\Big[2x^T(t)P_{2i}(A_i + B_i K_i)x(t) + \frac{1}{\varepsilon_1}x^T(t)P_{2i}B_i\tilde{M}_i\tilde{M}_i^T B_i^T P_{2i}x(t) \\
&\quad +\varepsilon_1\bar{e}(t)\tilde{I}^T\tilde{N}_i^T\tilde{N}_i\tilde{I}\bar{e}(t) + \frac{1}{\varepsilon_2}x^T(t)P_{2i}B_i\tilde{M}_i\tilde{M}_i^T B_i^T P_{2i}x(t) + \varepsilon_2 x^T(t)\tilde{N}_i^T\tilde{N}_i x(t) \\
&\quad +2x^T(t)P_{2i}\bar{K}_i\bar{e}(t) + \beta\bar{e}^T(t)P_{1i}\bar{e}(t) + \alpha x^T(t)P_{2i}x(t)\Big] \\
&\le \sum_{i=1}^{N}\sigma_i(t)\left(\begin{bmatrix} x(t) \\ \bar{e}(t)\end{bmatrix}^T\begin{bmatrix} \Lambda_{11i} & \Lambda_{12i} \\ \Lambda_{12i}^T & \Lambda_{22i}\end{bmatrix}\begin{bmatrix} x(t) \\ \bar{e}(t)\end{bmatrix}\right),
\end{aligned} \tag{2.39}
$$

where

$$
\begin{aligned}
\Lambda_{11i} &= (A_i + B_i K_i)^T P_{2i} + P_{2i}(A_i + B_i K_i) + \varepsilon_2\tilde{N}_i^T\tilde{N}_i + \alpha P_{2i} \\
&\quad +(\frac{1}{\varepsilon_1} + \frac{1}{\varepsilon_2})P_{2i}B_i\tilde{M}_i\tilde{M}_i^T B_i^T P_{2i}, \\
\Lambda_{12i} &= P_{2i}\bar{K}_i, \\
\Lambda_{22i} &= \varepsilon_1\tilde{I}^T\tilde{N}_i^T\tilde{N}_i\tilde{I} + \beta P_{1i}.
\end{aligned}
$$

Pre- and post-multiplying (2.32) by $diag\{P_{2i}, I, I\}$, one has

$$
\begin{bmatrix} \tilde{\Phi}_i' & P_{2i}B_i\tilde{M}_i & \tilde{N}_i^T \\ * & -(\frac{1}{\varepsilon_1} + \frac{1}{\varepsilon_2})^{-1}I & 0 \\ * & * & -\varepsilon_2^{-1}I \end{bmatrix} < 0, \tag{2.40}
$$

where $\tilde{\Phi}_i' = (A_i + B_i K_i)^T P_{2i} + P_{2i}(A_i + B_i K_i) + \alpha P_{2i}$, which implies $\Lambda_{11i} < 0$.

Define

$$\theta_1 = \min_{i=1,\ldots,N}\|\Pi_i\|; \tag{2.41}$$

$$\theta_2 = \min_{i=1,\ldots,N}\|\Lambda_{11i}\|; \tag{2.42}$$

$$\theta_3 = 2\min_{i=1,\ldots,N}\|\Lambda_{12i}\|; \tag{2.43}$$

$$\theta_4 = \min_{i=1,\ldots,N}\|\Lambda_{22i}\|; \tag{2.44}$$

Therefore, if the obtained results in Theorem 2.1 is feasible, we have

$$\dot{V}_{1i}(t, \bar{e}(t)) \leq -\theta_1 \|\bar{e}(t)\|^2, \tag{2.45}$$

and by Rayleigh's inequality, we have

$$x^T(t)\Lambda_{11i}x(t) \leq -\theta_2\|x(t)\|^2. \tag{2.46}$$

Select the constant scalar ρ satisfying the following condition:

$$\rho \geq \frac{\theta_3^2 + \theta_2\theta_4}{\theta_1\theta_2} > 0 \tag{2.47}$$

then we obtain

$$\begin{aligned}
&\dot{V}_i(t, x(t), \bar{e}(t)) + \alpha V_i(t, x(t), \bar{e}(t)) \\
&\leq -\theta_2\|x(t)\|^2 + 2\|x(t)\|\|\Lambda_{12i}\|\|\bar{e}(t)\| - \theta_1\rho\|\bar{e}(t)\|^2 + \theta_4\|\bar{e}(t)\|^2 \\
&\leq -\theta_2\|x(t)\|^2 + \theta_3\|x(t)\|\|\bar{e}(t)\| - \theta_1\rho\|\bar{e}(t)\|^2 + \theta_4\|\bar{e}(t)\|^2.
\end{aligned} \tag{2.48}$$

By (2.47), we can obtain that

$$\begin{aligned}
&\dot{V}_i(t, x(t), \bar{e}(t)) + \alpha V_i(t, x(t), \bar{e}(t)) \\
&\leq -\theta_2\|x(t)\|^2 + \sqrt{\theta_2(\theta_1\rho - \theta_4)}\|x(t)\|\|\bar{e}(t)\| - (\theta_1\rho - \theta_4)\|\bar{e}(t)\|^2 \\
&\leq -\left[(\sqrt{\theta_2}\|x(t)\| - \sqrt{\theta_1\rho - \theta_4}\|\bar{e}(t)\|)^2 + \sqrt{\theta_2(\theta_1\rho - \theta_4)}\|x(t)\|\|\bar{e}(t)\|\right]. \\
&< 0
\end{aligned} \tag{2.49}$$

From the condition (2.49), one can conclude that $\dot{V}_i(t, x(t), \bar{e}(t)) < -\alpha V_i(t, x(t), \bar{e}(t))$. According to Lemma 2.1 and condition (2.31), one can get that the closed-loop system (2.30) is asymptotically stable. The proof is completed. (□)

2.1.5 An Illustrative Example

In this section, a flight control switched system with of an F-18 aircraft is considered [27]. By using the body-axis coordinate system and the Newton's second law, the nonlinear model of the F-18 aircraft can be expressed by six-degree-of-freedom aircraft equations. For simplicity, we only analyze the longitudinal equations. The longitudinal axis system is shown in Fig. 2.1, where X and Z are the external forces and M is the moment, U and W are the coordinate components of the total translational velocity V_T, α_T is the attack angle and Θ is the pitch angle, Q represents the rational rate with the rotational inertia I_y. Then, the longitudinal equations of motion for the F-18 aircraft is modeled by

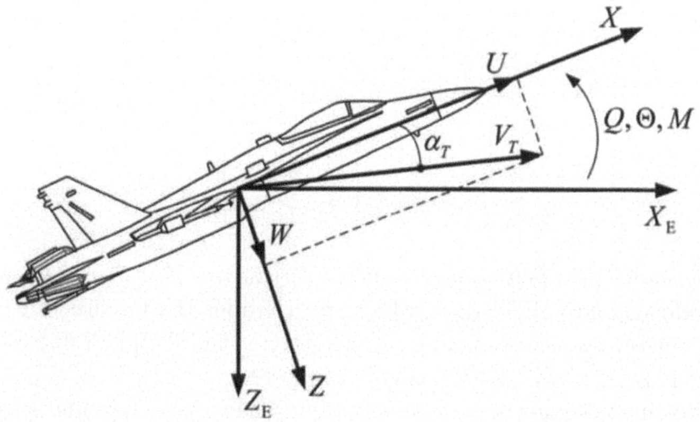

Fig. 2.1 Longitudinal axis system of F-18

$$\begin{cases} m\left[\dot{U} + QW + g\sin\Theta\right] = X; \\ m\left[\dot{W} - QU - g\cos\Theta\right] = Z; \\ \dot{Q}I_y = M; \end{cases} \tag{2.50}$$

where m is the aircraft mass and g is the acceleration of gravity. We consider the trimmed condition with zero rotational rates and side velocity by the assumption of straight and level coordinated flighting, which implies $Q_0 = 0$, $M_0 = 0$, and $W_0 = 0$. By the force decomposition, it follows on trimmed condition that

$$\begin{cases} mg\sin\Theta_0 = X_0; \\ -mg\cos\Theta_0 = Z_0; \end{cases} \tag{2.51}$$

Define the variables ν, ω of the translational velocities, q of the pitch angle rate, and δ of the pitch angle, then one has

$$U = U_0 + \nu,\ \Theta = \Theta_0 + \delta,\ W = \omega,\ Q = q,\ X = X_0 + \Delta X,$$
$$Z = Z_0 + \Delta Z,\ M = \Delta M. \tag{2.52}$$

Substituting (2.51) and (2.52) into (2.50), we can obtain

$$\begin{cases} m\left[\dot{\nu} + (g\cos\Theta_0)\delta\right] = \Delta X, \\ m\left[\dot{\omega} - qU_0 + (g\sin\Theta_0)\delta\right] = \Delta Z, \\ \dot{q}I_y = \Delta M, \end{cases} \tag{2.53}$$

where ΔX, ΔZ, and ΔM are linear functions with respect to ν and η. The speed may be considered as a constant in some short time period, which implies $\dot{\nu} = 0$. Therefore, it can be regarded that the response is only controlled by the angle of attack η and pitch angle rate q. Thus, the final longitudinal linear equations of the F-18 aircraft can be simplified into the following form:

$$\begin{bmatrix} \dot{\eta} \\ \dot{q} \end{bmatrix} = A_{long} \begin{bmatrix} \eta \\ q \end{bmatrix} + B_{long} \begin{bmatrix} \xi_E \\ \xi_{PTV} \end{bmatrix}, \tag{2.54}$$

where

$$A_{long} = \begin{bmatrix} Z_\eta \; Z_q \\ T_\eta \; T_q \end{bmatrix}, \; B_{long} = \begin{bmatrix} Z_E \; Z_{PTV} \\ T_E \; T_{PTV} \end{bmatrix}, \; \eta = \alpha_T - \alpha_0, \tag{2.55}$$

Z_η, Z_q, T_η, and T_q are longitudinal stability derivatives, Z_E, Z_{PTV}, T_E, and T_{PTV} are longitudinal control derivatives, which are determined by the structure of aircraft and flight conditions. The limitation of the elevator and the pitch thrust vectoring nozzle satisfy $\xi_E \in [-24°, 10.5°]$ and $\xi_{PTV} \in [-30°, 30°]$.

The following two pairs of parameters are chosen in a set of baseline aerodynamic data from wind tunnel and flight test data:

$$A_1 = A_{long}^{m8h12} = \begin{bmatrix} -1.562 \; 0.9862 \\ -14.96 \; -1.132 \end{bmatrix}, \; B_1 = B_{long}^{m8h12} = \begin{bmatrix} -0.2316 \; -0.04349 \\ -26.48 \quad -5.323 \end{bmatrix},$$

$$G_1 = G_{long}^{m8h12} = \begin{bmatrix} 0.5 \\ 0.5 \end{bmatrix}, \; C_1 = C_{long}^{m8h12} = \begin{bmatrix} 0.40 \; 0.2 \end{bmatrix};$$

$$A_2 = A_{long}^{m8h10} = \begin{bmatrix} -1.675 \; 0.9853 \\ -16.16 \; -1.212 \end{bmatrix}, \; B_2 = B_{long}^{m8h10} = \begin{bmatrix} -0.2449 \; -0.04649 \\ -28.34 \quad -5.742 \end{bmatrix},$$

$$G_2 = G_{long}^{m8h10} = \begin{bmatrix} 0.5 \\ 0.5 \end{bmatrix}, \; C_2 = C_{long}^{m8h10} = \begin{bmatrix} 0.34 \; 0.15 \end{bmatrix};$$

where A_{long}^{m8h12} means the longitudinal state matrix at Mach 0.8 and altitude 12 kft, and the other coefficient matrices have the same meanings.

If we select $\mu = 1.2$, $\alpha = 0.1$, and $\varepsilon = 0.5$, then the ADT $\tau_a = 2s$ satisfies the constraint (2.13). The uncertainty matrices are chosen as

$$M_1 = \begin{bmatrix} 0.01 \\ 0.01 \end{bmatrix}, \; M_2 = \begin{bmatrix} 0.02 \\ 0.02 \end{bmatrix}, \; N_1 = N_2 = 0.1.$$

By solving the conditions in Theorem 2.1, one can get

$$P_{11} = \begin{bmatrix} 3.4537 & -0.0347 & -1.2777 \\ -0.0347 & 0.3519 & 0.4369 \\ -1.2777 & 0.4369 & 2.272 \end{bmatrix},$$

$$P_{12} = \begin{bmatrix} 3.3995 & -0.0309 & -1.2512 \\ -0.0309 & 0.3337 & 0.4310 \\ -1.2512 & 0.4310 & 2.2225 \end{bmatrix},$$

$$Q_1 = \begin{bmatrix} -10.3053 \\ 0.7696 \\ -7.4708 \end{bmatrix}, \; Q_2 = \begin{bmatrix} -13.1215 \\ 0.3366 \\ -9.8479 \end{bmatrix},$$

$$\bar{L}_1 = \begin{bmatrix} -6.3216 \\ 13.2102 \\ -9.3808 \end{bmatrix}, \ \bar{L}_2 = \begin{bmatrix} -8.1753 \\ 15.9066 \\ -12.1183 \end{bmatrix}.$$

Then the non-fragile PIO gains in (2.4) can be calculated as follows:

$$L_1 = \begin{bmatrix} -6.3216 \\ 13.2102 \end{bmatrix}, \ L_2 = \begin{bmatrix} -8.1753 \\ 15.9066 \end{bmatrix}, \ H_1 = -9.3808, \ H_2 = -12.1183.$$

For simulation, a piecewise constant fault is considered, which is defined as

$$f(t) = \begin{cases} 0, & 0s \le t < 20s; \\ 10, & 20s \le t < 300s; \\ -10, & 300s \le t < 600s; \end{cases}$$

The switching signal $\sigma_i(t)$ is shown in Fig. 2.2, and the FE result is described in Fig. 2.3. One can get that the FE $\hat{f}(t)$ accurately converges to the real fault $f(t)$.

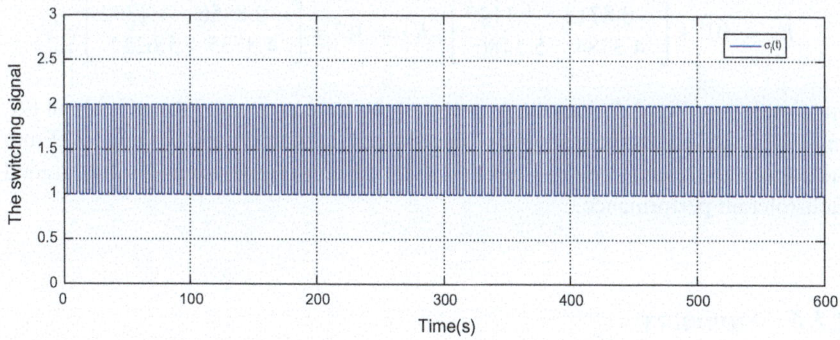

Fig. 2.2 The switching signal

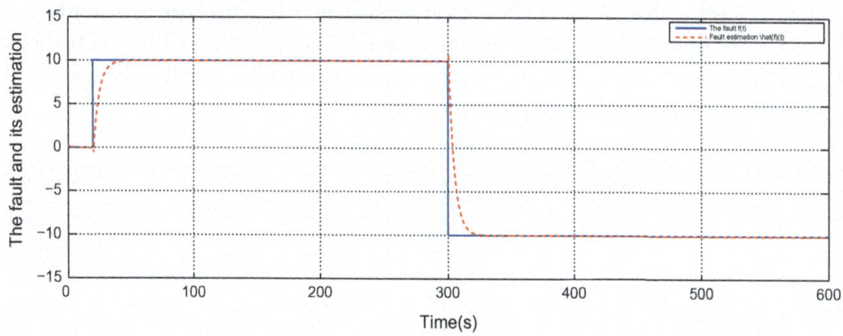

Fig. 2.3 The fault $f(t)$ and its estimation $\hat{f}(t)$

In the following, based on the online FE information, the non-fragile state feedback controller and the simulation results of FA are provided. We choose $\varepsilon_1 = \varepsilon_2 = 0.5$, the uncertain matrices are selected as

$$\tilde{M}_1 = \begin{bmatrix} 0.2 \\ 0.2 \end{bmatrix}, \ \tilde{M}_2 = \begin{bmatrix} 0.1 \\ 0.1 \end{bmatrix}, \ \tilde{N}_1 = \begin{bmatrix} 0.2 \ 0.2 \end{bmatrix}, \ \tilde{N}_2 = \begin{bmatrix} 0.1 \ 0.1 \end{bmatrix}.$$

By solving the conditions in Theorem 2.2, one can obtain

$$X_{11} = \begin{bmatrix} 8120.1 & 0 \\ 0 & 8120.1 \end{bmatrix}, X_{12} = \begin{bmatrix} 8120.1 & 0 \\ 0 & 8120.1 \end{bmatrix},$$
$$X_{21} = \begin{bmatrix} 1821.4 & -1547.3 \\ -1547.3 & 1822.1 \end{bmatrix}, X_{22} = \begin{bmatrix} 1806.0 & -1529.3 \\ -1529.3 & 1806.7 \end{bmatrix},$$
$$W_1 = 10^7 * \begin{bmatrix} 0.1376 & -0.6829 \\ -0.6829 & 3.3974 \end{bmatrix}, W_2 = 10^7 * \begin{bmatrix} 0.1429 & -0.7039 \\ -0.7039 & 3.4743 \end{bmatrix}.$$

Then the non-fragile controllers can be calculated as follows:

$$K_1 = 10^4 * \begin{bmatrix} -0.8715 & -1.1149 \\ 4.3384 & 5.5486 \end{bmatrix}, K_2 = 10^4 * \begin{bmatrix} -0.8856 & -1.1392 \\ 4.3735 & 5.6252 \end{bmatrix}.$$

For simulation, Fig. 2.4 gives the response of the state $x(t)$ with FA and without FA, and Fig. 2.5 describes the response of the output $y(t)$ with FA and without FA. From the simulation results, one can get the designed non-fragile controller has a good fault-tolerant performance.

2.1.6 Summary

In this section, we have studied the problems of non-fragile FE observer and fault-tolerant controller design for continuous-time switched systems with actuator faults. The designed observer and controller are both resilient to destabilizing fluctuations or quantization errors as well as a class of actuator faults. Finally, the proposed techniques are illustrated by a practical example.

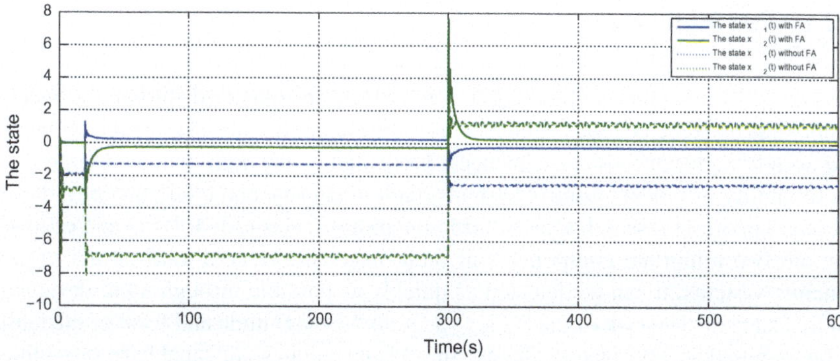

Fig. 2.4 The response of state x(t)

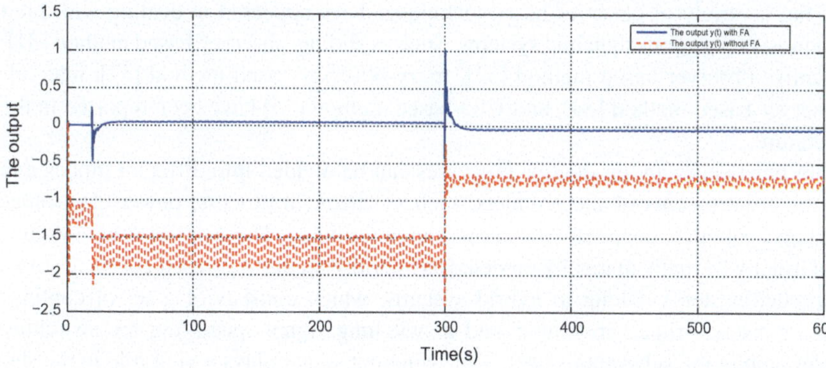

Fig. 2.5 The response of output y(t)

2.2 Non-fragile Fault Diagnosis with Actuator and Sensor Faults

In order to improve the fault diagnosis theory of switched systems, observer-based method is proposed to deal with the fault diagnosis issue subjected to the parameter uncertainty, unknown process disturbances, and unexpected faults. By employing PIO technique and the ADT method, an H_∞/H_- performance level is considered to ensure the worst case of robustness to the disturbance and sensitivity to the fault simultaneously. Moreover, the obtained fault diagnosis observer can give accurate FE online for the piecewise constant fault. Finally, a two-tank system example is introduced in the end to show the effectiveness of the proposed results.

2.2.1 Introduction

It is unavoidable for real engineering systems affected by outside disturbances, uncertainties and unexpected faults, which may tend the degrade of the system performance and even bring hazardous consequences. Motivated by the high-demand to improve the reliability of modern control systems, fault diagnosis, and tolerant control have become important research topics during the past decades [28–30]. FD and estimation are two important components of fault diagnosis. Once a fault occurs in the dynamic systems, it can be detected as quickly as possible through a pre-designed FD mechanism. The task of the FE is to acquire the magnitude and form of the fault as far as possible. Fortunately, the existing FE techniques are capable of providing exact information on the size of the faults that occur, thereby helping to reconstruct the fault signals. As such, FE is an important ingredient in improving active FTC. So far, a variety of observer-based techniques have appeared in dealing with fault diagnosis for various dynamic systems. Such as sliding-observer-based method [31], adaptive-observer-based method [32], fuzzy-observer-based method [33], interval-observer-based method [34], and UIO-based method [35] have been reported in the literature.

In practice, many productive processes can be divided into different modes due to the requirements of the products, such as chemical process, power electronics systems, robotics, cooperative control, and other fields. These engineering systems can usually be mathematically modeled as hybrid systems to give a theoretical study. Switched systems belong to hybrid systems, which consists of a set of continuous or discrete-time subsystems and a switching signal specifying the switching order among the subsystems [36]. The switching signal plays a vital role in the stability and performance analysis of the switched system. For the bygone decades, several switching signal techniques have been developed to ensure the stability of the switched systems with stable or unstable subsystems, such as maximum or minimum dwell-time method [12], ADT method [13], PDT method [37] etc. Based on the obtained switching techniques, the stability and performance analysis for switched systems have been in-depth investigated [38, 39].

Recently, observer-based fault diagnosis and FTC technology for the switched system has attracted more and more researchers' interests. In [40], a fault detector has been designed to guarantee the robustness to the disturbances for discrete-time switched systems subject to state delay. Under the case of asynchronous between the observer and the original system, the FD filter design method was proposed in [41]. For nonlinear switched stochastic systems, [42] proposed a fuzzy FD filter design programme. Based on the descriptor observer method, the problem of sensor FD has been studied in [43] for discrete-time switched systems. Sliding-mode observer-based approach, sensor FD, and estimation for switched systems was studied [44]. Actuator and sensor FE was, respectively, investigated in [45, 46] by utilizing UIO and reduced-order observer. PIO-based method, non-fragile actuator FE, and fault-tolerant controller design for switched systems are studied in [47]. Encouraged by these results, we are more interested in developing a PIO approach to attain multiple FD and FE for the switched system.

In this section, we extended the results in [47] of developing a fault diagnosis observer for the switched system subjected to the parameter uncertainty, unknown process disturbances, actuator faults, and sensor faults. Specifically, the multi-objective H_∞/H_- condition is exploited to attain the optimize robustness to the unknown input and sensitivity to the fault. Based on the obtained fault diagnosis observer, an accurate FE for the actuator and sensor faults can be provided. To sum up, the innovation points of this paper are summarized as follows:

- A multiple FD observer is designed for the switched system subjected to system uncertainties, unknown inputs, actuator, and sensor faults;
- By using ADT method, sufficient H_∞/H_- conditions are obtained to guarantee the robustness to the system uncertainty and unknown input and sensitivity to the fault;
- PIO-based FE algorithm is proposed, which can give an accurate actuator and sensor FE.

2.2.2 Problem Description

In this section, the following uncertain switched linear system are considered:

$$\begin{cases} \dot{x}(t) = \sum_{i=1}^{N} \sigma_i(t)(\tilde{A}_i x(t) + B_i u(t) + D_i d(t) + F_{1i} f(t)) \\ y(t) = \sum_{i=1}^{N} \sigma_i(t)(C_i x(t) + F_{2i} f(t)) \end{cases} \tag{2.56}$$

in which $x(t) \in \mathbb{R}^n$, $u(t) \in \mathbb{R}^p$, $d(t) \in \mathbb{R}^{n_d}$, and $y(t) \in \mathbb{R}^m$ are the state vector, the control input vector, the unknown input vector, and the output vector. $f(t) \in \mathbb{R}^{n_f}$ denotes the fault vector appearing in actuator and sensor meanwhile. N is the total number of the subsystems. $\sigma_i(t)$ is called switching signal and satisfies

$$\sigma_i(t) : [0, \infty) \to \{0, 1\}, \quad \sum_{i=1}^{N} \sigma_i(t) = 1. \tag{2.57}$$

It is noted that the switching signal is a piecewise constant function depending on the time t, and the ith mode will be activated when $\sigma_i(t) = 1$. A_i, B_i, C_i, D_i, F_{1i}, and F_{2i} are constant real matrices with appropriate dimensions. The uncertain system matrices $\tilde{A}_i$ satisfy

$$\tilde{A}_i = A_i + \Delta A_i \tag{2.58}$$

$$\Delta A_i = M_i F(t) N_i, \tag{2.59}$$

$$F^T(t)F(t) \leq I, \tag{2.60}$$

where $M_i \in \mathbb{R}^{n \times n_F}$ and $N_i \in \mathbb{R}^{n_{F'} \times n}$, $i = 1, \ldots, N$ are constant matrices, $F(t) \in \mathbb{R}^{n_F \times n_{F'}}$ represents uncertain matrix satisfying the condition in (2.60) and with Lebesgue measurable elements.

Remark 2.4 When $f(t) = f_a(t)$ and $F_{2i} = 0$, which implies only actuator fault occurs in the system. Correspondingly, when $f(t) = f_s(t)$ and $F_{1i} = 0$, which means only sensor fault occurs in the system. If the actuator fault and the sensor fault simultaneously appear in the system, we can define the fault signal as $f(t) = \left[f_a^T(t) \; f_s^T(t) \right]^T$, and the coefficients of the fault are decomposed as $F_{1i}(t) = \left[F_{a1i} \; F_{s1i} \right]$ and $F_{2i}(t) = \left[F_{a2i} \; F_{s2i} \right]$, where F_{a1i}, F_{a2i} and F_{s1i}, F_{s2i}, respectively, represent the coefficients of the actuator fault and the sensor fault with appropriate dimensions.

Remark 2.5 In practical applications, the occurrence of parameter uncertainties are unavoidable, which will affect the stability and performance of the dynamics if they are not treated properly. In this section, the uncertainty term is assumed to be norm-bounded. This form was first proposed in [21] and has been widely used to deal with non-fragile control issues [22].

For the main object of this section, the following assumptions are given:

- The pair of (A_i, C_i), $i = 1, 2, \ldots, N$, are detectable.
- The matrices F_{2i} are full column rank.
- The actuator fault is piecewise constant fault, i.e. $\dot{f}(t) = 0$.

Remark 2.6 In this section, we only focus on a class of constant faults or piecewise constant faults. From the perspective of practical engineering, this type of fault is very common. From this opinion, the consideration of this paper is general and practical. Moreover, FE for this kind of fault has been widely investigated and can be extended to other fault cases [20].

For simplicity, we rewrite the Eq. (2.56) as follows:

$$\begin{cases} \dot{x}(t) = \sum_{i=1}^{N} \sigma_i(t)(\tilde{A}_i x(t) + \tilde{B}_i \tilde{d}(t) + F_{1i} f(t)), s \\ y(t) = \sum_{i=1}^{N} \sigma_i(t)(C_i x(t) + F_{2i} f(t)), \end{cases} \tag{2.61}$$

where $\tilde{B}_i = \left[B_i \; D_i \right], \tilde{d}(t) = \left[u^T(t) \; d^T(t) \right]^T$.

The following fault diagnosis observer is employed to generate the residual signal:

$$\begin{cases} \dot{\hat{x}}(t) = \sum_{i=1}^{N} \sigma_i(t)(A_i\hat{x}(t) + B_iu(t) + L_i(y(t) - \hat{y}(t)) + F_{1i}\hat{f}(t)), \\ \hat{y}(t) = \sum_{i=1}^{N} \sigma_i(t)(C_i\hat{x}(t) + F_{2i}\hat{f}(t)), \\ \dot{\hat{f}}(t) = \sum_{i=1}^{N} \sigma_i(t)H_i(y(t) - \hat{y}(t)) - \hat{f}(t), \\ r(t) = y(t) - \sum_{i=1}^{N} \sigma_i(t)C_i\hat{x}(t), s \end{cases} \tag{2.62}$$

where $\hat{x}(t) \in \mathbb{R}^n$ and $\hat{f}(t) \in \mathbb{R}^{n_f}$ are, respectively, the estimation of the state and the fault of the system (2.56). $r(t) \in \mathbb{R}^m$ is the residual signal for late use to detect the fault. $L_i, H_i,\ i = 1, \ldots, N$ are the observer gains to be determined later.

Define

$$e(t) = x(t) - \hat{x}(t), \quad e_f(t) = f(t) - \hat{f}(t). \tag{2.63}$$

From (2.61) and (2.62), the derivative of the state error can be deduced as follows:

$$\begin{aligned} \dot{e}(t) &= \dot{x}(t) - \dot{\hat{x}}(t) \\ &= \sum_{i=1}^{N} \sigma_i(t)((A_i - L_iC_i)e(t) + (F_{1i} - L_iF_{2i})e_f(t) + \bar{D}_i\tilde{d}(t) + \Delta A_ix(t)), \end{aligned} \tag{2.64}$$

where $\bar{D}_i = \begin{bmatrix} 0 & D_i \end{bmatrix}$.

Then, the derivative of the fault error is given as

$$\begin{aligned} \dot{e}_f(t) &= \dot{f}(t) - \dot{\hat{f}}(t) \\ &= -\sum_{i=1}^{N} \sigma_i(t)(H_iC_ie(t) + H_iF_{2i}e_f(t)) + \dot{f}(t) + f(t) - e_f(t). \end{aligned} \tag{2.65}$$

In this section, our goal is to design an FD observer (2.61) such that the residual signal $r(t)$ is robust to the disturbance $d(t)$ and sensitive to the fault $f(t)$. Therefore, the FD problem can be formulated as an H_∞/H_- optimization: for some scalars $\gamma > 0$ and $\beta > 0$, such that

$$\sup_{d(t)\neq 0} \frac{\|r(t)\|_2}{\|d(t)\|_2} < \gamma, \ \ and \quad \inf_{f(t)\neq 0} \frac{\|r(t)\|_2}{\|f(t)\|_2} > \beta.$$

2.2.3 Main Results

In this subsection, an H_∞/H_- fault diagnosis observer is designed.

2.2.3.1 H_- Fault Sensitivity Conditions

Firstly, the H_- fault sensitivity conditions are addressed. By considering the augmented system is only affected by the influence of the fault on residual signal, and define $\xi(t) = \left[e^T(t)\ e_f^T(t)\ x^T(t)\right]^T$, then the following augmented systems can be obtained:

$$\begin{cases} \dot{\xi}(t) = \sum\limits_{i=1}^{N} \sigma_i(t)(\tilde{\mathscr{A}_i}\xi(t) + \tilde{F}_i f(t)), \\ r(t) = \sum\limits_{i=1}^{N} \sigma_i(t)(\tilde{C}_i\xi(t) + F_{2i} f(t)), \end{cases} \tag{2.66}$$

where

$$\tilde{\mathscr{A}_i} = \begin{bmatrix} A_i - L_i C_i & F_{1i} - L_i F_{2i} & \Delta A_i \\ -H_i C_i & -H_i F_{2i} - I & 0 \\ 0 & 0 & A_i + \Delta A_i \end{bmatrix}, \tilde{F}_i = \begin{bmatrix} 0 \\ I \\ F_{1i} \end{bmatrix}, \tilde{C}_i = \begin{bmatrix} C_i\ 0\ 0 \end{bmatrix}. \tag{2.67}$$

Theorem 2.3 *For given scalars $\beta > 0$, $\varepsilon_1 > 0$ and $\mu > 1$, if there exist positive definite matrices $P_{1i} > 0$, $P_{2i} > 0$, $P_{3i} > 0$, matrices $\bar{L}_i$ and $\bar{H}_i$, which satisfy*

$$P_{1i} \le \mu P_{1j}, \quad P_{2i} \le \mu P_{2j}, \quad P_{3i} \le \mu P_{3j} \tag{2.68}$$

$$\begin{bmatrix} \Phi_{11i} & \Phi_{12i} & 0 & -C_i^T F_{2i} & C_i^T & P_{1i} M_i & 0 \\ * & \Phi_{22i} & 0 & P_{2i} & 0 & 0 & 0 \\ * & * & \Phi_{33i} & P_{3i} F_{1i} & 0 & P_{3i} M_i & N_i^T \\ * & * & * & \beta^2 I - F_{2i}^T F_{2i} & 0 & 0 & 0 \\ * & * & * & * & -I & 0 & 0 \\ * & * & * & * & * & -\varepsilon_1 I & 0 \\ * & * & * & * & * & * & -\varepsilon_1^{-1} I \end{bmatrix} < 0, \tag{2.69}$$

where $\Phi_{11i} = sym(P_{1i}A_i - \bar{L}_i C_i) + \alpha P_{1i}$, $\Phi_{12i} = P_{1i}F_{1i} - \bar{L}_i F_{2i} - C_i^T \bar{H}_i^T$, $\Phi_{22i} = sym(-\bar{H}_i F_{2i} - P_{2i}) + \alpha P_{2i}$, $\Phi_{33i} = sym(P_{3i}A_i) + \alpha P_{3i}$, *then, the system (2.66) is GUAS for any arbitrary switching signal satisfying (2.13) and has an H_- performance index β. Under this case, the FD observer parameters of (2.62) can be designed as $L_i = P_{1i}^{-1}\bar{L}_i$, $H_i = P_{2i}^{-1}\bar{H}_i$.*

Proof Firstly, we establish the stability of system (2.66). When $f(t) = 0$, we define the following switched Lyapunov function:

$$V_i(t,\xi(t)) = \xi^T(t)\left(\sum_{i=1}^{N}\sigma_i(t)P_i\right)\xi(t), \tag{2.70}$$

where $P_i = diag\{P_{1i}, P_{2i}, P_{3i}\}$, $P_{1i} > 0$, $P_{2i} > 0$, $P_{3i} > 0$. By substituting (2.66) into (2.70), one has

$$\begin{aligned}
&\dot{V}_i(t,\xi(t)) + \alpha V_i(t,\xi(t)) \\
&= \dot{\xi}^T(t)\left(\sum_{i=1}^{N}\sigma_i(t)P_i\right)\xi(t) + \xi^T(t)\left(\sum_{i=1}^{N}\sigma_i(t)P_i\right)\dot{\xi}(t) \\
&+\xi^T(t)\left(\sum_{i=1}^{N}\sigma_i(t)\alpha P_i\right)\xi(t) \\
&= \sum_{i=1}^{N}\sigma_i(t)\xi^T(t)\Phi_i\xi(t),
\end{aligned}$$

where $\Phi_i = [\Phi_{ai}, \Phi_{bi}, \Phi_{ci}]$,

$$\begin{aligned}
\Phi_{ai} &= \begin{bmatrix} sym(P_{1i}A_i - \bar{L}_iC_i) + \alpha P_{1i} \\ P_{1i}F_{1i} - \bar{L}_iF_{2i} - C_i^T\bar{H}_i^T \\ P_{1i}\Delta A_i \end{bmatrix}, \\
\Phi_{bi} &= \begin{bmatrix} P_{1i}F_{1i} - \bar{L}_iF_{2i} - C_i^T\bar{H}_i^T \\ sym(-\bar{H}_iF_{2i} - P_{2i}) + \alpha P_{2i} \\ 0 \end{bmatrix}, \\
\Phi_{ci} &= \begin{bmatrix} P_{1i}\Delta A_i \\ 0 \\ sym(P_{3i}(A_i + \Delta A_i)) + \alpha P_{3i} \end{bmatrix}.
\end{aligned} \tag{2.71}$$

On the other hand,

$$\Phi_i = \Phi_{1i} + \bar{M}_i^T F(t)\bar{N}_i + \bar{N}_i^T F^T(t)\bar{M}_i, \tag{2.72}$$

where $\Phi_{1i} = [\Phi_{di}, \Phi_{ei}, \Phi_{fi}]$

$$\begin{aligned}
\Phi_{di} &= \begin{bmatrix} sym(P_{1i}A_i - \bar{L}_iC_i) + \alpha P_{1i} \\ P_{1i}F_{1i} - \bar{L}_iF_{2i} - C_i^T\bar{H}_i^T \\ 0 \end{bmatrix}, \\
\Phi_{ei} &= \begin{bmatrix} P_{1i}F_{1i} - \bar{L}_iF_{2i} - C_i^T\bar{H}_i^T \\ sym(-\bar{H}_iF_{2i} - P_{2i}) + \alpha P_{2i} \\ 0 \end{bmatrix},
\end{aligned}$$

$$\Phi_{fi} = \begin{bmatrix} 0 \\ 0 \\ sym(P_{3i}A_i) + \alpha P_{3i} \end{bmatrix},$$

$$\bar{M}_i = \begin{bmatrix} M_i^T P_{1i} & 0 & M_i^T P_{3i} \end{bmatrix}, \quad \bar{N}_i = \begin{bmatrix} 0 & 0 & N_i \end{bmatrix}.$$

By Lemma 2.2, one can get that there exists a scalar $\varepsilon_1 > 0$ such that

$$\Phi_i \leq \Phi_{1i} + \varepsilon_1^{-1}\bar{M}_i^T\bar{M}_i + \varepsilon_1\bar{N}_i^T\bar{N}_i. \tag{2.73}$$

From (2.69), which means

$$\Phi_{1i} + \varepsilon_1^{-1}\bar{M}_i^T\bar{M}_i + \varepsilon_1\bar{N}_i^T\bar{N}_i < 0. \tag{2.74}$$

Equation (2.74) combines with the condition (2.68) and Lemma 2.1, which implies the system (2.66) is asymptotically stable.

Consider the system (2.66), we introduce the following performance index:

$$J = \int_0^t (\beta^2 f^T(t)f(t) - r^T(t)r(t))dt \tag{2.75}$$

one can get

$$\begin{aligned} J &= \int_0^t (\beta^2 f^T(t)f(t) - r^T(t)r(t) + \dot{V}_i(t, \xi(t)))dt - V_i(t, \xi(t)) \\ &\leq \int_0^t (\beta^2 f^T(t)f(t) - r^T(t)r(t) + \dot{V}_i(t, \xi(t)) \\ &+\alpha V_i(t, \xi(t)))dt - V_i(t, \xi(t)). \end{aligned} \tag{2.76}$$

By (2.69), which means

$$\beta^2 f^T(t)f(t) - r^T(t)r(t) + \dot{V}_i(t, \xi(t)) + \alpha V_i(t, \xi(t)) < 0. \tag{2.77}$$

Then we have $J < 0$, which means $\|r(t)\|_2 > \beta\|f(t)\|_2$ when $t \to \infty$. The proof is completed. (□)

2.2.3.2 H_∞ Robustness Conditions

In the following, the H_∞ robustness conditions are addressed in this subsection. To this end, we define $\xi(t) = \begin{bmatrix} e^T(t) & e_f^T(t) & x^T(t) \end{bmatrix}^T$, the augmented system is first rewritten in the following fault-free form:

$$\begin{cases} \dot{\xi}(t) = \sum_{i=1}^{N} \sigma_i(t)(\tilde{\mathscr{A}_i}\xi(t) + \tilde{D}_i d(t)), \\ r(t) = \sum_{i=1}^{N} \sigma_i(t)\tilde{C}_i\xi(t), \end{cases} \tag{2.78}$$

where

$$\tilde{\mathscr{A}_i} = \begin{bmatrix} A_i - L_iC_i & F_{1i} - L_iF_{2i} & \Delta A_i \\ -H_iC_i & -H_iF_{2i} - I & 0 \\ 0 & 0 & A_i + \Delta A_i \end{bmatrix}, \tilde{D}_i = \begin{bmatrix} \bar{D}_i \\ 0 \\ \bar{B}_i \end{bmatrix}, \tilde{C}_i = \begin{bmatrix} C_i \; 0 \; 0 \end{bmatrix}. \tag{2.79}$$

Theorem 2.4 *For given scalars $\gamma > 0$, $\varepsilon_2 > 0$ and $\mu > 1$, if there exist positive definite matrices $P_{1i} > 0$, $P_{2i} > 0$, $P_{3i} > 0$, matrices $\bar{L}_i$ and $\bar{H}_i$, which satisfy*

$$P_{1j} \leq \mu P_{1i}, \quad P_{2j} \leq \mu P_{2i}, \quad P_{3j} \leq \mu P_{3i} \tag{2.80}$$

$$\begin{bmatrix} \Phi_{11i} & \Phi_{12i} & 0 & P_{1i}\bar{D}_i & C_i^T & P_{1i}M_i & 0 \\ * & \Phi_{22i} & 0 & P_{2i} & 0 & 0 & 0 \\ * & * & \Phi_{33i} & P_{3i}\bar{B}_i & 0 & P_{3i}M_i & N_i^T \\ * & * & * & -\gamma^2 I & 0 & 0 & 0 \\ * & * & * & * & -I & 0 & 0 \\ * & * & * & * & * & -\varepsilon_2 I & 0 \\ * & * & * & * & * & * & -\varepsilon_2^{-1} I \end{bmatrix} < 0, \tag{2.81}$$

where $\Phi_{11i} = sym(P_{1i}A_i - \bar{L}_iC_i) + \alpha P_{1i}$, $\Phi_{12i} = P_{1i}F_{1i} - \bar{L}_iF_{2i} - C_i^T\bar{H}_i^T$, $\Phi_{22i} = sym(-\bar{H}_iF_{2i} - P_{2i}) + \alpha P_{2i}$, $\Phi_{33i} = sym(P_{3i}A_i) + \alpha P_{3i}$, then, the system (2.78) is GUAS for any arbitrary switching signal satisfying (2.13) and has an H_∞ performance index γ. Under this case, the FD observer parameters of (2.62) can be designed as $L_i = P_{1i}^{-1}\bar{L}_i$, $H_i = P_{2i}^{-1}\bar{H}_i$.

Proof According to similar proof line of Theorem 2.3, the conclusion can be easily derived.

Consider the system (2.78), we define the following performance index:

$$J = \int_0^t (r^T(t)r(t) - \gamma^2 d^T(t)d(t))dt \tag{2.82}$$

one can get

$$
\begin{aligned}
J &= \int_0^t (r^T(t)r(t) - \gamma^2 d^T(t)d(t) + \dot{V}_i(t,\xi(t)))dt - V_i(t,\xi(t)) \\
&\leq \int_0^t (r^T(t)r(t) - \gamma^2 d^T(t)d(t) + \dot{V}_i(t,\xi(t)) \\
&+ \alpha V_i(t,\xi(t)))dt - V_i(t,\xi(t)).
\end{aligned} \tag{2.83}
$$

By (2.81), which means

$$
r^T(t)r(t) - \gamma^2 d^T(t)d(t) + \dot{V}_i(t,\xi(t)) + \alpha V_i(t,\xi(t)) < 0. \tag{2.84}
$$

Then we have $J < 0$, which means $\|r(t)\|_2 < \gamma\|d(t)\|_2$ when $t \to \infty$. The proof is completed. (□)

2.2.3.3 Residual Evaluation

According to ordinary technical approach, the root mean square value (RMSV) of the residual signal is widely used in threshold computation, which can measure the average energy of a signal over a time interval. Then, we give the residual evaluation function as follows:

$$
J(r(t)) = \|r\|_{RMSV} = \left(\frac{1}{T}\int_0^T r^T(t)r(t)dt\right)^{1/2} \tag{2.85}
$$

Due to the determination of the evaluation function, the threshold value is also determined, the threshold value J_{th} can be represented as

$$
J_{th} = \sup_{d(t)\in L_2[0,\infty),\, f=0} J(r(t)). \tag{2.86}
$$

Therefore, FD can be implemented through the following decision logic:

$$
\|r\| \geq J_{th} \Rightarrow with\ faults \Rightarrow alarm, \tag{2.87}
$$

$$
\|r\| < J_{th} \Rightarrow fault\ free. \tag{2.88}
$$

Remark 2.7 In the Theorems 2.3 and 2.4, one can get that there are N LMIs should be simultaneously solved for each condition. Moreover, the iteration conditions in (2.68) and (2.80) further increase the calculation amount. The above reasons may lead to a computational complexity of the proposed methods.

Remark 2.8 In this section, a PIO is borrowed as a fault diagnosis observer. Under this technology, multiple FD and FE problems are solved. Compared with other observer-based fault diagnosis methods, it has the advantages of easy implementation and wide application in practical engineering systems.

2.2.4 An Illustrative Example

To verify the effectiveness of the designed fault diagnosis observer, a two-tank liquid-level control system in the Sect. 1.2.8 is employed to verify the proposed techniques, which is shown in the Fig. 1.10. The system parameters are given as follows:

$$A_1 = 10^{-4} * \begin{bmatrix} -7 & 7 \\ 11 & -11 \end{bmatrix}, A_2 = 10^{-4} * \begin{bmatrix} -13 & 7 \\ 11 & -11 \end{bmatrix}, A_3 = 10^{-4} * \begin{bmatrix} -7 & 7 \\ 11 & -44 \end{bmatrix},$$
$$B_1 = B_2 = B_3 = \begin{bmatrix} 0.2 \\ 0.1 \end{bmatrix} F_{11} = F_{12} = F_{13} = \begin{bmatrix} 2 \\ 3 \end{bmatrix}, C_1 = C_2 = C_3 = \begin{bmatrix} 1 & 2 \end{bmatrix},$$
$$F_{21} = F_{22} = F_{23} = 0.2.$$

We set $\mu = 1.002$ and $\varepsilon = 0.2$. By applying the results in Theorem 2.3, the optimizing performance indices $\alpha = 0.001$ and $\beta = 0.5$ with the parameters of the fault diagnosis observer L_i and H_i are obtained as follows:

$$\bar{L}_1 = 10^7 * \begin{bmatrix} 4.6576 \\ 9.2737 \end{bmatrix}, \quad \bar{L}_2 = 10^7 * \begin{bmatrix} 4.7085 \\ 9.3855 \end{bmatrix}, \bar{L}_3 = 10^7 * \begin{bmatrix} 4.7160 \\ 9.3021 \end{bmatrix},$$
$$\bar{H}_1 = 3.6718 * 10^8, \bar{H}_2 = 3.5738 * 10^8, \quad \bar{H}_3 = 3.6924 * 10^8.$$

Furthermore, one can get the ADT value is $\tau_a^* = \frac{ln\mu}{\alpha} = 1.9980$. The switching signal is described in Fig. 2.6, which satisfies the ADT condition in (2.13). The outside disturbance signal is set as a white noise, which is shown in Fig. 2.7. Firstly, a constant fault $f_1(t)$ is given by

$$f(t) = \begin{cases} 2; & 20s \le t < 60s \\ 0; & others \end{cases}$$

which is described in Fig. 2.8. The contrasting results of the residual signal $r(t)$ affected by the fault $f(t)$ and fault free is shown in Fig. 2.9. Figure 2.10 gives the state response of the residual evaluation function $J(r(t))$ with fault and fault free.

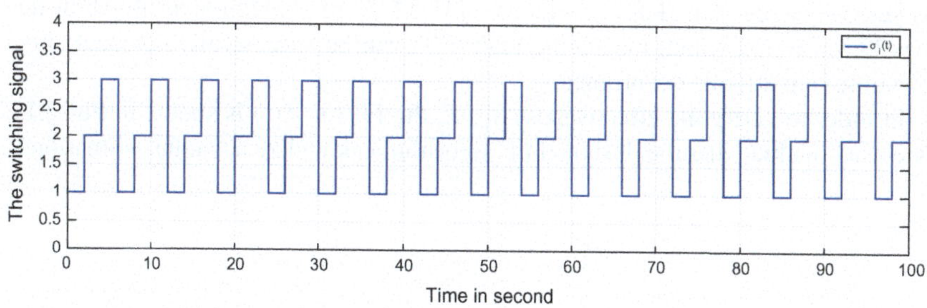

Fig. 2.6 The switching signal $\sigma_i(t)$

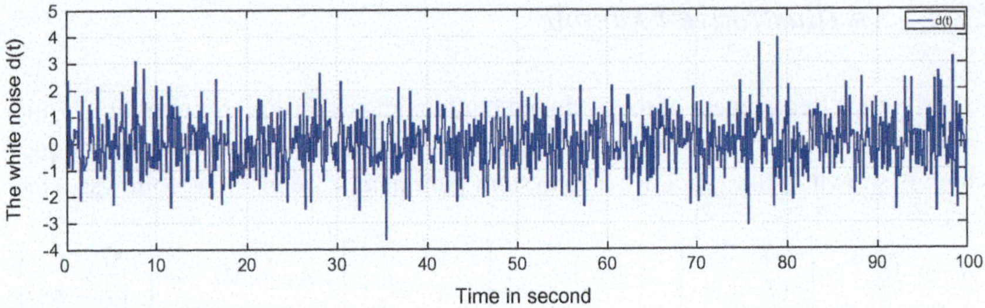

Fig. 2.7 The white noise $d(t)$

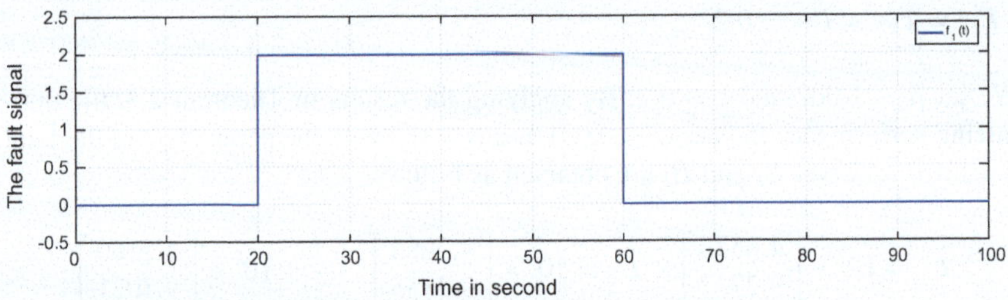

Fig. 2.8 The fault signal $f(t)$

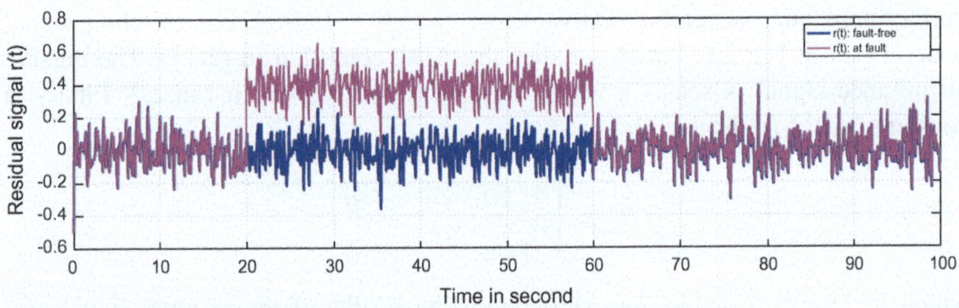

Fig. 2.9 The Residual signal $r(t)$

By selecting a predefined threshold $J_{th} = 131.5246$, the simulation results show that $J(r(23.8)) > J_{th}$, which means the fault $f(t)$ can be successfully detected about four time step after its occurrence.

In order to verify the effectiveness of FE, the FE for $f(t)$ is shown in Fig. 2.11. One can conclude that the designed FE algorithm can realize accurate estimation.

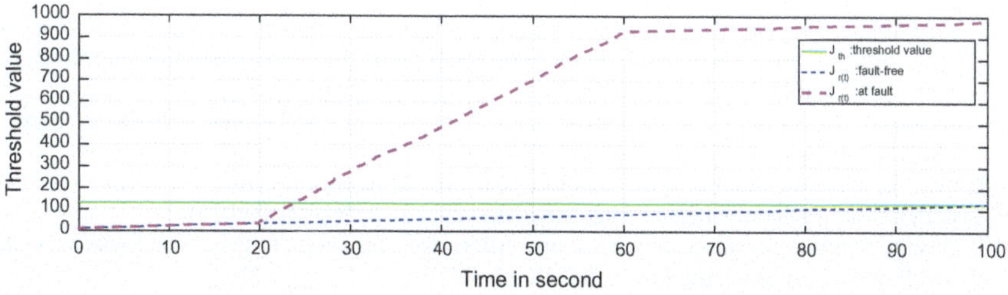

Fig. 2.10 Threshold value J_{th}

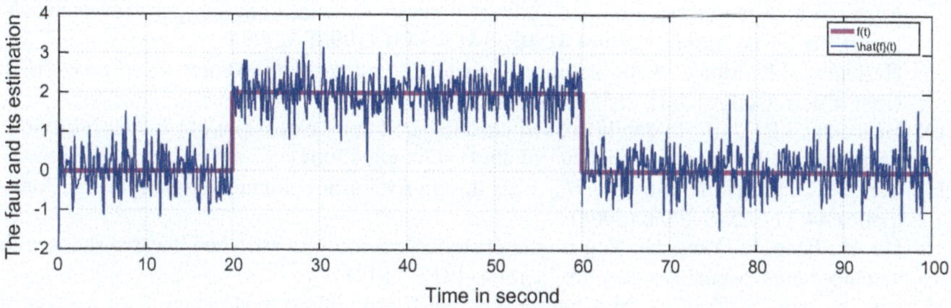

Fig. 2.11 Threshold value $f(t)$ and its estimation $\hat{f}(t)$

2.2.5 Summary

The main purpose of this section is to address the design of a fault diagnosis observer for switched systems. Based on switched Lyapunov function method and ADT technique, sufficient conditions are obtained to construct a desired FD observer. At the same time, an efficient FE algorithm is provided.

References

1. He, X., Wang, Z., Qin, L., Zhou, D.: Active fault-tolerant control for an internet-based networked three-tank system. IEEE Trans. Control Syst. Technol. **24**(6), 2150–2157 (2016)
2. Van, M., Ge, S., Ren, H.: Robust fault-tolerant control for a class of second-order nonlinear systems using an adaptive third-order sliding mode control. IEEE Trans. Syst. Man Cybern. Syst. **47**(2), 221–228 (2017)
3. Zhang, L., Chen, M., Wu, Q., Wu, B.: FTC for uncertain networked control systems with induced delays and actuator saturation. IEEE Access **4**, 6574–6584 (2016)
4. Yadegar, M., Afshar, A., Meskin, N.: "Fault-tolerant control of non-linear systems based on adaptive virtual actuator. IET Control Theory Appl. **11**(9), 1371–1379 (2017)
5. Yang, H., Jiang, B., Cocquempot, V.: Qualitative fault tolerance analysis for a class of hybrid systems. Nonlinear Anal. Hybrid Syst. **2**(3), 846–861 (2008)
6. Du, D., Tan, Y., Zhang, Y.: Dynamic output feedback FTCler design for discrete-time switched systems with actuator fault. Nonlinear Anal. Hybrid Syst. **16**, 93–103 (2015)

7. Pavlichkov, S.S., Dashkovskiy, S.N., Pang, C.K.: Uniform stabilization of nonlinear systems with arbitrary switchings and dynamic uncertainties. IEEE Trans. Autom. Control **62**(5), 2207–2222 (2017)
8. Xiang, W.: Necessary and sufficient condition for stability of switched uncertain linear systems under dwell-time constraint. IEEE Trans. Autom. Control **61**(11), 3619–3624 (2016)
9. Wu, L., Yang, R., Shi, P., Su, X.: Stability analysis and stabilization of 2-D switched systems under arbitrary and restricted switchings. Automatica **59**, 206–215 (2015)
10. Xiang, W.: On equivalence of two stability criteria for continuous-time switched systems with dwell time constraint. Automatica **54**, 36–40 (2015)
11. Briat, C.: Convex lifted conditions for robust image-stability analysis and image-stabilization of linear discrete-time switched systems with minimum dwell-time constraint. Automatica **50**(3), 976–983 (2014)
12. Morse, A.S.: Supervisory control of families of linear set-point controllers part I: exact matching. IEEE Trans. Autom. Control **41**(10), 1413–1431 (1996)
13. Hespanha, J.P., Morse, A.S.: Stability of switched systems with average dwell time. IEEE Conf. Decis. Control **3**(3), 2655–2660 (2002)
14. Hespanha, J.P.: Uniform stability of switched linear systems extensions of Lasalles invariance principle. IEEE Trans. Autom. Control **49**(4), 470–482 (2004)
15. Yang, G., Che, W.: Non-fragile H_∞ filter design for linear continuous-time systems. Automatica **44**(11), 2849–2856 (2008)
16. Hu, H., Jiang, B., Yang, H.: Non-fragile H_2 reliable control for switched linear systems with actuator faults. Signal Process. **93**(7), 1804–1812 (2013)
17. Liu, Y., Niu, Y., Zou, Y.: Non-fragile observer-based sliding mode control for a class of uncertain switched systems. J. Frankl. Inst. **351**(2), 952–963 (2014)
18. Wang, D., Shi, P., Wang, W., Karimi, H.R.: Non-fragile H_∞ control for switched stochastic delay systems with application to water quality process. Int. J. Robust Nonlinear Control **24**(11), 1677–1693 (2014)
19. Gao, Z., Liu, X., Chen, Q.: Unknown input observer-based robust fault estimation for systems corrupted by partially decoupled disturbances. IEEE Trans. Ind. Electron. **63**(4), 2537 (2016)
20. Zhang, K., Jiang, B., Cocquempot, V.: Robust fault estimation observer design with finite-time convergence specification. IET Control Theory Appl. **11**(1), 1–9 (2017)
21. Petersen, I.R.: A stabilization algorithm for a class of uncertain linear systems. Syst. Control Lett. **8**(4), 351–357 (1987)
22. Li, F., Shi, P., Wu, L., Zhang, X.: Fuzzy-model-based D-stability and nonfragile control for discrete-time descriptor systems with multiple delays. IEEE Trans. Fuzzy Syst. **22**(4), 1019–1025 (2014)
23. Chang, X., Yang, G.: Nonfragile H_∞ filter design for T-S fuzzy systems in standard form. IEEE Trans. Ind. Electron. **61**(7), 3448–3458 (2014)
24. Gahinet, P., Apkarian, P.: An linear matrix inequality approach to H_∞ control. Int. J. Robust Nonlinear Control **4**, 422–448 (1994)
25. Zhou, L., Ho, D.W.C., Zhai, G.: Stability analysis of switched linear singular systems. Automatica **49**, 1481–1487 (2013)
26. Lien, C.H.: Delay-dependent stability criteria for uncertain neutral systems with multiple time-varying delays via LMI approach. IEE Proc. Control Theory Appl. **152**(6), 707–714 (2005)
27. Adams, R.J., Buffington, J.M., Sparks, A.G., Banda, S.S.: Robust Multivariable Flight Control. Springer, New York, NY, USA (2012)
28. Merzouki, R., Samantaray, A.K., Pathak, P.M., Bouamama, B.O.: Model-Based Fault Diagnosis and FTC. Springer, Berlin (2012)
29. Jiang, B., Zhang, K., Shi, P.: Integrated fault estimation and accommodation design for discrete-time Takagi-Sugeno fuzzy systems with actuator faults. IEEE Trans. Fuzzy Syst. **19**(2), 291–304 (2011)
30. Su, Q., Fan, Z., Zhang, D., Li, J.: Finite-time fault detection filtering for switched singular systems with all modes unstable: an ADT approach. Int. J. Control Autom. Syst. **17**(8), 2026–2036 (2019)

31. Yin, S., Gao, H., Qiu, J., Kaynak, O.: Descriptor reduced-order sliding mode observers design for switched systems with sensor and actuator faults. Automatica **76**, 282–292 (2017)
32. Xie, X., Yue, D., Hu, S.: Fault estimation observer design of discrete-time nonlinear systems via a joint real-time scheduling law. IEEE Trans. Syst. ManCybern. Syst. **47**(7), 1451–1463 (2017)
33. Liu, G., Cao, Y., Chang, X.: Fault detection observer design for fuzzy systems with local nonlinear models via fuzzy Lyapunov function. Int. J. Control Autom. Syst. **15**(5), 2233–2242 (2017)
34. Guo, S., Zhu, F.: Acuator fault detection and interval reconstruction based on interval observers. IFAC PapersOnLine **50**(1), 5061–5066 (2017)
35. Chen, H., Du, D., Zhu, D., Yang, Y.: UIO-based fault estimation and accommodation for nonlinear switched systems. Int. J. Control Autom. Syst. **17**(2), 435–444 (2019)
36. Xiao, X., Park, J.H., Zhou, L., Lu, G.: Event-triggered control of discrete-time switched linear systems with network transmission delays. Automatica **111**, (2020). https://doi.org/10.1016/j.automatica.2019.108585
37. Zhang, L., Zhuang, S., Shi, P.: Non-weighted quasi-time-dependent H_∞ filtering for switched linear systems with persistent dwell-time. Automatica **54**, 201–209 (2015)
38. Daafouz, J., Riedinger, P., Iung, C.: Stability analysis and control synthesis for switched systems: a switched Lyapunov function approach. IEEE Trans. Autom. Control **47**(11), 1883–1887 (2002)
39. Zhang, L., Shi, P.: Stability, l_2-gain and asynchronous H_∞ control of discrete-time switched systems with average dwell time. IEEE Trans. Autom. Control **54**(9), 2192–2199 (2009)
40. Wang, D., Wang, W., Shi, P.: Robust fault detection for switched linear systems with state delays. IEEE Trans. Syst. Man Cybern. B Cybern. **39**(3), 800–805 (2009)
41. Du, D., Jiang, B., Shi, P., Karimi, H.R.: Fault detection for continues-time switched systems under asynchronous switching. Int. J. Robust Nonlinear Control **24**(11), 1694–1706 (2014)
42. Su, X., Shi, P., Wu, L., Song, Y.: Fault detection filtering for nonlinear switched stochastic systems. IEEE Trans. Autom. Control **61**(5), 1310–1315 (2016)
43. Du, D.: Fault detection for discrete-time linear systems based on descriptor observer approach. Appl. Math. Comput. **293**(15), 575–585 (2017)
44. Li, J., Pan, K., Zhang, D., Su, Q.: Robust fault detection and estimation observer design for switched systems. Nonlinear Anal. Hybrid Syst. **34**, 30–42 (2019)
45. Du, D., Cocquempot, V., Jiang, B.: Robust fault estimation observer design for switched systems with unknown input. Appl. Math. Comput. **348**, 70–83 (2019)
46. Han, J., Liu, X., Wei, X., Hu, X., Zhang, H.: Reduced-order observer based fault estimation and fault-tolerant control for switched stochastic systems with actuator and sensor faults. ISA Trans. **88**, 91–101 (2019)
47. Du, D., Cocquempot, V.: Non-fragile actuator fault estimation and accommodation for switched systems. IFAC-PapersOnLine **51**(24), 110–116 (2018)

Chapter 3
Reduced-Order Observer-Based FEA for Continuous-Time Switched Systems

This section investigates the problems of actuator FEA for discrete-time switched systems with state delay. By using reduced-order observer method and switched Lyapunov function technique, an FE algorithm is achieved for the discrete-time switched system with actuator fault and state delay. Then based on the obtained online FE information, a switched dynamic output feedback controller is employed to compensate for the effect of faults by stabilizing the closed-loop systems. Finally, an example is proposed to illustrate the obtained results.

3.1 Introduction

In modern complex systems, dependability is as important as performances. Once a fault (sensor, actuator, or component failures) occurred in the system, the system's behavior may be drastically decreased, ranging from performance degradation to instability. Making a general survey of the related literature, FTC technique is an efficient method in order to reach the system objectives, or if this turns to be impossible, to achievable new objectives to avoid catastrophic behaviors. Therefore, FTC has been the subject of intensive investigations over the past two decades, and a number of achievements have been obtained, which can be found in several excellent papers, please refer to [1–5] and the references therein.

As for FTC, which can be divided into two main approaches: passive and active. While passive fault tolerance considers system faults as a special kind of uncertainties [6], active fault tolerance is based on FDI and accommodation technique [7]. Though FDI can give information about whether there exist faults occurring [8], the magnitude of the fault cannot be precisely provided. Therefore, the next step

D. Du et al., *Observer-Based Fault Diagnosis and Fault-Tolerant Control for Switched Systems*, Studies in Systems, Decision and Control 323,
https://doi.org/10.1007/978-981-15-9073-3_3

is to search for an efficient approach to estimate the magnitude of the fault, which is called fault estimate or fault reconstruction [9]. Finally, using the estimated fault information, a fault-tolerant controller can be designed to compensate for the effect of the fault. From the above discussion, it can be seen that FE plays a very important role in active FTC. Therefore, the study of FE has become a hot research topic owing to its importance in active FTC. During the past decade, various effective methods, such as sliding-mode observer approach using equivalent output injection signal [10], adaptive technique [11], learning method based on neural network [12], and so on, have been developed to realize FE. Generally speaking, adaptive technique is usually used to obtain actuator fault information for the continuous case. Another commonly employed technique to realize FE is based on descriptor observer technique, but it is usually utilized to estimate sensor fault [13]. Recently, reduced-order observer method is proposed to solve actuator FE for discrete-time systems, this motives us to investigate this issue for discrete-time switched systems.

Switched system is a class of hybrid systems, which consists of a finite number of subsystems and an associated switching signal governing the switching among them. Many real-world process and systems can be modeled as switched systems, including chemical processes, computer-controlled systems, switched circuits, and so on. Therefore, it has been intensively investigated in the past decades [14–20]. Compared with fruitful stability and stabilization results for switched systems, fault diagnosis and FTC achievements are relatively few. In [21, 22], FD for switched system was separately investigated for continuous case and discrete case. [13] investigated sensor FEA approaches for continuous-time switched systems, and [23, 24] separately studied actuator FEA approaches for continuous-time switched systems with constant time-delay case and time-varying delay case. However, to the best of our knowledge, the problems of actuator FE and compensation for discrete-time switched systems with state delay are still under research, which motivate us to study this meaningful and challenging topic.

Based on the above discussion, our objective of this section is to analyze and develop a general framework of FEA for discrete-time switched systems with actuator faults and state delay. The advantages of this section are shown in the following aspects: firstly, the actuator FE algorithm for discrete-time switched systems with state delay is resolved by using reduced-order observer approach; secondly, a switched dynamic output feedback controller is designed to compensate for the effect of actuator faults; thirdly, all the obtained efficient conditions are formulated in the form of LMI technique, which can be easily solved; finally, a simulation example is provided to illustrate the effectiveness of the proposed results.

3.2 Problem Statements and Preliminaries

Consider the following discrete-time switched linear system with state delay:

$$
\begin{aligned}
x(t+1) &= \sum_{i=1}^{N}\delta_i(t)A_i x(t) + \sum_{i=1}^{N}\delta_i(t)A_{di}x(t-d) + \sum_{i=1}^{N}\delta_i(t)B_i(u(t)+f(t)) \\
&\quad + \sum_{i=1}^{N}\delta_i(t)D_i\omega(t) && (3.1)\\
y(t) &= Cx(t) && (3.2)\\
z(t) &= \sum_{i=1}^{N}\delta_i(t)C_{2i}x(t), && (3.3)
\end{aligned}
$$

where $x(t) \in \mathbb{R}^n$ is the state vector, $u(t) \in \mathbb{R}^m$ is the control input vector, $y(t) \in \mathbb{R}^p$ is the measurable output vector, $z(t) \in \mathbb{R}^l$ is the controlled output, $f(t) \in \mathbb{R}^m$ represents the additive actuator fault, $\omega(t) \in \mathbb{R}^r$ is the disturbance which is assumed to belong to $l_2[0, \infty)$, and the positive integer d denotes the known state delay. The vector $\delta_i(t)$ is called switching signal, which specifies which subsystem will be activated at the discrete time t, where

$$
\delta_i(t) : Z^+ = \{0, 1, 2 \ldots\} \to \{0, 1\}, \quad \sum_{i=1}^{N}\delta_i(t) = 1, \quad \forall t \in Z^+.
$$

At an arbitrary continuous time t, $\delta_i(t)$ is dependent on t or $x(t)$, or both, or other switching rules. Similar to the switched signal adopted in [25], we assume that the sequence of subsystems in switching signal $\delta_i(t)$ is a priori, that is, its instantaneous value is available in real time. Matrices A_i, A_{di}, B_i, C, C_{2i}, and D_i are constant matrices with appropriate dimensions.

For the purpose of this note, we give the following lemma and assumption:

Lemma 3.1 ([26]) *The equilibrium 0 of*

$$
x(t+1) = f(x(t)) \tag{3.4}
$$

is globally uniformly asymptotically stable if there is a function $V : \mathbb{Z}^+ \times \mathbb{R}^n \to \mathbb{R}$ such that
(i)V is a positive definite function, decrescent, and radially unbounded;
(ii) $\Delta V(t, x(t)) = V(t+1, x(t+1)) - V(t, x(t))$ is negative definite along the solutions of (3.4).

Assumption 4 It is supposed that matrices B_i and C are of full rank, i.e., $rank(B_i) = m$ and $rank(C) = p$, and the pairs (A_i, B_i) and (A_i, C) are, respectively, controllable and observable.

Since the matrix C is of full column rank, there always exist a matrix $C^{\perp} \in \mathbb{R}^{(n-p)\times n}$ such that $\begin{bmatrix} C^{\perp} \\ C \end{bmatrix} \in \mathbb{R}^{n\times n}$ is a nonsingular matrix.

3.3 FE Design

3.3.1 State Transformation

We construct nonsingular matrix $T = \begin{bmatrix} C^{\perp}_{(n-p)\times n} \\ C_{p\times n} \end{bmatrix}^{-1}$, which satisfies

$$CT = \begin{bmatrix} 0_{p\times(n-p)} & I_p \end{bmatrix}.$$

Then define the following state transformation:

$$x(t) = T\bar{x}(t). \tag{3.5}$$

By using the condition in (3.5), the switched systems (3.1) and (3.2) can be transformed as follows:

$$\begin{bmatrix} \bar{x}_1(t+1) \\ \bar{x}_2(t+1) \end{bmatrix} = \sum_{i=1}^{N} \delta_i(t) \begin{bmatrix} A_{11i} & A_{12i} \\ A_{21i} & A_{22i} \end{bmatrix} \begin{bmatrix} \bar{x}_1(t) \\ \bar{x}_2(t) \end{bmatrix} + \sum_{i=1}^{N} \delta_i(t) \begin{bmatrix} A_{11di} & A_{12di} \\ A_{21di} & A_{22di} \end{bmatrix} \begin{bmatrix} \bar{x}_1(t-d) \\ \bar{x}_2(t-d) \end{bmatrix}$$
$$+ \sum_{i=1}^{N} \delta_i(t) \begin{bmatrix} B_{1i} \\ B_{2i} \end{bmatrix} (u(t) + f(t)) + \sum_{i=1}^{N} \delta_i(t) \begin{bmatrix} D_{1i} \\ D_{2i} \end{bmatrix} \omega(t), \tag{3.6}$$

$$y(t) = \begin{bmatrix} 0_{p\times(n-p)} & I_p \end{bmatrix} \begin{bmatrix} \bar{x}_1(t) \\ \bar{x}_2(t) \end{bmatrix}, \tag{3.7}$$

where

$$\bar{x}(t) = \begin{bmatrix} \bar{x}_1(t) \\ \bar{x}_2(t) \end{bmatrix}, \quad T^{-1}A_iT = \begin{bmatrix} A_{11i} & A_{12i} \\ A_{21i} & A_{22i} \end{bmatrix}, \quad T^{-1}A_{di}T = \begin{bmatrix} A_{11di} & A_{12di} \\ A_{21di} & A_{22di} \end{bmatrix},$$
$$T^{-1}B_i = \begin{bmatrix} B_{1i} \\ B_{2i} \end{bmatrix}, \quad T^{-1}D_i = \begin{bmatrix} D_{1i} \\ D_{2i} \end{bmatrix}, \quad CT = \begin{bmatrix} 0_{p\times(n-p)} & I_p \end{bmatrix}.$$

Then, the system (3.6) and (3.7) can be rewritten as

$$\bar{x}_1(t+1) = \sum_{i=1}^{N} \delta_i(t)A_{11i}\bar{x}_1(t) + \sum_{i=1}^{N} \delta_i(t)A_{12i}y(t) + \sum_{i=1}^{N} \delta_i(t)A_{11di}\bar{x}_1(t-d)$$

$$+\sum_{i=1}^{N}\delta_i(t)A_{12di}y(t-d)+\sum_{i=1}^{N}\delta_i(t)B_{1i}u(t)$$
$$+\sum_{i=1}^{N}\delta_i(t)B_{1i}f(t)+\sum_{i=1}^{N}\delta_i(t)D_{1i}\omega(t), \tag{3.8}$$
$$y(t+1)=\sum_{i=1}^{N}\delta_i(t)A_{21i}\bar{x}_1(t)+\sum_{i=1}^{N}\delta_i(t)A_{22i}y(t)$$
$$+\sum_{i=1}^{N}\delta_i(t)A_{21di}\bar{x}_1(t-d)+\sum_{i=1}^{N}\delta_i(t)A_{22di}y(t-d)$$
$$+\sum_{i=1}^{N}\delta_i(t)B_{2i}u(t)+\sum_{i=1}^{N}\delta_i(t)B_{2i}f(t)+\sum_{i=1}^{N}\delta_i(t)D_{2i}\omega(t). \tag{3.9}$$

By introducing the virtual input $\eta(t)$ and $\rho(t)$:

$$\eta(t)=\sum_{i=1}^{N}\delta_i(t)A_{12i}y(t)+\sum_{i=1}^{N}\delta_i(t)A_{12di}y(t-d)+\sum_{i=1}^{N}\delta_i(t)B_{1i}u(t) \tag{3.10}$$
$$\rho(t)=y(t+1)-\sum_{i=1}^{N}\delta_i(t)A_{22i}y(t)-\sum_{i=1}^{N}\delta_i(t)A_{22di}y(t-d)$$
$$-\sum_{i=1}^{N}\delta_i(t)B_{2i}u(t) \tag{3.11}$$

one can get

$$\bar{x}_1(t+1)=\sum_{i=1}^{N}\delta_i(t)A_{11i}\bar{x}_1(t)+\sum_{i=1}^{N}\delta_i(t)A_{11di}\bar{x}_1(t-d)+\eta(t)$$
$$+\sum_{i=1}^{N}\delta_i(t)B_{1i}f(t)+\sum_{i=1}^{N}\delta_i(t)D_{1i}\omega(t) \tag{3.12}$$
$$\rho(t)=\sum_{i=1}^{N}\delta_i(t)A_{21i}\bar{x}_1(t)+\sum_{i=1}^{N}\delta_i(t)A_{21di}\bar{x}_1(t-d)+\sum_{i=1}^{N}\delta_i(t)B_{2i}f(t)$$
$$+\sum_{i=1}^{N}\delta_i(t)D_{2i}\omega(t). \tag{3.13}$$

3.3.2 Reduced-Order Observer Design

For the dynamics (3.12) and (3.13), we construct the following reduced-order switched state observer:

$$\begin{aligned}\hat{\bar{x}}_1(t+1) &= \sum_{i=1}^{N}\delta_i(t)A_{11i}\hat{\bar{x}}_1(t) + \sum_{i=1}^{N}\delta_i(t)A_{11di}\hat{\bar{x}}_1(t-d) + \eta(t)\\ &\quad + \sum_{i=1}^{N}\delta_i(t)B_{1i}\hat{f}(t) - \sum_{i=1}^{N}\delta_i(t)G_i(\hat{\rho}(t)-\rho(t)),\end{aligned} \tag{3.14}$$

$$\hat{\rho}(t) = \sum_{i=1}^{N}\delta_i(t)A_{21i}\hat{\bar{x}}_1(t) + \sum_{i=1}^{N}\delta_i(t)A_{21di}\hat{\bar{x}}_1(t-d) + \sum_{i=1}^{N}\delta_i(t)B_{2i}\hat{f}(t), \tag{3.15}$$

$$\hat{f}(t+1) = \hat{f}(t) - \sum_{i=1}^{N}\delta_i(t)F_i(\hat{\rho}(t)-\rho(t)), \tag{3.16}$$

where $\hat{\bar{x}}(t) \in \mathbb{R}^{n-p}$ is the reduced-order observer state, $\hat{\rho}(t) \in \mathbb{R}^{p}$ is the reduced-order observer output, $\hat{f}(t) \in \mathbb{R}^{m}$ is the fault estimation of the fault $f(t)$, G_i and F_i are reduced-order observer gain matrices to be designed.

Let

$$e(t) = \hat{\bar{x}}_1(t) - \bar{x}_1(t), \quad e_f(t) = \hat{f}(t) - f(t) \tag{3.17}$$

then the error dynamics are given as follows:

$$\begin{aligned}e(t+1) &= \sum_{i=1}^{N}\delta_i(t)(A_{11i}-G_iA_{21i})e(t) + \sum_{i=1}^{N}\delta_i(t)(A_{11di}-G_iA_{21di})e(t-d)\\ &\quad + \sum_{i=1}^{N}\delta_i(t)(B_{1i}-G_iB_{2i})e_f(t) + \sum_{i=1}^{N}\delta_i(t)(G_iD_{2i}-D_{1i})\omega(t)\end{aligned} \tag{3.18}$$

$$\begin{aligned}e_f(t+1) &= \hat{f}(t+1) - f(t+1)\\ &= \hat{f}(t) - \sum_{i=1}^{N}\delta_i(t)F_iA_{21i}e(t) - \sum_{i=1}^{N}\delta_i(t)F_iA_{21di}e(t-d)\\ &\quad - \sum_{i=1}^{N}\delta_i(t)F_iB_{2i}e_f(t) + \sum_{i=1}^{N}\delta_i(t)F_iD_{2i}\omega(t) - f(t+1)\\ &= \hat{f}(t) - f(t) - \sum_{i=1}^{N}\delta_i(t)F_iA_{21i}e(t) - \sum_{i=1}^{N}\delta_i(t)F_iA_{21di}e(t-d)\end{aligned}$$

$$
\begin{aligned}
&-\sum_{i=1}^{N}\delta_i(t)F_iB_{2i}e_f(t)+\sum_{i=1}^{N}\delta_i(t)F_iD_{2i}\omega(t)-(f(t+1)-f(t))\\
&=-\sum_{i=1}^{N}\delta_i(t)F_iA_{21i}e(t)-\sum_{i=1}^{N}\delta_i(t)F_iA_{21di}e(t-d)+\sum_{i=1}^{N}\delta_i(t)F_iD_{2i}\omega(t)\\
&+\sum_{i=1}^{N}\delta_i(t)(I_m-F_iB_{2i})e_f(t)-\Delta f(t)
\end{aligned}\tag{3.19}
$$

where $\Delta f(t)=f(t+1)-f(t)$. Combining the dynamics in (3.18) and (3.19), one can get the following augmented system:

$$
\begin{aligned}
\bar{e}(t+1)=&\sum_{i=1}^{N}\delta_i(t)(\overline{A}_{1i}-\overline{G}_i\overline{A}_{2i})\bar{e}(t)+\sum_{i=1}^{N}\delta_i(t)(\overline{A}_{1di}-\overline{G}_i\overline{A}_{2di})\bar{e}(t-d)\\
&+\sum_{i=1}^{N}\delta_i(t)(\overline{G}_i\overline{D}_{1i}-\overline{D}_{2i})v(t),
\end{aligned}\tag{3.20}
$$

where

$$
\begin{aligned}
&\bar{e}(t)=\begin{bmatrix}e(t)\\ e_f(t)\end{bmatrix},\ v(t)=\begin{bmatrix}\omega(t)\\ \Delta f(t)\end{bmatrix},\ \overline{A}_{1i}=\begin{bmatrix}A_{11i} & B_{1i}\\ 0_{m\times(n-p)} & I_m\end{bmatrix},\ \overline{A}_{2i}=\begin{bmatrix}A_{21i} & B_{2i}\end{bmatrix},\\
&\overline{A}_{1di}=\begin{bmatrix}A_{11di} & 0_{(n-p)\times m}\\ 0_{m\times(n-p)} & 0_{m\times m}\end{bmatrix},\ \overline{A}_{2di}=\begin{bmatrix}A_{21di} & 0_{(n-p)\times m}\end{bmatrix},\ \overline{D}_{1i}=\begin{bmatrix}D_{2i} & 0_{p\times m}\end{bmatrix},\\
&\overline{D}_{2i}=\begin{bmatrix}D_{1i} & 0_{(n-p)\times m}\\ 0_{m\times d} & I_m\end{bmatrix},\ \overline{G}_i=\begin{bmatrix}G_i\\ F_i\end{bmatrix}.
\end{aligned}\tag{3.21}
$$

The objective of this section is to determine the observer gains in order to guarantee asymptotic convergence of the estimated state in (3.14) to the state in (3.12). That is to say, the estimated error $\bar{e}(t+1)$ tends to zero when $t\to\infty$.

In the following, a reduced-order observer design method under an H_∞ performance index is proposed to achieve robust FE for discrete-time switched systems with state delay.

Theorem 3.1 *Given a constant $\gamma_1>0$, if there exist positive definite matrices P_i, Q_i, nonsingular matrices Ω, and matrix H_i, for any $i,j,l\in\{1,2,\ldots,N\}$, such that*

$$
\begin{bmatrix}
P_j-(\Omega+\Omega^T) & \Omega^T\overline{A}_{1i}-H_i\overline{A}_{2i} & \Omega^T\overline{A}_{1di}-H_i\overline{A}_{2di} & H_i\overline{D}_{1i}-\Omega^T\overline{D}_{2i} & 0\\
* & -P_i+Q_i & 0 & 0 & \bar{I}_m\\
* & * & -Q_l & 0 & 0\\
* & * & * & -\gamma_1^2 I_{r+m} & 0\\
* & * & * & * & -I_m
\end{bmatrix}<0\tag{3.22}
$$

where $\bar{I}_m = \begin{bmatrix} 0_{(n-p)\times m} & I_m \end{bmatrix}^T$*.Then the augmented switched system (3.20) is stable with* H_∞ *performance index* γ_1*, and the reduced-order FE observer is given by*

$$\overline{G}_i = \Omega^{-T} H_i. \tag{3.23}$$

Proof Suppose that (3.22) holds, then it is easy to see from (3.22) that

$$(P_j - \Omega)^T P_j^{-1} (P_j - \Omega) \geq 0, \tag{3.24}$$

which implies that

$$P_j - (\Omega + \Omega^T) \geq -\Omega^T P_j^{-1} \Omega. \tag{3.25}$$

Combining the condition in (3.23), (3.22) can be transformed into the following formation:

$$\begin{bmatrix} P_j^{-1} & \overline{A}_{1i} - \overline{G}_i \overline{A}_{2i} & \overline{A}_{1di} - \overline{G}_i \overline{A}_{2di} & \overline{G}_i \overline{D}_{1i} - \overline{D}_{2i} & 0 \\ * & -P_i + Q_i & 0 & 0 & \bar{I}_m \\ * & * & -Q_l & 0 & 0 \\ * & * & * & -\gamma_1^2 I_{r+m} & 0 \\ * & * & * & * & -I_m \end{bmatrix} < 0. \tag{3.26}$$

Consider the following switched Lyapunov function:

$$V(t) = \bar{e}^T(t) \left(\sum_{i=1}^{N} \delta_i(t) P_i \right) \bar{e}(t) + \sum_{s=t-d}^{t-1} \bar{e}^T(s) \left(\sum_{i=1}^{N} \delta_i(s) Q_i \right) \bar{e}(s) \tag{3.27}$$

and define $\Delta V(t) = V(t+1) - V(t)$. Then along the error dynamics (3.20),

$$\begin{aligned} \Delta V(t) &= \bar{e}^T(t+1) \left(\sum_{i=1}^{N} \delta_i(t+1) P_i \right) \bar{e}(t+1) + \sum_{s=t-d+1}^{t} \bar{e}^T(s) \left(\sum_{i=1}^{N} \delta_i(s) Q_i \right) \bar{e}(s) \\ &\quad - \bar{e}^T(t) \left(\sum_{i=1}^{N} \delta_i(t) P_i \right) \bar{e}(t) - \sum_{s=t-d}^{t-1} \bar{e}^T(s) \left(\sum_{i=1}^{N} \delta_i(s) Q_i \right) \bar{e}(s) \\ &= \bar{e}^T(t+1) \left(\sum_{i=1}^{N} \delta_i(t+1) P_i \right) \bar{e}(t+1) - \bar{e}^T(t) \left(\sum_{i=1}^{N} \delta_i(t) P_i \right) \bar{e}(t) \\ &\quad + \bar{e}^T(t) \left(\sum_{i=1}^{N} \delta_i(t) Q_i \right) \bar{e}(t) - \bar{e}^T(t-d) \left(\sum_{i=1}^{N} \delta_i(t-d) Q_i \right) \bar{e}(t-d). \end{aligned} \tag{3.28}$$

It follows that for any nonzero vector $\bar{e}(t)$ and the particular case $\delta_i(t) = 1, \delta_{k\neq i}(t) = 0, \delta_j(t+1) = 1, \delta_{k\neq j}(t+1) = 0$, and $\delta_l(t-d) = 1, \delta_{k\neq l}(t-d) = 0$, we have

$$\Delta V(t) = \vartheta^T(t)\left(\begin{bmatrix} (\overline{A}_{1i} - \overline{G}_i\overline{A}_{2i})^T \\ (\overline{A}_{1di} - \overline{G}_i\overline{A}_{2di})^T \\ (\overline{G}_i\overline{D}_{1i} - \overline{D}_{2i})^T \end{bmatrix} P_j \left[\, \overline{A}_{1i} - \overline{G}_i\overline{A}_{2i} \;\; \overline{A}_{1di} - \overline{G}_i\overline{A}_{2di} \;\; \overline{G}_i\overline{D}_{1i} - \overline{D}_{2i} \,\right] + \begin{bmatrix} -P_i + Q_i & 0 & 0 \\ 0 & -Q_l & 0 \\ 0 & 0 & 0 \end{bmatrix}\right)\vartheta(t), \tag{3.29}$$

where $\vartheta(t) = \left[\, \bar{e}^T(t) \;\; \bar{e}^T(t-d) \;\; v^T(t) \,\right]^T$.

Then from the condition in (3.28), one can get that $\Delta V(t) < 0$, which concludes from Lemma 1 that augmented switched system (3.20) is stable.

Let

$$J_{1K} = \sum_{t=0}^{K-1} \left[e_f^T(t)e_f(t) - \gamma_1^2 v^T(t)v(t) \right] \tag{3.30}$$

where K is an arbitrary positive integer. For any nonzero $v(t) \in l_2[0, \infty)$ and zero initial condition $\bar{e}(t) = 0$, one has

$$\begin{aligned} J_{1K} &= \sum_{t=0}^{K-1} \left[e_f^T(t)e_f(t) - \gamma_1^2 v^T(t)v(t) + \Delta V(t) \right] - V(K) \\ &\le \sum_{t=0}^{K-1} \left(\bar{I}_m\bar{I}_m^T\bar{e}^T(t)\bar{e}(t) - \gamma_1^2 v^T(t)v(t) + \Delta V(t) \right). \end{aligned} \tag{3.31}$$

It is noted that

$$\begin{aligned} &\bar{I}_m\bar{I}_m^T\bar{e}^T(t)\bar{e}(t) - \gamma_1^2 v^T(t)v(t) + \Delta V(t) \\ &= \vartheta^T(t)\left(\begin{bmatrix} (\overline{A}_{1i} - \overline{G}_i\overline{A}_{2i})^T \\ (\overline{A}_{1di} - \overline{G}_i\overline{A}_{2di})^T \\ (\overline{G}_i\overline{D}_{1i} - \overline{D}_{2i})^T \end{bmatrix} P_j \left[\, \overline{A}_{1i} - \overline{G}_i\overline{A}_{2i} \;\; \overline{A}_{1di} - \overline{G}_i\overline{A}_{2di} \;\; \overline{G}_i\overline{D}_{1i} - \overline{D}_{2i} \,\right] \right. \\ &\quad \left. + \begin{bmatrix} -P_i + Q_i + \bar{I}_m\bar{I}_m^T & 0 & 0 \\ 0 & -Q_l & 0 \\ 0 & 0 & -\gamma_1^2 I_{r+m} \end{bmatrix}\right)\vartheta(t). \end{aligned} \tag{3.32}$$

By Schur complement formula, the inequality is equivalent to the following inequality:

$$\begin{bmatrix} P_j^{-1} & \overline{A}_{1i} - \overline{G}_i\overline{A}_{2i} & \overline{A}_{1di} - \overline{G}_i\overline{A}_{2di} & \overline{G}_i\overline{D}_{1i} - \overline{D}_{2i} \\ * & -P_i + Q_i + \bar{I}_m\bar{I}_m^T & 0 & 0 \\ * & * & -Q_l & 0 \\ * & * & * & -\gamma_1^2 I_{r+m} \end{bmatrix} < 0. \tag{3.33}$$

It follows from (3.33) and Schur complement formula that

$$e_f^T(t)e_f(t) - \gamma_1^2 \nu^T(t)\nu(t) + \Delta V(t) < 0, \tag{3.34}$$

which implies, for any K, $J_{1K} < 0$. Then, one has that for any nonzero $\nu(t) \in l_2[0, \infty)$, $\|e_f(t)\|_2 < \gamma_1 \|\nu(t)\|_2$. (□)

3.3.3 FE Algorithm Design

Based on the obtained results in the Sect. 3.2, we construct the following augmented system:

$$\begin{aligned}\begin{bmatrix} \hat{\hat{x}}_1(t+1) \\ \hat{f}(t+1) \end{bmatrix} &= \sum_{i=1}^{N} \delta_i(t) \begin{bmatrix} A_{11i} & B_{1i} \\ 0_{m\times(n-p)} & I_m \end{bmatrix} \begin{bmatrix} \hat{\hat{x}}_1(t) \\ \hat{f}(t) \end{bmatrix} \\ &+ \sum_{i=1}^{N} \delta_i(t) \begin{bmatrix} A_{11di} & 0_{(n-p)\times m} \\ 0_{m\times(n-p)} & 0_m \end{bmatrix} \begin{bmatrix} \hat{\hat{x}}_1(t-d) \\ \hat{f}(t-d) \end{bmatrix} \\ &- \sum_{i=1}^{N} \delta_i(t) \begin{bmatrix} G_i \\ F_i \end{bmatrix} (\hat{\rho}(t) - \rho(t)) + \begin{bmatrix} I_{n-p} \\ 0_{m\times(n-p)} \end{bmatrix} \eta(t), \end{aligned} \tag{3.35}$$

Combining (3.15) and (3.21), one has

$$\begin{aligned}\begin{bmatrix} \hat{\hat{x}}_1(t+1) \\ \hat{f}(t+1) \end{bmatrix} &= \sum_{i=1}^{N} \delta_i(t)(\overline{A}_{1i} - \overline{G}_i \overline{A}_{2i}) \begin{bmatrix} \hat{\hat{x}}_1(t) \\ \hat{f}(t) \end{bmatrix} + \sum_{i=1}^{N} \delta_i(t)(\overline{A}_{1di} - \overline{G}_i \overline{A}_{2di}) \begin{bmatrix} \hat{\hat{x}}_1(t-d) \\ \hat{f}(t-d) \end{bmatrix} \\ &+ \sum_{i=1}^{N} \delta_i(t) \overline{G}_i \rho(t) + \overline{I}\eta(t), \end{aligned} \tag{3.36}$$

where $\overline{I} = \begin{bmatrix} I_{n-p} \\ 0_{m\times(n-p)} \end{bmatrix}$. Substituting (3.10) and (3.11) into (3.36), one has

$$\begin{aligned}\begin{bmatrix} \hat{\hat{x}}_1(t+1) \\ \hat{f}(t+1) \end{bmatrix} &= \sum_{i=1}^{N} \delta_i(t)(\overline{A}_{1i} - \overline{G}_i \overline{A}_{2i}) \begin{bmatrix} \hat{\hat{x}}_1(t) \\ \hat{f}(t) \end{bmatrix} + \sum_{i=1}^{N} \delta_i(t)(\overline{A}_{1di} - \overline{G}_i \overline{A}_{2di}) \begin{bmatrix} \hat{\hat{x}}_1(t-d) \\ \hat{f}(t-d) \end{bmatrix} \\ &+ \sum_{i=1}^{N} \delta_i(t) \overline{G}_i y(t+1) + \sum_{i=1}^{N} \delta_i(t)(\overline{I} A_{12i} - \overline{G}_i A_{22i}) y(t) \\ &+ \sum_{i=1}^{N} \delta_i(t)(\overline{I} A_{12di} - \overline{G}_i A_{22di}) y(t-d) \\ &+ \sum_{i=1}^{N} \delta_i(t)(\overline{I} B_{1i} - \overline{G}_i B_{2i}) u(t). \end{aligned} \tag{3.37}$$

Define the following invariable:

$$\zeta(t+1) = \begin{bmatrix} \hat{\bar{x}}_1(t+1) \\ \hat{f}(t+1) \end{bmatrix} - \sum_{i=1}^{N} \delta_i(t)\overline{G}_i y(t+1). \tag{3.38}$$

Then, it follows from (3.37) that

$$\begin{aligned} \zeta(t+1) = & \sum_{i=1}^{N} \delta_i(t)(\overline{A}_{1i} - \overline{G}_i\overline{A}_{2i})\zeta(t) + \sum_{i=1}^{N} \delta_i(t)(\overline{A}_{1di} - \overline{G}_i\overline{A}_{2di})\zeta(t-d) \\ & + \sum_{i=1}^{N} \delta_i(t)(\overline{I}A_{12i} - \overline{G}_i A_{22i})y(t) + \sum_{i=1}^{N} \delta_i(t)(\overline{I}A_{12di} - \overline{G}_i A_{22di})y(t-d) \\ & + \sum_{i=1}^{N} \delta_i(t)(\overline{I}B_{1i} - \overline{G}_i B_{2i})u(t). \end{aligned} \tag{3.39}$$

Furthermore, it follows from that

$$\begin{bmatrix} \hat{\bar{x}}_1(t) \\ \hat{f}(t) \end{bmatrix} = \zeta(t) - \sum_{i=1}^{N} \delta_i(t-1)\overline{G}_i y(t), \tag{3.40}$$

where $\sum_{i=1}^{N} \delta_i(t-1)$ means the subsystem of time $t-1$ is activated.

3.4 FA Design

On the basis of the obtained online FE information, we design a fault-tolerant controller to guarantee the stability in the presence of faults. Since the state $x(t)$ is unavailable directly, we utilize the switched dynamical output feedback controller scheme to construct the fault-tolerant controller. For the discrete-time switched system (3.1)–(3.3), we consider the following dynamic output feedback controller:

$$\hat{x}(t+1) = \sum_{i=1}^{N} \delta_i(t)A_{Ki}\hat{x}(t) + \sum_{i=1}^{N} \delta_i(t)B_{Ki}y(t) \tag{3.41}$$

$$u(t) = \sum_{i=1}^{N} \delta_i(t)C_{Ki}\hat{x}(t) - \hat{f}(t), \tag{3.42}$$

where $\hat{x}(t) \in \mathbb{R}^n$ is the controller state vector, A_{Ki}, B_{Ki}, and C_{Ki} are controller gains to be determined later.

Substituting (3.2) into (3.41), one has

$$\hat{x}(t+1) = \sum_{i=1}^{N} \delta_i(t) A_{Ki}\hat{x}(t) + \sum_{i=1}^{N} \delta_i(t) B_{Ki} C x(t) \tag{3.43}$$

Then substituting $u(t)$ into (3.1), one has

$$\begin{aligned} x(t+1) &= \sum_{i=1}^{N} \delta_i(t) A_i x(t) + \sum_{i=1}^{N} \delta_i(t) A_{di} x(t-d) + \sum_{i=1}^{N} \delta_i(t) B_i C_{Ki}\hat{x}(t) \\ &\quad + \sum_{i=1}^{N} \delta_i(t) B_i (f(t) - \hat{f}(t)) + \sum_{i=1}^{N} \delta_i(t) D_i \omega(t) \\ &= \sum_{i=1}^{N} \delta_i(t) A_i x(t) + \sum_{i=1}^{N} \delta_i(t) A_{di} x(t-d) + \sum_{i=1}^{N} \delta_i(t) B_i C_{Ki}\hat{x}(t) \\ &\quad - \sum_{i=1}^{N} \delta_i(t) B_i e_f(t) + \sum_{i=1}^{N} \delta_i(t) D_i \omega(t). \end{aligned} \tag{3.44}$$

It follows that

$$\xi(t+1) = \sum_{i=1}^{N} \delta_i(t)\widetilde{A}_i \xi(t) + \sum_{i=1}^{N} \delta_i(t)\widetilde{A}_{di} H\xi(t-d) + \sum_{i=1}^{N} \delta_i(t)\widetilde{B}_i \mu(t) \tag{3.45}$$

$$z(t) = \sum_{i=1}^{N} \delta_i(t)\widetilde{C}_i \xi(t), \tag{3.46}$$

where

$$\begin{aligned} &\xi(t) = \begin{bmatrix} x(t) \\ \hat{x}(t) \end{bmatrix},\ \mu(t) = \begin{bmatrix} \omega(t) \\ e_f(t) \end{bmatrix},\ \widetilde{A}_i = \begin{bmatrix} A_i & B_i C_{Ki} \\ B_{Ki} C & A_{Ki} \end{bmatrix},\ \widetilde{A}_{di} = \begin{bmatrix} A_{di} \\ 0 \end{bmatrix}, \\ &\widetilde{B}_i = \begin{bmatrix} D_i & -B_i \\ 0 & 0 \end{bmatrix},\ \widetilde{C}_i = \begin{bmatrix} C_{2i} & 0 \end{bmatrix},\ H = \begin{bmatrix} I & 0 \end{bmatrix}. \end{aligned} \tag{3.47}$$

3.5 H_∞ Performance Analysis

Lemma 3.2 *Given a constant $\gamma_2 > 0$, the augmented switched systems (3.45)–(3.46) is stable with H_∞ performance γ_2, if there exist positive matrices $\bar{P}$ and $\bar{Q}$, such that the following LMI holds:*

$$\begin{bmatrix} -\bar{P} & \bar{P}\widetilde{A}_i & \bar{P}\widetilde{A}_{di} & \bar{P}\widetilde{B}_i \\ * & -\bar{P}+H^T\bar{Q}H+\widetilde{C}_i^T\widetilde{C}_i & 0 & 0 \\ * & * & -\bar{Q} & 0 \\ * & * & * & -\gamma_2^2 I \end{bmatrix} < 0. \tag{3.48}$$

Proof Firstly, we establish the stability of augmented switched system (3.45). When $\mu(t)=0$, (3.45) becomes

$$\xi(t+1)=\sum_{i=1}^{N}\delta_i(t)\widetilde{A}_i\xi(t)+\sum_{i=1}^{N}\delta_i(t)\widetilde{A}_{di}H\xi(t-d) \tag{3.49}$$

Define Lyapunov function:

$$V(t)=\xi^T(t)\bar{P}\xi(t)+\sum_{s=t-d}^{t-1}\xi^T(s)H^T\bar{Q}H\xi(s). \tag{3.50}$$

Then along the error dynamics (3.49),

$$\begin{aligned} \Delta V(t) &= \xi^T(t+1)\bar{P}\xi(t+1)+\sum_{s=t-d+1}^{t}\xi^T(s)H^T\bar{Q}H\xi(s)-\xi^T(t)\bar{P}\xi(t) \\ &\quad -\sum_{s=t-d}^{t-1}\xi^T(s)H^T\bar{Q}H\xi(s) \\ &= \xi^T(t+1)\bar{P}\xi(t+1)-\xi^T(t)\bar{P}\xi(t)+\xi^T(t)H^T\bar{Q}H\xi(t) \\ &\quad -\xi^T(t-d)H^T\bar{Q}H\xi(t-d) \end{aligned} \tag{3.51}$$

This with the relationship

$$x(t-d)=H\xi(t-d) \tag{3.52}$$

which implies

$$\Delta V(t)=\tilde{\xi}^T(t)\left(\begin{bmatrix}\widetilde{A}_i^T \\ \widetilde{A}_{di}^T\end{bmatrix}\bar{P}\left[\widetilde{A}_i\ \widetilde{A}_{di}\right]+\begin{bmatrix}-\bar{P}+H^T\bar{Q}H & 0 \\ 0 & -\bar{Q}\end{bmatrix}\right)\tilde{\xi}(t),$$

where $\tilde{\xi}(t)=\left[\xi^T(t)\ x^T(t-d)\right]^T$, which follows from (3.48) and Schur complete lemma that the system (3.49) is stable.

Let

$$J_{2K}=\sum_{t=0}^{K-1}\left(z^T(t)z(t)-\gamma_2^2\mu^T(t)\mu(t)\right), \tag{3.53}$$

where K is an arbitrary positive integer. For any nonzero $\mu(t) \in [0, \infty)$ and zero initial condition $\xi(t) = 0$, one has

$$J_{2K} = \sum_{t=0}^{K-1} \left(z^T(t)\xi(t) - \gamma_2^2 \mu^T(t)\mu(t) + \Delta V(t)\right) - V(K)$$
$$\leq \sum_{t=0}^{K-1} \left(z^T(t)z(t) - \gamma_2^2 \mu^T(t)\mu(t) + \Delta V(t)\right). \tag{3.54}$$

It is noted that

$$z^T(t)z(t) - \gamma_2^2 \mu^T(t)\mu(t) + \Delta V(t)$$
$$= \bar{z}^T(t)\left(\begin{bmatrix} \widetilde{A}_i^T \\ \widetilde{A}_{di}^T \\ \widetilde{B}_i^T \end{bmatrix} \bar{P}\left[\widetilde{A}_i \ \widetilde{A}_{di} \ \widetilde{B}_i\right] + \begin{bmatrix} -\bar{P} + H^T\bar{Q}H + \widetilde{C}_i^T\widetilde{C}_i & 0 & 0 \\ 0 & -\bar{Q} & 0 \\ 0 & 0 & -\gamma_2^2 I \end{bmatrix}\right)\bar{z}(t),$$

where $\bar{z}(t) = \left[\xi^T(t) \ x^T(t-d) \ \mu^T(t)\right]^T$.

It follows from (3.48) and Schur complement formula that

$$z^T(t)z(t) - \gamma_2^2 \mu^T(t)\mu(t) + \Delta V(t) < 0, \tag{3.55}$$

which implies for any K, $J_{2K} < 0$. Then, one has that for any nonzero $\nu(t) \in [0, \infty)$, $\|z(t)\|_2 < \gamma_2 \|\mu(t)\|_2$. (□)

Based on Lemma 3.2, a dynamic output feedback controller design method is developed in the following theorem.

Theorem 3.2 *Given a constant $\gamma_2 > 0$, then there exists a dynamic output feedback controller in the form of (3.41) and (3.42) such that the augmented switched system (3.45)–(3.46) is stable with H_∞ performance γ_2 if there exist positive definite matrices X, Y, matrices Ω_i, M_i, N_i, such that the following linear matrix inequalities*

$$\begin{bmatrix} \Lambda_{11} & \Lambda_{12} & \Lambda_{13} & 0 & 0 \\ * & \Lambda_{22} & 0 & \Lambda_{24} & \bar{H} \\ * & * & -\gamma_2^2 I & 0 & 0 \\ * & * & * & -I & 0 \\ * & * & * & * & -\bar{Q}^{-1} \end{bmatrix} < 0 \tag{3.56}$$

holds, for all $i \in \{1, 2, \ldots, N\}$ and some given positive matrix $\bar{Q}$, where

$$\Lambda_{11} = \begin{bmatrix} -Y & -I \\ -I & -X \end{bmatrix}, \Lambda_{12} = \begin{bmatrix} A_iY + B_iN_i & A_i & A_{di} \\ \Omega_i & XA_i + M_iC & XA_{di} \end{bmatrix}, \Lambda_{13} = \begin{bmatrix} D_i & -B_i \\ XD_i & -XB_i \end{bmatrix},$$

$$\Lambda_{22} = \begin{bmatrix} -Y & -I & 0 \\ -I & -X & 0 \\ 0 & 0 & -\bar{Q} \end{bmatrix}, \Lambda_{24} = \begin{bmatrix} Y^TC_{2i}^T \\ C_{2i}^T \\ 0 \end{bmatrix}, \bar{H} = \begin{bmatrix} Y \\ I \\ 0 \end{bmatrix}.$$

In this case, a desired stable dynamic output feedback H_∞ controller is given the parameters as follows:

$$\begin{aligned} A_{Ki} &= S^{-1}(\Omega_i - XA_iY - M_iCY - XB_iN_i)W^{-T}, \\ B_{Ki} &= S^{-1}M_i, \\ C_{Ki} &= N_iW^{-T}, \end{aligned} \tag{3.57}$$

where S and W are any nonsingular matrices satisfying

$$SW^T = I - XY. \tag{3.58}$$

Proof Suppose that the inequalities (3.56) holds, which implies

$$\begin{bmatrix} -Y & -I \\ -I & -X \end{bmatrix} < 0, \tag{3.59}$$

which, by the Schur complement formula, gives that $X - Y^{-1} > 0$. Therefore, $I - XY$ is nonsingular, hence there exist nonsingular matrices S and W such that (3.58) holds.

Now, we introduce the following nonsingular matrices:

$$\Pi_1 = \begin{bmatrix} Y & I \\ W^T & 0 \end{bmatrix}, \ \Pi_2 = \begin{bmatrix} I & X \\ 0 & S^T \end{bmatrix}. \tag{3.60}$$

Let

$$\bar{P} = \Pi_2\Pi_1^{-1} \tag{3.61}$$

then

$$\bar{P} = \begin{bmatrix} X & S \\ S^T & \Xi \end{bmatrix}, \tag{3.62}$$

where

$$\Xi = W^{-1}Y(X - Y^{-1})YW^{-T} > 0. \tag{3.63}$$

Observe that

$$\Xi - S^TX^{-1}S = S^T(YX - I)^{-1}(Y - X^{-1})(XY - I)^{-1}S > 0. \tag{3.64}$$

Then we have $\bar{P} > 0$.

Considering the notations in (3.47) with , and A_{Ki}, B_{Ki}, and C_{Ki} in (3.57), we can verify that the matrix inequality in (3.56) can be rewritten as

$$\begin{bmatrix} -\Pi_2^T \bar{P}^{-1} \Pi_2 & \Pi_2^T \widetilde{A}_i \Pi_1 & \Pi_2^T \widetilde{A}_{di} & \Pi_2^T \widetilde{B}_i & 0 & 0 \\ * & -\Pi_1^T \bar{P} \Pi_1 + \Pi_1^T \widetilde{C}_i & 0 & 0 & \Pi_1^T \widetilde{C}_i^T & \Pi_1^T H^T \\ * & * & -\bar{Q} & 0 & 0 & 0 \\ * & * & * & -\gamma_2^2 I & 0 & 0 \\ * & * & * & * & -I & 0 \\ * & * & * & * & * & -\bar{Q}^{-1} \end{bmatrix} < 0. \quad (3.65)$$

Pre- and post-multiplying this inequality by $diag\{\Pi_2^{-T}, \Pi_1^{-T}, I, I, I, I\}$ and its transpose, respectively, which, by the Schur complement, the result then follows from Lemma 3.1. (□)

Remark 3.1 For the sake of clarity, only a class of actuator fault case is considered in this section. However, note that such additive actuator faults considered in this section can be readily extended to general class of additive faults [27]. Meanwhile, the additive faults representation is more general than the multiplicative ones, which can be modeled as additive actuator ones [3].

Remark 3.2 The time delay is assumed to be constant in this section. Moreover, by using the technique in [28], the results may be extended to time-varying case.

Remark 3.3 For the existence of the terms A_{di}, switched Lyapunov function is not suitable in designing dynamic output feedback controller. A common Lyapunov function is employed to achieve the results. This maybe result in some conservatism.

3.6 An Illustrate Example

Consider the discrete-time switched linear system (3.1)–(3.3) consisting of three subsystems described by

$$A_1 = \begin{bmatrix} -0.5 & 0.3 \\ 0.1 & 1 \end{bmatrix}, A_2 = \begin{bmatrix} 0.2 & -0.1 \\ 1 & 0.3 \end{bmatrix}, A_3 = \begin{bmatrix} 1 & -0.2 \\ 2 & 0.05 \end{bmatrix}, A_{d1} = \begin{bmatrix} 0.1 & 0.3 \\ 0 & 1 \end{bmatrix},$$
$$A_{d2} = \begin{bmatrix} -0.2 & 0.1 \\ 0 & 0.3 \end{bmatrix}, A_{d3} = \begin{bmatrix} 0.2 & 1 \\ 0 & 0.05 \end{bmatrix}, B_1 = \begin{bmatrix} 0.1 \\ 0.2 \end{bmatrix}, B_2 = \begin{bmatrix} 0.1 \\ 0.2 \end{bmatrix}, B_3 = \begin{bmatrix} 0.1 \\ 0.2 \end{bmatrix},$$
$$D_1 = \begin{bmatrix} 0.1 \\ 0.2 \end{bmatrix}, D_2 = \begin{bmatrix} 0.1 \\ 0.2 \end{bmatrix}, D_3 = \begin{bmatrix} 0.1 \\ 0.2 \end{bmatrix}, C_{11} = C_{12} = C_{13} = \begin{bmatrix} 1 & 2 \end{bmatrix},$$
$$C_{21} = C_{22} = C_{23} = \begin{bmatrix} 2 & 1 \end{bmatrix}.$$

The state transformation matrix can be chosen as $T = \begin{bmatrix} 1 & 0 \\ -0.5 & 0.5 \end{bmatrix}$. It follows that the augmented matrices in (3.21) can be constructed as

$$\overline{A}_{11} = \begin{bmatrix} -0.65 & 0.1 \\ 0 & 1 \end{bmatrix}, \overline{A}_{12} = \begin{bmatrix} 0.25 & 0.1 \\ 0 & 1 \end{bmatrix}, \quad \overline{A}_{13} = \begin{bmatrix} 1.1 & 0.1 \\ 0 & 1 \end{bmatrix},$$
$$\overline{A}_{21} = \begin{bmatrix} 0.15 & 0.5 \end{bmatrix}, \quad \overline{A}_{22} = \begin{bmatrix} -0.05 & 0.5 \end{bmatrix}, \quad \overline{A}_{23} = \begin{bmatrix} -0.1 & 0.5 \end{bmatrix},$$
$$\overline{A}_{1d1} = \begin{bmatrix} -0.05 & 0 \\ 0 & 0 \end{bmatrix}, \quad \overline{A}_{1d2} = \begin{bmatrix} -0.25 & 0 \\ 0 & 0 \end{bmatrix}, \quad \overline{A}_{1d3} = \begin{bmatrix} -0.3 & 0 \\ 0 & 0 \end{bmatrix},$$
$$\overline{A}_{2d1} = \begin{bmatrix} 0.15 & 0 \end{bmatrix}, \quad \overline{A}_{2d2} = \begin{bmatrix} 0.05 & 0 \end{bmatrix}, \quad \overline{A}_{2d3} = \begin{bmatrix} 0.5 & 0 \end{bmatrix},$$
$$\overline{D}_{11} = \begin{bmatrix} 0.5 & 0 \end{bmatrix}, \quad \overline{D}_{12} = \begin{bmatrix} 0.5 & 0 \end{bmatrix}, \quad \overline{D}_{13} = \begin{bmatrix} 0.5 & 0 \end{bmatrix},$$
$$\overline{D}_{21} = \begin{bmatrix} 0.1 & 0 \\ 0 & 1 \end{bmatrix}, \quad \overline{D}_{22} = \begin{bmatrix} 0.1 & 0 \\ 0 & 1 \end{bmatrix}, \quad \overline{D}_{23} = \begin{bmatrix} 0.1 & 0 \\ 0 & 1 \end{bmatrix}.$$

By solving the condition (3.22) in Theorem 3.1, one can get the optimal value $\gamma_1 = 1.82$ with reduced-order observer parameters as follows:

$$\overline{G}_1 = \begin{bmatrix} 0.4401 \\ -0.0039 \end{bmatrix}, \overline{G}_2 = \begin{bmatrix} 0.1357 \\ 0.0007 \end{bmatrix}, \overline{G}_3 = \begin{bmatrix} -0.0551 \\ 0.0026 \end{bmatrix}.$$

Then, the positive definite matrix Q is set up as $\begin{bmatrix} 1 & 0 \\ 0 & 2 \end{bmatrix}$, by solving the conditions in Theorem 3.2, one obtains the minimum attenuation value $\gamma_2 = 2.16$ with the dynamic output feedback controller parameters as follows:

$$A_{K1} = \begin{bmatrix} -0.2997 & 2.9905 \\ 0.1125 & 2.2276 \end{bmatrix}, \quad A_{K2} = \begin{bmatrix} -0.1028 & 1.5334 \\ 0.0709 & 1.6935 \end{bmatrix},$$
$$A_{K3} = \begin{bmatrix} 0.1191 & 0.1245 \\ 0.1377 & 4.0849 \end{bmatrix}, \quad B_{K1} = \begin{bmatrix} -0.8266 \\ -0.4915 \end{bmatrix}, \quad B_{K2} = \begin{bmatrix} -0.6498 \\ -0.3593 \end{bmatrix},$$
$$B_{K3} = \begin{bmatrix} -0.4300 \\ -0.9382 \end{bmatrix}, \quad C_{K1} = \begin{bmatrix} 0.4399 & -11.6092 \end{bmatrix},$$
$$C_{K2} = \begin{bmatrix} -1.2777 & -0.5005 \end{bmatrix}, \quad C_{K3} = \begin{bmatrix} -3.0098 & 5.7662 \end{bmatrix}.$$

For simulation, an actuator constant fault signal $f_1(t)$ is set up as

$$f_1(t) = \begin{cases} 0, & 0 \leq t \leq 20 \\ 2. & 20 < t \leq 100 \end{cases}$$

If the external disturbance is set as a step signal, Fig. 3.1 depicts the fault $f_1(t)$ (solid line) and its estimate $\hat{f}_1(t)$ (dotted line). An actuator time-varying fault signal $f_2(t)$ is set up as

$$f_2(t) = \begin{cases} 0, & 0 \leq t \leq 20 \\ sin(t) + 2. & 20 < t \leq 100 \end{cases}$$

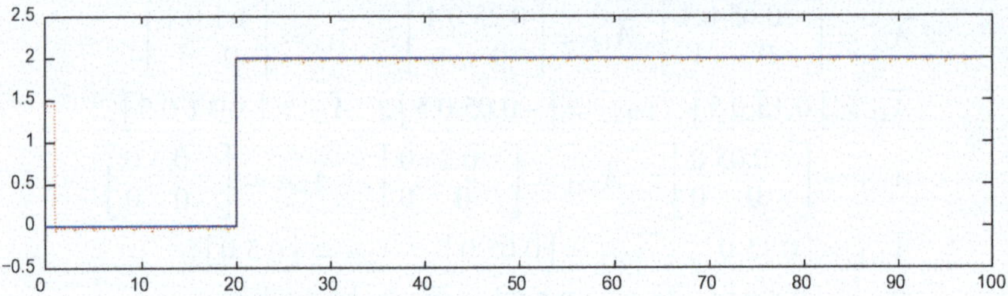

Fig. 3.1 Fault $f_1(t)$ (solid line) and its estimate $\hat{f}_1(t)$ (dotted line)

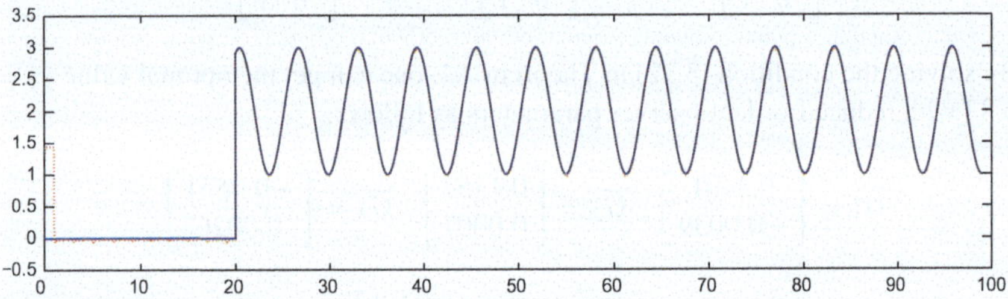

Fig. 3.2 Fault $f_2(t)$ (solid line) and its estimate $\hat{f}_2(t)$ (dotted line)

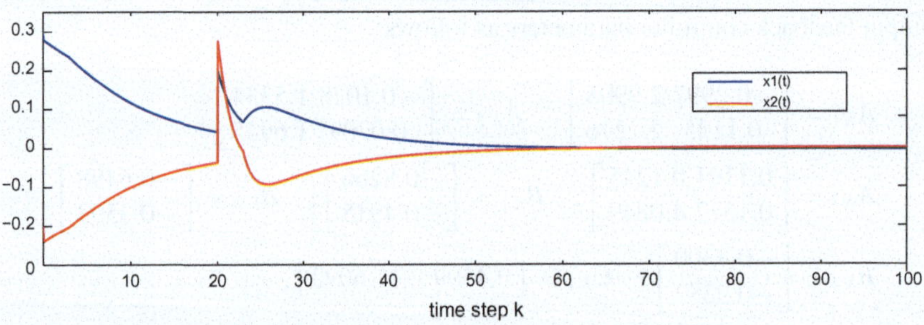

Fig. 3.3 State x(t) with fault accommodation

Figure 3.2 depicts the fault $f_2(t)$ (solid line) and its estimate $\hat{f}_2(t)$ (dotted line). It follows from Figs. 3.1 and 3.2 that the proposed reduced-order switched observer has good performance to estimate the actuator fault. Therefore, the simulation results exhibit well estimate operation. Figure 3.3 shows the simulation result of state response with fault accommodation.

3.7 Summary

By utilizing the reduced-order rank observer method and switched Lyapunov function technique, FE and compensation for discrete-time switched system with state delay are studied in this work, and an efficient reduced-order rank observer design is obtained in the form of LMI. Through a dynamic output feedback controller fault compensation, an fault-tolerant operation is realized. Finally, simulation results are presented to verify the effectiveness of the proposed method.

References

1. Berdjag, D., Cocquempot, V., Christophe, C., Shumesky, A., Zhirabok, A.: Algebraic approach for model decomposition: application to fault detection and isolation in discrete-event systems. Int. J. Appl. Math. Comput. Sci. **21**(1), 109–125 (2011)
2. Gholami, M., Cocquempot, V., Henrik, S., Thomas, B.: Passive FTC of piecewise affine systems based on H_∞ synthesis. In: IFAC Workshop Series, pp. 28–35 (2011)
3. Frank, P.M., Ding, S.X., Koppen-Seliger, B.: Current developments in the theory of FDI. In: Proc. IFAC Safeprocess, Budapest, Hungary, pp. 16–27 (2000)
4. Feng, J., Wang, S., Zhao, Q.: Closed-loop design of fault detection for networked non-linear systems with mixed delays and packet losses. IET Control Theory Appl. **7**(6), 858–868 (2013)
5. Wang, S., Feng, J., Zhang, H.: Robust FTC for a class of networked control systems with state delay and stochastic actuator failures. In: International Journal of Adaptive Control and Signal Processing (2013)
6. Tao, G., Chen, S., Joshi, S.M.: An adaptive actuator failure compensation controller using output feedback. IEEE Trans. Automat. Control **47**(3), 506–511 (2002)
7. Zhang, K., Jiang, B., Shi, P.: Fast fault estimation and accommodation for dynamical systems. IET Control Theory Appl. **3**(2), 189–199 (2009)
8. Zhong, M., Li, S., Zhao, Y.: Robust H_∞ fault detection for uncertain LDTV systems using krein space approach. Int. J. Innov. Comput. Inf. Control **9**(4), 1637–1649 (2013)
9. Tabatabaeipour, S.M., Bak, T.: Robust observer-based fault estimation and accommodation of discrete-time piecewise linear systems. J. Frankl. Inst. **351**(1), 277–295 (2014)
10. Edwards, C., Spurgeon, S.K., Patton, R.J.: Sliding mode observers for fault detection and isolation. Automatica **36**(4), 541–553 (2000)
11. Tan, C., Tao, G., Qi, R.: A discrete-time parameter estimation based adaptive actuator failure compensation control scheme. Int. J. Control **86**(2), 276–289 (2013)
12. Polycarpou, M.M.: Fault accommodation of a class of multivariable nonlinear dynamical systems using a learning approach. IEEE Trans. Automat. Control **46**(5), 736–742 (2001)
13. Du, D., Jiang, B., Shi, P.: Sensor fault estimation and compensation for time-delay switched systems. Int. J. Syst. Sci. **43**(4), 629–640 (2012)
14. Zhao, J., Hill, D.J.: Passivity and stability of switched systems: a multiple storate function method. Syst. Control Lett. **57**(2), 158–164 (2008)
15. Vesely, V., Rosinova, D.: Robust MPC controller design for switched systems using multi-parameter dependent Lyapunov function. Int. J. Innov. Comput. Inf. Control **10**(1), 269–280 (2014)
16. Wu, L., Zheng, W.: Weighted H_∞ model reduction for linear switched systems with time-varying delay. Automatica **45**(1), 186–193 (2009)
17. Attia, S.B., Salhi, S., Ksouri, M.: Static switched output feedback stabilization for linear discrete-time switched systems. Int. J. Innov. Comput. Inf. Control **8**(5), 3203–3213 (2012)

18. Zhang, L., Gao, H.: Asynchronously switched control of switched linear systems with average dwell time. Automatica **46**, 953–958 (2010)
19. Zhang, L., Shi, P., Boukas, E., Wang, C.: H_∞ model reduction for uncertain switched linear discrete-time systems. Automatica **44**, 2944–2949 (2008)
20. Cheng, D., Guo, L., Lin, Y., Wang, Y.: Stabilization of switched linear systems. IEEE Trans. Automat. Control **50**(5), 661–666 (2005)
21. Wang, D., Wang, W., Shi, P.: Robust fault detection for switched linear systems with state delays. IEEE Trans. Syst. Man Cybern. B: Cybern. **39**(3), 800–805 (2009)
22. Du, D., Jiang, B., Shi, P.: Fault detection for discrete-time switched systems with intermittent measurements. Int. J. Control **85**(1), 78–87 (2012)
23. Du, D., Jiang, B., Shi, P.: Active fault-tolerant control for switched systems with time delay. Int. J. Adapt. Control Signal Process. **25**(5), 466–480 (2011)
24. Du, D., Jiang, B., Shi, P.: Fault estimation and accommodation for switched systems with time-varying delay. Int. J. Control Autom. Systems **9**(3), 442–451 (2011)
25. Daafouz, J., Riedinger, P., Iung, C.: Stability analysis and control synthesis for switched systems: a switched Lyapunov function approach. IEEE Trans. Autom. Control **47**(11), 1883–1887 (2002)
26. Vidyasagar, M.: Nonlinear Systems Analysis. Prentice-Hall International Editions, Englewood Cliffs, NJ (1993)
27. Patton R.J., Chen, J., Lopez-Toribio, C.J.: Fuzzy observers for nonlinear dynamic systems fault diagnosis. In: Proc. 37th IEEE Conf. Decis. Control, Tampa, FL, pp. 84–89 (1998)
28. Du, D., Jiang, B., Zhou, S.: Delay-dependent robust stabilisation of uncertain discrete-time switched systems with time-varying state delay. Int. J. Syst. Sci. **39**(3), 305–313 (2008)

Chapter 4
Interval-Observer-Based Robust FE for Discrete-Time Switched Systems

In this section, the problem of robust FE observer for discrete-time switched systems with unknown input is investigated. Firstly, let the state vector and fault vector of the initial system into the state vector of the augmented system, which can obtain the augmented system. Next, based on the P-radius technique, an interval observer is designed to estimate the state and the fault of the augmented system which can ensure the error system is stable. Then, some assumptions are given in order to simplify the calculation, and the LMI technique and Schur Complement Lemma are used to solve the observer. Finally, an interval hull computation approximation algorithm of FE is obtained for discrete-time switched systems. Compared with the region-based method, this method has more precise boundary and higher efficiency.

4.1 Introduction

In the past decades, because of the increasing demands for all kinds of dynamic systems, the problem of fault diagnosis has been paid more attention. At present, there are many literature on the research of fault diagnosis [1–4]. Fault diagnosis is divided into FD and FE. The purpose of FD is to detect the fault as soon as possible, while the task of FE is to locate the system fault and obtain more information about it. The observer-based method has been proved to be one of the most popular research methods.

In recent years, lots of observer-based methods for the issue of FE have been developed, such as sliding-mode observer [5], fuzzy observer [6], adaptive observer [7], etc. The FE technique based on sliding-mode observer can produce bounded estimation for both constant and time-varying faults. The issues FE and FTC(FTC) for fuzzy systems are investigated in [8]. While the adaptive-observer-based FE

D. Du et al., *Observer-Based Fault Diagnosis and Fault-Tolerant Control for Switched Systems*, Studies in Systems, Decision and Control 323,
https://doi.org/10.1007/978-981-15-9073-3_4

technique can only achieve asymptotical estimation for constant faults. However, we know that faults and disturbances are inevitable in dynamic systems. In order to reduce the influence of faults and disturbances, the method to decouple residual from interference can eliminate the influence of disturbances. The UIO technique is an effective method for disturbance decoupling. For example, in [9] a robust fault diagnosis observer was designed for the autonomous spacecraft by the UIO technique. Similarly, in [10], an FE observer for the actuator of a wind turbine system was designed by using the UIO technique. It is worth noting that the UIO-based technique mentioned in the references are limited by some existing conditions, which motivates us to search for a feasible technique to decouple the partial disturbances.

On the other hand, the switching system has attracted wide attention because of its extensive application in practical engineering systems [11, 12]. The switching system is a hybrid system composed of several subsystems and a switching signal specifying the switching sequence among them. According to switching characteristics, switching signals can be divided into arbitrary switching and constrained switching. ADT is a typical kind of constrained switching signal, which was firstly proposed in [13]. In [14], ADT switching has been proven to be more general and less conservative.

In this section, we mainly focus on designing an UIO-based robust FE observer for discrete-time switched systems with unknown input. The main contributions of this section can be summarized in the following aspects:

- an UIO-based robust FE observer is designed for discrete-time switched systems. The FE is completely robust to the unknown input disturbance;
- efficient conditions of robust FE observer design for the switched system with measurement disturbance are obtained;
- the proposed FE algorithm is suitable for all kinds of FE.

4.2 The Preliminaries

Definition 4.1 ([15]) An unitary box $\mathbf{B}^n \in \mathbb{R}^n$ is a box composed by $(x_1, \ldots, x_n)^T$, $x_i \in [a_i, b_i], a_i \leq b_i, i = 1, 2, \ldots, n$.

Definition 4.2 ([15]) For given the center vector $p \in \mathbb{R}^n$ and the generator matrix $H \in \mathbb{R}^{n\times m}$, $m \geq n$, an m-zonotope $\mathscr{X}$ is the set $\mathscr{X} = \langle p, H\rangle = p \oplus H\mathbf{B}^m = \{p + Hx : x \in \mathbf{B}^m\}$,

where $\oplus$ denotes the Minkowski sum, which is usually used between the sets:

$$\mathscr{X} \oplus \mathscr{Y} = \{x + y : x \in \mathscr{X}, y \in \mathscr{Y}\}. \tag{4.1}$$

According to [15], the following properties are hold for zonotopes.

Property 4.1 *For given* $p_1 \in \mathbb{R}^n$, $p_2 \in \mathbb{R}^n$ *and the generator matrix* $H_1 \in \mathbb{R}^{n\times m}$, $H_2 \in \mathbb{R}^{n\times m}$,

$$\langle p_1, H_1\rangle \oplus \langle p_2, H_2\rangle = \langle p_1 + p_2, [H_1, H_2]\rangle. \tag{4.2}$$

Property 4.2 *For given the center vector $p \in \mathbb{R}^n$ and the generator matrix $H \in \mathbb{R}^{n\times m}$, $K \in \mathbb{R}^{m\times n}$,*

$$K\langle p, H\rangle = \langle Kp, KH\rangle. \tag{4.3}$$

Property 4.3 *For interval hulls $Box(S_1) = [\bar{a}_1, \underline{a}_1]$ and $Box(S_2) = [\bar{a}_2, \underline{a}_2]$, the following equation satisfies*

$$\begin{aligned} &Box(S_1 \oplus S_2)\\ &= Box(S_1) \oplus Box(S_2)\\ &= [\bar{a}_1, \underline{a}_1] \oplus [\bar{a}_2, \underline{a}_2] = [\bar{a}_1 + \bar{a}_2, \bar{a}_2 + \underline{a}_2] \end{aligned}$$

Definition 4.3 ([16]) For a zonotope $\mathscr{Z} = \langle p, H\rangle \subset \mathbb{R}^n$, its interval hull is defined as $\mathscr{Z} \subseteq Box(\mathscr{Z}) = [\underline{a}, \bar{a}]$, where

$$\begin{cases} \underline{a} = p_i - \sum_{j=1}^{m} |H_{ij}| \\ \bar{a} = p_i + \sum_{j=1}^{m} |H_{ij}|, i = 1, 2, \ldots, n. \end{cases}$$

Definition 4.4 ([15]) The P-radius of an m-zonotope $\mathscr{Z} = p \oplus H\mathbf{B}^m$ is defined as $\mathscr{L}(k) = \max_{z\in\mathscr{Z}} \|z - p\|_P^2$, where $P = P^T > 0$ is a symmetric and positive definite matrix.

4.3 Problem Description

Consider the following discrete-time switched system:

$$\begin{cases} x(t+1) = \sum\limits_{i=1}^{N} \alpha_i(t)(A_i x(t) + B_i u(t) + D_{di} d(t) + D_{\omega i}\omega(t) + F_{1i} f(t)) \\ y(t) = \sum\limits_{i=1}^{N} \alpha_i(t)(C_i x(t) + D_{\upsilon i}\upsilon(t) + F_{2i} f(t)), \end{cases} \tag{4.4}$$

where $x(t) \in \mathbb{R}^{n_x}$ is the state vector, $u(t) \in \mathbb{R}^{n_u}$ is the known input vector, $y(t) \in \mathbb{R}^{n_y}$ is the measured output vector; $d(t) \in \mathbb{R}^{n_d}$ is the independence Gaussian white noise disturbance, $\omega(t) \in \mathbb{R}^{n_\omega}$ is the unknown input vector, $\upsilon(t) \in \mathbb{R}^{n_\upsilon}$ is the measurement noise vector, and $f(t) \in \mathbb{R}^{n_f}$ is the fault vector.

It is assumed that the initial state and the perturbations are bounded: $x(0) \in \mathscr{X}_0 = \langle p_0, H_0\rangle$, $d(t) \in \mathscr{D} = \langle 0, I_{n_d}\rangle$, $\omega(t) \in \mathscr{W} = \langle 0, I_{n_\omega}\rangle$, $\upsilon(t) \in \mathscr{V} = \langle 0, I_{n_\upsilon}\rangle$, where $p_0 \in \mathbb{R}^{n_x}$ and $H_0 \in \mathbb{R}^{n_x\times r}$ are known constant vector and matrix, $I_{n_d} \in \mathbb{R}^{n_d\times n_d}$, $I_{n_\omega} \in \mathbb{R}^{n_\omega\times n_\omega}$, and $I_{n_\upsilon} \in \mathbb{R}^{n_\upsilon\times n_\upsilon}$ denote identity matrices with proper dimensions. $A_i, B_i, C_i, D_{di}, D_{\omega i}, D_{\upsilon i}, F_{1i}$, and F_{2i} are matrices with appropriate dimensions.

Assumption 5 The matrices D_{di} are full column rank.

In order to realize FE, an augmented vector of the state and the fault are constructed as

$$\bar{x}(t) = \begin{bmatrix} x(t) \\ f(t) \end{bmatrix} \tag{4.5}$$

then the system (4.4) can be transformed to the following augmented system:

$$\begin{cases} E\bar{x}(t+1) = \sum_{i=1}^{N} \alpha_i(t)(\bar{A}_i\bar{x}(t) + \bar{B}_i u(t) + \bar{D}_{di}d(t) + \bar{D}_{\omega i}\omega(t)), \\ y(t) = \sum_{i=1}^{N} \alpha_i(t)(\bar{C}_i\bar{x}(t) + D_{\upsilon i}\upsilon(t)), \end{cases} \tag{4.6}$$

where

$$E = \begin{bmatrix} I_{n_x} & 0 \\ 0 & 0 \end{bmatrix}, \bar{A}_i = \begin{bmatrix} A_i & F_{1i} \\ 0 & 0 \end{bmatrix}, \bar{B}_i = \begin{bmatrix} B_i \\ 0 \end{bmatrix},$$
$$\bar{D}_{di} = \begin{bmatrix} D_{di} \\ 0 \end{bmatrix}, \bar{D}_{\omega i} = \begin{bmatrix} D_{\omega i} \\ 0 \end{bmatrix}, \bar{C}_i = \begin{bmatrix} C_i & F_{2i} \end{bmatrix}.$$

In order to realize FE, the following UIO is employed to estimate the state of the augmented system (4.6):

$$\begin{cases} \hat{\bar{x}}(t+1) = \sum_{i=1}^{N} \alpha_i(t)(J_i\bar{A}_i\hat{\bar{x}}(t) + J_i\bar{B}_i u(t) + L_i(y(t) - \hat{y}(t)) + R_i y(t+1)) \\ \hat{y}(t) = \sum_{i=1}^{N} \alpha_i(t)\bar{C}_i\hat{\bar{x}}(t)), \end{cases} \tag{4.7}$$

where $\hat{\bar{x}}(t) \in \mathbb{R}^{n_x+n_f}$ is the estimation of $\bar{x}(t)$, $\hat{y}(t) \in \mathbb{R}^{n_y}$ is the output of the observer. $J_i \in \mathbb{R}^{n_x \times n_x}$, $L_i \in \mathbb{R}^{n_x \times n_y}$, and $R_i \in \mathbb{R}^{n_x \times n_y}$ are the observer parameters to be determined later. The initial state estimation $\hat{\bar{x}}(0)$ is set as $\hat{\bar{x}}(0) = p_0$.

Define $e(t) = \bar{x}(t) - \hat{\bar{x}}(t)$, by (4.4) and (4.7), one has $e(0) \in \Omega_0 = \langle 0, H_0 \rangle$. For the goal of this issue, the observer parameters J_i and R_i of are supposed to satisfy the following assumption:

$$J_i E + R_i \bar{C}_i = I. \tag{4.8}$$

Based on (4.6)–(4.8), one has

$$\begin{aligned} e(t+1) &= (J_i\bar{A}_i - L_i\bar{C}_i)e(t) + J_i\bar{D}_{di}d(t) + J_i\bar{D}_{\omega i}\omega(t) \\ &\quad - L_i D_{\upsilon i}\upsilon(t) - R_i D_{\upsilon i}\upsilon(t+1). \end{aligned} \tag{4.9}$$

If the following conditions hold:

$$J_i \bar{D}_{di} = 0 \tag{4.10}$$

then the error system (4.9) reduces to

$$e(t+1) = (J_i \bar{A}_i - L_i \bar{C}_i)e(t) + J_i \bar{D}_{\omega i}\omega(t) - L_i D_{\upsilon i}\upsilon(t) - R_i D_{\upsilon i}\upsilon(t+1). \tag{4.11}$$

Remark 4.1 By letting $J_i \bar{D}_{di} = 0$, one can see that the independence Gaussian white noise disturbance $d(t)$ is successfully decoupled from the system.

On the other hand, from (4.8) and (4.10), one can get

$$\begin{bmatrix} J_i & R_i \end{bmatrix} \begin{bmatrix} E & \bar{D}_{di} \\ \bar{C}_i & 0 \end{bmatrix} = \begin{bmatrix} I & 0 \end{bmatrix} \tag{4.12}$$

then a set of feasible solution of J_i and R_i can be obtained by

$$\begin{bmatrix} J_i & R_i \end{bmatrix} = \begin{bmatrix} I & 0 \end{bmatrix} \Psi_i^{\dagger} + \partial(I - \Psi_i \Psi_i^{\dagger}), \tag{4.13}$$

where

$$\Psi_i = \begin{bmatrix} E & \bar{D}_{di} \\ \bar{C}_i & 0 \end{bmatrix}. \tag{4.14}$$

$\Psi_i^{\dagger}$ represents the Moore–Penrose inverse, and ∂ is an arbitrary matrix with more extra design freedom degree. From the error system (4.11), one can conclude that the reachable set of $e(t+1)$ can be obtained from $e(t)$, $\omega(t)$, and $\upsilon(t)$ recursively by using the properties of zonotopes in (4.2) and (4.3). Since $e(0)$, $\omega(t)$, and $\upsilon(t)$ are bounded in centered zonotopes, $e(t)$ is also bounded in a zonotope denoted by $\Omega = \langle 0, H(k) \rangle$, where $H(k) \in \mathbb{R}^{(n_x+n_f)\times r}$. According to (4.11), the generator matrix $H(k)$ of the zonotope Ω can be obtained by

$$H_e(k) = \begin{bmatrix} (J_i \bar{A}_i - L_i \bar{C}_i)H_e(k-1) & J_i D_{\omega i} & -L_i D_{\upsilon i} & -R_i D_{\upsilon i} \end{bmatrix}. \tag{4.15}$$

For the sake of minimizing the effects by the disturbances on the state estimation, the observer gains L_i need to be designed such that the size of Ω is minimized. By the Definition 4.4, the P-radius of the zonotope Ω is defined as

$$\mathscr{L}(k) = \max_{z \in \mathbf{B}^r} \| H_e(k) z \|_P^2. \tag{4.16}$$

If there exist a scalar $0 < \gamma < 1$ such that $\mathscr{L}(k) < \gamma \mathscr{L}(k-1)$, the size of Ω is decreasing. Due to the outside disturbance and noise, this condition is hard to verify. A relaxation of this condition can be

$$\mathscr{L}(k) = \max_{z \in \mathbf{B}^r} \| H_e(k) z \|_P^2 + \delta, \tag{4.17}$$

where δ is a positive constant which represents the max influence from the unknown disturbance and measurement noise. Under the condition (4.17), one can get $\mathscr{L}(k)$ is bounded. If γ is smaller or $tr(P)$ is bigger, the size of Ω is smaller.

In the following theorem, the desired observer gains L_i are given which can guarantee the size of Ω is smaller, i.e. $e(t) \to 0$ with $t \to \infty$.

Theorem 4.1 *For a given scalar $0 < \gamma < 1$, matrices J_i and R_i, if there exist a positive definite matrix P, matrices W_i and G_i such that the following optimization conditions can be solved:*

$$\min_{\gamma, P, G_i, W_i} -tr(P) \tag{4.18}$$

$$s.t. \begin{bmatrix} P - G_i - G_i^T & G_i J_i \bar{A}_i - W_i \bar{C}_i & G_i J_i \bar{D}_{\omega i} & -W_i D_{\upsilon i} & -G_i R_i D_{\upsilon i} \\ * & -\gamma P & 0 & 0 & 0 \\ * & * & -\bar{D}_{\omega i}^T \bar{D}_{\omega i} & 0 & 0 \\ * & * & * & -D_{\upsilon i}^T D_{\upsilon i} & 0 \\ * & * & * & * & -D_{\upsilon i}^T D_{\upsilon i} \end{bmatrix} < 0 \tag{4.19}$$

then the error system (4.11) is bounded stable. Under this case, the observer gains L_i can be calculated by $L_i = G_i^{-1} W_i$.

Proof Define a zonotope size P-radius as follows:

$$\mathscr{L}(k) = \max_{z \in \mathbf{B}^r} \| H_e(k) z \|_P^2, r = n_x + n_f + n_\omega + 2n_\upsilon. \tag{4.20}$$

If there exist two scalars $0 < \gamma < 1$ and $\delta > 0$, such that the following condition holds:

$$\mathscr{L}(k) < \gamma \mathscr{L}(k-1) + \delta \tag{4.21}$$

then the size of the zonotope will be decreased.

Define the generator matrix formation as

$$H_e(k) = \left[(J_i \bar{A}_i - L_i \bar{C}_i) H_e(k-1) \;\; J_i D_{\omega i} \;\; -L_i D_{\upsilon i} \;\; -R_i D_{\upsilon i} \right] \tag{4.22}$$

then by (4.11), the max influence value δ of the unknown input can be chosen as

$$\delta = \max_{\beta \in \mathbf{B}^{n_\omega}} \| \bar{D}_{\omega i} \beta \|_2^2 + \max_{\eta \in \mathbf{B}^{n_\upsilon}} \| D_{\upsilon i} \eta \|_2^2 + \max_{\lambda \in \mathbf{B}^{n_\upsilon}} \| D_{\upsilon i} \lambda \|_2^2. \tag{4.23}$$

By (4.21) and (4.23), one has

$$\max_{z \in \mathbf{B}^r} \|H_e(k)z\|_P^2 < \gamma \max_{\hat{z} \in \mathbf{B}^{n_x+n_f}} \|H_e(k-1)\hat{z}\|_P^2 + \max_{\beta \in \mathbf{B}^{n_\omega}} \|\bar{D}_{\omega i}\beta\|_2^2 + \max_{\eta \in \mathbf{B}^{n_\upsilon}} \|D_{\upsilon i}\eta\|_2^2 + \max_{\lambda \in \mathbf{B}^{n_\upsilon}} \|D_{\upsilon i}\lambda\|_2^2, \tag{4.24}$$

which is equivalent to the following inequality:

$$\|H_e(k)z\|_P^2 - \gamma\|H_e(k-1)\hat{z}\|_P^2 - \|\bar{D}_{\omega i}\beta\|_2^2 - \|D_{\upsilon i}\eta\|_2^2 - \|D_{\upsilon i}\lambda\|_2^2 < 0. \tag{4.25}$$

Furthermore, (4.25) can be written as

$$\begin{aligned} &z^T H_e(k)^T P H_e(k) z - \gamma \hat{z}^T H_e(k-1)^T P H_e(k-1)\hat{z} \\ &-\beta^T \bar{D}_{\omega i}^T \bar{D}_{\omega i}\beta - \eta^T D_{\upsilon i}^T D_{\upsilon i}\eta - \lambda^T D_{\upsilon i}^T D_{\upsilon i}\lambda < 0. \end{aligned} \tag{4.26}$$

Define

$$z = \begin{bmatrix} \rho \\ \beta \\ \eta \\ \lambda \end{bmatrix}, \rho = H_e(k-1)\hat{z} \tag{4.27}$$

then (4.25) can be transformed as

$$z^T(\Theta_i^T P \Theta_i + \Lambda_i)z < 0, \tag{4.28}$$

where

$$\Theta_i = \begin{bmatrix} J_i\bar{A}_i - L_i\bar{C}_i & J_i D_{\omega i} & -L_i D_{\upsilon i} & -R_i D_{\upsilon i} \end{bmatrix}, \tag{4.29}$$

$$\Lambda_i = \begin{bmatrix} -\gamma P & 0 & 0 & 0 \\ * & -\bar{D}_{\omega i}^T \bar{D}_{\omega i} & 0 & 0 \\ * & * & -D_{\upsilon i}^T D_{\upsilon i} & 0 \\ * & * & * & -D_{\upsilon i}^T D_{\upsilon i} \end{bmatrix}. \tag{4.30}$$

By Schur Complement Lemma, (4.28) is equivalent to

$$\begin{bmatrix} -P^{-1} & \Theta_i \\ \Theta_i^T & \Lambda_i \end{bmatrix} < 0. \tag{4.31}$$

On the other hand, for any invertible matrix G_i, the inequality $(P - G_i)^T P^{-1}(P - G_i) < 0$ hold. Define

$$W_i = G_i L_i. \tag{4.32}$$

Then one can conclude that (4.19) implies (4.28) holds. The proof is completed. (□)

4.4 The Analysis of the FE Error

In this section, an interval hull computation approximation algorithm of FE is obtained for discrete-time switched systems, which can prove that the obtained algorithm has more accurate boundary and higher efficiency than the zonotope-based technique. For system (4.11), under the assumptions that the initial state, system disturbance, and measured noise are belong to the corresponding zonotopes, we have

$$\Omega_{t+1} = (J_i\bar{A}_i - L_i\bar{C}_i)\Omega_t \oplus J_i\bar{D}_{\omega i}\mathscr{W}_t \oplus (-L_iD_{vi})\mathscr{V}_t \oplus (-R_iD_{vi})\mathscr{V}_{t+1}, \quad (4.33)$$

$$\begin{aligned}\Omega_{t+1} = {} & (J_i\bar{A}_i - L_i\bar{C}_i)^k\Omega_t \oplus \bigoplus_{j=0}^{k-1}(J_i\bar{A}_i - L_i\bar{C}_i)^j J_i\bar{D}_{\omega i}\mathscr{W}_t \\ & \oplus \bigoplus_{j=0}^{k-1}(J_i\bar{A}_i - L_i\bar{C}_i)^j(-L_iD_{vi})\mathscr{V}_t \\ & \oplus \bigoplus_{j=0}^{k-1}(J_i\bar{A}_i - L_i\bar{C}_i)^j(-R_iD_{vi})\mathscr{V}_{t+1}. \end{aligned} \quad (4.34)$$

Based on Property 4.3, the interval hull of Ω_{t+1} can be obtained by

$$\begin{aligned}Box(\Omega_{t+1}) = {} & Box((J_i\bar{A}_i - L_i\bar{C}_i)^k\Omega_0) \oplus \bigoplus_{j=0}^{k-1} Box((J_i\bar{A}_i - L_i\bar{C}_i)^j J_i\bar{D}_{\omega i}\mathscr{W}_t) \\ & \oplus \bigoplus_{j=0}^{k-1} Box((J_i\bar{A}_i - L_i\bar{C}_i)^j(-L_iD_{vi})\mathscr{V}_t) \\ & \oplus \bigoplus_{j=0}^{k-1} Box((J_i\bar{A}_i - L_i\bar{C}_i)^j(-R_iD_{vi})\mathscr{V}_{t+1}). \end{aligned} \quad (4.35)$$

Remark 4.2 The form of the fault can be arbitrary. That is to say, there is no limitation on the formation of the fault. The method proposed here can be realized to estimate them accurately.

4.5 Summary

In this section, an UIO-based robust FE observer is designed for discrete-time switched systems with unknown input. By using P-radius technique, an FE algorithm is proposed to achieve robust FE. Our designed method is general in the sense that the same design procedure is applied for any fault vector.

References

1. Marseglia, G.R., Raimondo, D.M.: Active fault diagnosis: a multi-parametric approach. Automatica **79**, 223–230 (2017)
2. Yao, L., Qin, J., Wang, H., Jiang, B.: Design of new fault diagnosis and FTC scheme for non-Gaussian singular stochastic distribution systems. Automatica **48**(9), 2305–2313 (2012)
3. Du, D.: Fault detection for discrete-time systems based on descriptor observer approach. Appl. Math. Comput. **293**, 575–585 (2017)
4. Xiong, J., Chang, X., Yi, X.: Design of robust nonfragile fault detection filter for uncertain dynamic systems with quantization. Appl. Math. Comput. **338**, 774–788 (2018)
5. Yan, X., Edwards, C.: Robust decentralized actuator fault detection and estimation for large-scale systems using a sliding mode observer. Int. J. Control **81**(4), 591–606 (2007)
6. Zhang, K., Jiang, B., Shi, P., Xu, J.: Analysis and design of robust H_∞ fault estimation observer with finite-frequency specifications for discrete-time fuzzy systems. IEEE Trans. Fuzzy Syst. **45**(7), 1225–1235 (2015)
7. Jiang, B., Staroswiecki, M., Cocquempot, V.: Fault accommodation for nonlinear dynamic systems. IEEE Trans. Autom. Control **51**(9), 1578–1583 (2006)
8. Huang, S., Yang, G.: FTCler design for T-S fuzzy systems with time-varying delay and actuator faults: a k-step fault-estimation approach. IEEE Trans. Fuzzy Syst. **22**(6), 1526–1540 (2014)
9. Fonod, R., Henry, D., Charbonnely, C., Bornschleglz, E.: A class of nonlinear unknown input observer for fault diagnosis: application to fault tolerant control of an autonomous spacecraft. In: Proc. UKACC Int. Conf. Control, Loughborogh, U.K., pp. 19–24 (2014)
10. Sun, X., Patton, R.: Robust actuator multiplicative fault estimation with unknown input decoupling for a wind turbine system. In: Proc. Conf. Control Fault-Tolerant Syst, Nice, France, pp. 263–268 (2013)
11. Daafouz, J., Riedinger, P., Iung, C.: Stability analysis and control synthesis for switched systems: a switched Lyapunov function approach. IEEE Trans. Autom. Control **47**(11), 1883–1887 (2002)
12. Johansson, M., Rantzer, A.: Computation of piecewise quadratic Lyapunov functions for hybrid systems. IEEE Trans. Autom. Control **43**(4), 555–559 (1998)
13. Hespanha, J.P., Morse, A.S.: Stability of switched systems with average dwell time. IEEE Conf. Decis. Control **3**(3), 2655–2660 (2002)
14. Hespanha, J.P.: Uniform stability of switched linear systems extensions of Lasalles invariance principle. IEEE Trans. Autom. Control **49**(4), 470–482 (2004)
15. HieuLe, V.T., Stoica, C., Alamo, T., Camacho, E.F., Dumur, D.: Zonotopic guaranteed state estimation for uncertain systems. Automatica **49**, 3418–3424 (2013)
16. Tang, W., Wang, Z., Shen, Y., Rodrigues, M., Theilliol, D.: Observer and interval hull computation fault detection based on multi-objective observer and interval hull computation. In: 10th Safeprocess. Poland (2018)

Chapter 5
Adaptive Observer-Based Fault Diagnosis for Continuous-Time Switched Systems

5.1 FE for Actuator and Sensor Faults

This section investigates the problem of FE for a continuous-time switched system with actuator and sensor faults. Firstly, based on the ADT method and switched Lyapunov function, an FE observer is designed for the augmented switched system, which can ensure that the error system is asymptotically stable with an H_∞ performance index. Next, an algorithm is proposed to realize FE accurately. Finally, an example is given to illustrate the effectiveness of the designed methods.

5.1.1 Introduction

In recent decades, with the increasing demands of safety and reliability of various dynamic systems, fault diagnosis has attracted more and more attention. However, in practical terms, the system fault is inevitable, this not only brings security risks to the system but also causes huge economic losses. Therefore, as a common and effective method to deal with the system fault, fault diagnosis has been widely developed and applied in recent years [1–5]. The fault diagnosis technique consists of FD and FE. The focus of FD is to detect the fault as soon as possible, and the focus of FE is to describe the shape of the system's fault. FE can obtain more information than FD.

For the issue of FE, a variety of methods have appeared in recent years, among which the observer-based method is the most popular, such as adaptive observer [6], sliding-mode observer [7], fuzzy observer [8], etc. The technique of FE based on adaptive observer can achieve the asymptotic estimation of constant faults, while the technique of FE based on sliding-mode observer can produce the bounded estimation of constant faults, and both of them can realize bounded estimation for time-varying faults. In [9], an FE method based on UIO was proposed, which combined the FE

D. Du et al., *Observer-Based Fault Diagnosis and Fault-Tolerant Control for Switched Systems*, Studies in Systems, Decision and Control 323,
https://doi.org/10.1007/978-981-15-9073-3_5

observer and fault-tolerant controller. In order to keep a balance between estimation accuracy and estimation cost, an adjustable dynamic observer design method is proposed in [10]. Based on the output information about the system, the PIO is applied in [11], to reconstruct the actuator fault. In [12], it is proposed that the outputs and differentials of the PIO are both considered to be known and can be utilized. However, if there is measurement interference or sensor fault in the system, traditional observer cannot be directly applied, measure disturbances or sensor fault may be amplified by the observer gain. In [13, 14], many observer design methods are mentioned, among which generalized observer design method has been widely applied which includes the state of the original system and the sensor fault. It can be seen from the above results that the system model plays an important role. However, it is very difficult to know all the information of the dynamic system in practice. In addition, it should be noted that when the actuator fault is time-varying with slow speed, the PIO can obtain better estimation performance. On the contrary, if the fault changes frequently, the estimation performance may be affected by the fault differential.

On the other aspect, the control and synthesis of the switched system has attracted more and more attention due to its extensive applications in practical engineering systems. The concept of ADT firstly appeared in [15], which was one of the constrained switching signals and has been widely used in fault diagnosis and fault-tolerant control.

In this section, we mainly focus on designing a FE observer for the continuous-time switched system. The main contributions of this section can be summarized in the following aspects:

- Based on the adaptive approximation technique, a new adaptive dynamic PIO is designed to estimate system states, actuator and sensor faults, and measurement disturbance. In dynamic PIO, output and output differential information are used. However, the previously assumed measurement output differential is measurable.
- The designed FE observer satisfies the H_∞ performance is gradually stable.
- The proposed FE algorithm is suitable for various FE.

5.1.2 Model Description and Preliminaries

In this section, the following switched system is considered:

$$\begin{cases} \dot{x}(t) = \sum\limits_{i=1}^{N} \sigma_i(t)(A_i x(t) + B_i u(t) + D_i d(t) + F_i f_a(t)) \\ y(t) = \sum\limits_{i=1}^{N} \sigma_i(t)(C_i x(t) + W_i w(t) + G_i f_s(t)) \end{cases} \tag{5.1}$$

where $x(t) \in \mathbb{R}^n$, $u(t) \in \mathbb{R}^{n_u}$, and $y(t) \in \mathbb{R}^{n_y}$ are the system state, the control input and the measurement output. $f_a(t) \in \mathbb{R}^{n_p}$, $f_s(t) \in \mathbb{R}^{n_s}$, $d(t) \in \mathbb{R}^{n_d}$ and $w(t) \in \mathbb{R}^{n_w}$ denote the actuator fault, the sensor fault, the outside disturbance and the measurement disturbance. $\sigma_i(t)$ is called a switching signal and satisfies

$$\sigma_i(t) : [0, \infty) \to \{0, 1\}, \quad \sum_{i=1}^{N} \sigma_i(t) = 1 \tag{5.2}$$

It is noted that the switching signal is a piecewise constant function depending on the time t, and the $i - th$ mode will be activated when $\sigma_i(t) = 1$. A_i, B_i, C_i, D_i, F_i, W_i, and G_i are constant real matrices with appropriate dimensions.

Definition 5.1 ([16]) For any switching signal $\sigma_i(t)$ and any $t_2 > t_1 > 0$, let $N_{\sigma_i(t)}$ (t_1, t_2) denote the number of switchings $\sigma_i(t)$ on an interval (t_1, t_2). If

$$N_{\sigma_i(t)}(t_1, t_2) \leq N_0 + \frac{t_2 - t_1}{\tau_a}$$

holds for a given $N_0 \geq 0$ and $\tau_a > 0$, then the constant τ_a is called the average dwell-time and N_0 is the chattering bound.

The following lemmas are used in this section:

Lemma 5.1 ([17]) *Consider the continuous-time switched system* $\dot{x}(t) = f_{\sigma_i(t)}$ $(x(t))$*, and let* $\alpha > 0$*,* $\mu > 1$ *be given constants. Suppose that there exist* $\mathscr{C}^1$ *functions* $V_{\sigma_i(t)} : \mathbb{R}^n \to \mathbb{R}$ *and two class* $\mathscr{K}_\infty$ *functions* κ_1 *and* κ_2 *such that*

$$\kappa_1(|x(t)|) \leq V_{\sigma_i(t)}(x(t)) \leq \kappa_2(|x(t)|)$$

$$\dot{V}_{\sigma_i(t)}((t)) \leq -\alpha V_{\sigma_i(t)}(x(t))$$

and $\forall(\sigma_i(t_l) = i, \sigma_i(t_l^-) = j) \in N \times N, i \neq j,$

$$V_i(x(t)) \leq \mu V_j(x(t))$$

then the switched system $\dot{x}(t) = f_{\sigma_i(t)}(x(t))$ *is GUAS for any switching signal with ADT*

$$\tau_a > \tau_a^* = \frac{\ln \mu}{\alpha}. \tag{5.3}$$

Note that there is sensor fault and outside disturbance in the system, then the classic Luenberger observer may amplify the effect of the sensor fault $f_s(t)$ and the outside disturbance $w(t)$. Fortunately, augmented state observer can avoid this problem.

Remark 5.1 In practical applications, the uncertainty of parameters is common, we assume that the uncertain parameters are bounded and has been implemented and applied to the problem. For the main object of this section, the following assumptions are given:

Assumption 6 The matrices W_i and G_i are full column rank.

For simplicity, we make the following simplification: Let $\eta(t) = [w^T(t), f_s^T(t)]^T$, $\eta(t) \in \mathbb{R}^{n_\eta}$. Then the output equation can be written as

$$y(t) = \sum_{i=1}^{N} \sigma_i(t)(C_i x(t) + \bar{W}_i \eta(t)) \tag{5.4}$$

where $\bar{W}_i = [W_i, G_i]$. Let $\bar{x}(t) = [x^T(t), \eta^T(t)]^T$. Based on (5.4), then we have

$$\begin{cases} J\dot{\bar{x}}(t) = \sum_{i=1}^{N} \sigma_i(t)(\bar{A}_i\bar{x}(t) + \bar{B}_i u(t) + \bar{D}_i d(t) + \bar{F}_i f_a(t) + L\bar{W}_i\eta(t)) \\ y(t) = \sum_{i=1}^{N} \sigma_i(t)(\bar{C}_i\bar{x}(t)) = \sum_{i=1}^{N} \sigma_i(t)(\bar{\bar{C}}_i\bar{x}(t) + \bar{W}_i\eta(t)) \end{cases} \tag{5.5}$$

where

$$J = \begin{bmatrix} I_n & 0 \\ 0 & 0_{n_y \times n_\eta} \end{bmatrix}, \ \bar{A}_i = \begin{bmatrix} A_i & 0 \\ 0 & -\bar{W}_i \end{bmatrix}, \ \bar{B}_i = \begin{bmatrix} B_i \\ 0_{n_y \times n_u} \end{bmatrix}, \ \bar{D}_i = \begin{bmatrix} D_i \\ 0_{n_y \times n_d} \end{bmatrix},$$
$$\bar{F}_i = \begin{bmatrix} F_i \\ 0_{n_y \times n_p} \end{bmatrix}, L = \begin{bmatrix} 0_{n \times n_y} \\ I_{n_y} \end{bmatrix}, \ \bar{C}_i = \begin{bmatrix} C_i & \bar{W}_i \end{bmatrix}, \ \bar{\bar{C}}_i = \begin{bmatrix} C_i & 0_{n_y \times n_\eta} \end{bmatrix}.$$

According to (5.4), it can be found that

$$L\bar{W}_i\eta(t) = L(y(t) - \sum_{i=1}^{N} \sigma_i(t)\bar{\bar{C}}_i\bar{x}(t))$$

thus,

$$J\dot{\bar{x}}(t) = \sum_{i=1}^{N} \sigma_i(t)((\bar{A}_i - L\bar{\bar{C}}_i)\bar{x}(t) + \bar{B}_i u(t) + \bar{D}_i d(t) + \bar{F}_i f_a(t) + Ly(t)) \tag{5.6}$$

we add $L\dot{y}(t)$ to both sides of (5.6), then we have

$$\begin{aligned} J_0\dot{\bar{x}}(t) = \sum_{i=1}^{N} \sigma_i(t)((\bar{A}_i - L\bar{\bar{C}}_i)\bar{x}(t) + \bar{B}_i u(t) \\ + \bar{D}_i d(t) + \bar{F}_i f_a(t) + Ly(t) + L\dot{y}(t)) \end{aligned} \tag{5.7}$$

where

$$J_0 = J + L\bar{C}_i = \begin{bmatrix} I_n & 0_{n\times n_\eta} \\ C_i & \bar{W}_i \end{bmatrix}.$$

Because $\bar{W}_i$ is full rank, its left inverse exists. Let $\bar{W}_i^{\dagger}$ represent one left inverse matrix of $\bar{W}_i$ which means $\bar{W}_i^{\dagger}\bar{W}_i = I_{n_\eta}$. Thus, there is matrix

$$J_0^{\dagger} = \begin{bmatrix} I_n & 0_{n\times n_y} \\ -\bar{W}_i^{\dagger}C_i & \bar{W}_i^{\dagger} \end{bmatrix}.$$

such that $J_0^{\dagger}J_0 = I_{n_{\bar{x}}}$. Based on above analysis, we have

$$\begin{aligned}\dot{\bar{x}}(t) &= \sum_{i=1}^{N}\sigma_i(t)(J_0^{\dagger}(\bar{A}_i - L\bar{\bar{C}}_i)\bar{x}(t) + J_0^{\dagger}\bar{B}_i u(t) \\ &\quad + J_0^{\dagger}\bar{D}_i d(t) + J_0^{\dagger}\bar{F}_i f_a(t) + J_0^{\dagger}Ly(t) + J_0^{\dagger}L\dot{y}(t)) \\ &= \sum_{i=1}^{N}\sigma_i(t)(\bar{\bar{A}}_i\bar{x}(t) + \bar{\bar{B}}_i u(t) + \bar{\bar{D}}_i d(t) + \bar{\bar{F}}_i f_a(t) + L_O y(t) + L_O\dot{y}(t))\end{aligned} \tag{5.8}$$

where, $\bar{\bar{A}}_i = J_0^{\dagger}(\bar{A}_i - L\bar{\bar{C}}_i)$, $\bar{\bar{B}}_i = J_0^{\dagger}\bar{B}_i$, $\bar{\bar{D}}_i = J_0^{\dagger}\bar{D}_i$, $\bar{\bar{F}}_i = J_0^{\dagger}\bar{F}_i$, $L_O = J_0^{\dagger}L$. Note that the system (5.8) contains $\dot{y}(t)$, in many cases, we know measurement output $y(t)$ rather than $\dot{y}(t)$. In order to avoid this problem, a new variable $z(t)$ is introduced.

Let

$$\begin{cases} z(t) = \bar{x}(t) - L_O y(t) \\ y(t) = \sum_{i=1}^{N}\sigma_i(t)(\bar{C}_i\bar{x}(t)) \end{cases} \tag{5.9}$$

then we can get $\bar{x}(t) = z(t) + L_O y(t)$. Thus, the system (5.9) can be rewritten as

$$\begin{cases} \dot{z}(t) = \dot{\bar{x}}(t) - L_O\dot{y}(t) \\ \quad = \sum_{i=1}^{N}\sigma_i(t)(\bar{\bar{A}}_i\bar{x}(t) + \bar{\bar{B}}_i u(t) + \bar{\bar{D}}_i d(t) + \bar{\bar{F}}_i f_a(t) + L_O y(t)) \\ \quad = \sum_{i=1}^{N}\sigma_i(t)(\bar{\bar{A}}_i z(t) + \bar{\bar{B}}_i u(t) + \bar{\bar{D}}_i d(t) + \bar{\bar{F}}_i f_a(t) + \bar{\bar{A}}_i L_O y(t) + L_O y(t)) \\ y(t) = \sum_{i=1}^{N}\sigma_i(t)(\bar{C}_i\bar{x}(t)) \end{cases} \tag{5.10}$$

For the system (5.10), the following FE observer is considered: where $\hat{z}(t) \in \mathbb{R}^{n_{\bar{x}}}$ and $\upsilon \in \mathbb{R}^{n_p}$ represent the observer state, $\hat{z}(t)$, $\hat{x}(t)$, $\hat{\bar{x}}(t)$, $\hat{f}_a(t)$, $\hat{y}(t)$ represent the estimations of $z(t)$, $x(t)$, $\bar{x}(t)$, $f_a(t)$ and $y(t)$, δ is an adjustable parameter, $K_{1i} \in \mathbb{R}^{n_{\bar{x}}\times n_y}$ and $K_{2i} \in \mathbb{R}^{n_p\times n_y}$ denote the observer gain matrices to be designed.

$$\begin{cases} \dot{\hat{z}}(t) = \sum_{i=1}^{N} \sigma_i(t)((\bar{\bar{A}}_i \hat{z}(t) + \bar{\bar{B}}_i u(t) + \bar{\bar{F}}_i \hat{f}_a(t) + \bar{\bar{A}}_i L_O y(t) + L_O y(t) \\ \qquad + K_{1i}(y(t) - \hat{y}(t))) \\ \dot{v}(t) = K_{2i}(y(t) - \hat{y}(t)) \\ \hat{\bar{x}}(t) = \sum_{i=1}^{N} \sigma_i(t)(\hat{z}(t) + L_O y(t)) \\ \hat{f}_a(t) = \sum_{i=1}^{N} \sigma_i(t)(v(t) + \delta(F_i^T C_i^T)(y(t) - \hat{y}(t))) \\ \hat{y}(t) = \sum_{i=1}^{N} \sigma_i(t)(\bar{C}_i \hat{\bar{x}}(t)) \end{cases} \tag{5.11}$$

Define

$$\begin{cases} e(t) = \bar{x}(t) - \hat{\bar{x}}(t), \\ e_f(t) = f_a(t) - \hat{f}_a(t), \\ e_z(t) = z(t) - \hat{z}(t), \end{cases} \tag{5.12}$$

According (5.9) and (5.11), we have $e(t) = e_z(t)$, then

$$\begin{aligned} \dot{e}(t) &= \dot{e}_z(t) = \dot{z}(t) - \dot{\hat{z}}(t) \\ &= \sum_{i=1}^{N} \sigma_i(t)((\bar{\bar{A}}_i - K_{1i}\bar{C}_i)e(t) + \bar{\bar{D}}_i d(t) + \bar{\bar{F}}_i e_f(t)) \end{aligned} \tag{5.13}$$

then, the derivative of the fault error is given as

$$\begin{aligned} \dot{e_f}(t) &= \dot{f}_a(t) - \dot{\hat{f}}_a(t) \\ &= \dot{f}_a(t) - \dot{v}(t) - \delta F_i^T C_i^T (\dot{y}(t) - \dot{\hat{y}}(t)) \\ &= \dot{f}_a(t) - K_{2i}\bar{C}_i e(t) - \delta F_i^T C_i^T \bar{C}_i \dot{e}(t) \end{aligned} \tag{5.14}$$

If we define $\xi(t) = [e^T(t), e_f^T(t)]^T$, then we have

$$J_1 \dot{\xi}(t) = (\tilde{A}_i - \bar{K}_i \tilde{C}_i)\xi(t) + \tilde{D}_i \bar{d}(t) \tag{5.15}$$

where

$$\begin{aligned} J_1 &= \begin{bmatrix} I_{n_{\bar{x}}} & 0_{n_{\bar{x}} \times n_p} \\ \delta F_i^T C_i^T \bar{C}_i & I_{n_p} \end{bmatrix}, \ \tilde{A}_i = \begin{bmatrix} \bar{\bar{A}}_i & \bar{\bar{F}}_i \\ 0_{n_p \times n_{\bar{x}}} & 0_{n_p \times n_p} \end{bmatrix}, \ \tilde{D}_i = \begin{bmatrix} \bar{\bar{D}}_i & 0_{\bar{x} \times n_p} \\ 0_{n_p \times n_d} & I_{n_p} \end{bmatrix}, \\ \bar{d}(t) &= \begin{bmatrix} d(t) \\ \dot{f}_a(t) \end{bmatrix}, \bar{K}_i = \begin{bmatrix} K_{1i} \\ K_{2i} \end{bmatrix}, \ \tilde{C}_i = [\, \bar{C}_i \ 0_{n_y \times n_p} \,]. \end{aligned} \tag{5.16}$$

We define $J_1^\dagger J_1 = I_{n_{\bar{x}} + n_p}$, then based on (5.15), we have

$$\dot{\xi}(t) = J_1^{\dagger}(\tilde{A}_i - \bar{K}_i\tilde{C}_i)\xi(t) + J_1^{\dagger}\tilde{D}_i\bar{d}(t) \tag{5.17}$$

5.1.3 Main Results

5.1.3.1 Robust Performance Analysis

In this subsection, the robust condition of the disturbance signal $\bar{d}(t)$ to the residual signal $\xi(t)$ will be established.

Theorem 5.1 *For given scalars $\gamma > 0$, $\mu > 1$ and $\alpha > 0$, for the case that $\bar{d}(t) = 0$: the error system (5.17) is asymptotically stable with the H_∞ performance index γ, such that*

$$\sup_{\bar{d}(t)=0} \frac{\|\xi(t)\|_2}{\|\bar{d}(t)\|_2} < \gamma. \tag{5.18}$$

For the case that $\bar{d}(t) \neq 0$, if there exist positive definite matrices $P_i = P_i^T > 0$ and matrices Q_i such that

$$P_i \leq \mu P_j \tag{5.19}$$

$$\begin{bmatrix} sym(P_i J_1^{\dagger}\tilde{A}_i - Q_i\tilde{C}_i) + \alpha P_i + I_{n_{\bar{x}}+n_p} & P_i J_1^{\dagger}\tilde{D}_i \\ * & -\gamma^2 I_{n_{\bar{d}}} \end{bmatrix} < 0 \tag{5.20}$$

then for the switching signal $\sigma_i(t)$ with ADT constraint (5.3), the system (5.17) is asymptotically stable with H_∞ performance index γ. Under this case, the observer gain matrixs of (5.11) can be designed as $\bar{K}_i = [K_{1i}, K_{2i}]^T = J_1 P_i^{\dagger} Q_i$.

Proof Firstly, we establish the stability of system (5.17). When $\bar{d}(t) = 0$, we define the following switched Lyapunov function:

$$V_i(t, \xi(t)) = \xi^T(t)\left(\sum_{i=1}^{N}\sigma_i(t)P_i\right)\xi(t) \tag{5.21}$$

where $P_i > 0$, then along the trajectory of system (5.17), one has

$$\begin{aligned} &\dot{V}_i(t, \xi(t)) + \alpha V_i(t, \xi(t)) \\ &= \dot{\xi}^T(t)\left(\sum_{i=1}^{N}\sigma_i(t)P_i\right)\xi(t) + \xi^T(t)\left(\sum_{i=1}^{N}\sigma_i(t)P_i\right)\dot{\xi}(t) \\ &\quad + \xi^T(t)\left(\sum_{i=1}^{N}\sigma_i(t)\alpha P_i\right)\xi(t) \end{aligned}$$

$$= \sum_{i=1}^{N} \sigma_i(t)\xi^T(t)(sym(P_i J_1^{\dagger} \tilde{A}_i - Q_i \tilde{C}_i) + \alpha P_i)\xi(t)$$

we define the following performance index:

$$J = \int_0^t (\xi^T(t)\xi(t) - \gamma^2 \bar{d}^T(t)\bar{d}(t))dt$$

by (5.18), which means

$$\xi^T(t)\xi(t) - \gamma^2 \bar{d}^T(t)\bar{d}(t) < 0$$

then we have $J < 0$, which means $\|\xi(t)\| < \gamma\|\bar{d}(t)\|$ when $t \to \infty$.

When $\bar{d}(t) \neq 0$, we set

$$\begin{aligned} \varphi_i(t) &= \dot{V}_i(t, \xi(t)) + \alpha V_i(t, \xi(t)) + \xi^T(t)\xi(t) - \gamma^2 \bar{d}^T(t)\bar{d}(t) \\ &= \sum_{i=1}^{N} \sigma_i(t) \left[\xi^T(t) \ \bar{d}^T(t) \right] \begin{bmatrix} \phi & P_i J_1^{\dagger} \tilde{D}_i \\ ** & -\gamma^2 I_{n_{\bar{d}}} \end{bmatrix} \begin{bmatrix} \xi(t) \\ \bar{d}(t) \end{bmatrix} \end{aligned} \tag{5.22}$$

where $\phi = sym(P_i J_1^{\dagger} \tilde{A}_i - Q_i \tilde{C}_i) + \alpha P_i + I_{n_{\bar{x}}+n_p}$. Thus if (5.20) holds and Lemma 5.1, one can get the $\phi < 0$. In the following, the proof of H_∞ performance index will be given. Consider the system (5.17), we define the following performance index:

$$J = \int_0^t (\xi^T(t)\xi(t) - \gamma^2 \bar{d}^T(t)\bar{d}(t))dt \tag{5.23}$$

one can get

$$\begin{aligned} J &= \int_0^t (\xi^T(t)\xi(t) - \gamma^2 \bar{d}^T(t)\bar{d}(t) + \dot{V}_i(t, \xi(t)))dt - V_i(t, \xi(t)) \\ &\leq \int_0^t (\xi^T(t)\xi(t) - \gamma^2 \bar{d}^T(t)\bar{d}(t) + \dot{V}_i(t, \xi(t)) \\ &\quad + \alpha V_i(t, \xi(t)))dt - V_i(t, \xi(t)) \end{aligned} \tag{5.24}$$

by (5.20), which means

$$\xi^T(t)\xi(t) - \gamma^2 \bar{d}^T(t)\bar{d}(t) + \dot{V}_i(t, \xi(t)) + \alpha V_i(t, \xi(t)) < 0 \tag{5.25}$$

Then we have $J < 0$, which means $\|\xi(t)\| < \gamma\|\bar{d}(t)\|$ when $t \to \infty$. The proof is completed. (□)

Algorithm:

Step 1: Construct the augmented system (5.6);

Step 2: Calculate the gains K_{1i}, K_{2i} and solve the unknown matrices P_i and Q_i by the conditions (5.18), (5.19) and (5.20) in the Theorem 5.1;
Step 3: Implement the observer (5.11) to obtain the state estimation $\dot{\hat{z}}(t)$, and then we can get the state estimation $\hat{\bar{x}}(t)$ and the FE $\hat{f}_a(t)$;
Step 4: Since the state estimation of the system has been obtained in step 3, the actuator FE $\hat{f}_s(t)$ and measurement disturbance estimation $\hat{w}(t)$ will obtained by the following equations:

$$\hat{f}_s(t) = \begin{bmatrix} 0_n & 0 & I_{n_s} \end{bmatrix} \hat{\bar{x}}(t) \tag{5.26}$$

$$\hat{w}(t) = \begin{bmatrix} 0_n & I_{n_w} & 0 \end{bmatrix} \hat{\bar{x}}(t). \tag{5.27}$$

5.1.4 An Illustrative Example

Consider the system (5.1) with the following parameters:

$$A_1 = \begin{bmatrix} -1.35 & -0.98 \\ 17.1 & -1.85 \end{bmatrix}, A_2 = \begin{bmatrix} -1.87 & -0.98 \\ 12.6 & -2.63 \end{bmatrix}, A_3 = \begin{bmatrix} -0.65 & -0.31 \\ 0.07 & 0.85 \end{bmatrix},$$
$$B_1 = \begin{bmatrix} 1.1 & 0.2 \\ 0.2 & 3 \end{bmatrix}, B_2 = \begin{bmatrix} 1 & 0.5 \\ 0.2 & 1.2 \end{bmatrix}, B_3 = \begin{bmatrix} 2 & 0.3 \\ -1 & 2 \end{bmatrix}, C_1 = \begin{bmatrix} 1 & 0.2 \\ -1.1 & 2 \end{bmatrix},$$
$$C_2 = \begin{bmatrix} -1 & 0.2 \\ 0.3 & 1 \end{bmatrix}, C_3 = \begin{bmatrix} -2 & 0.1 \\ 1.5 & 2 \end{bmatrix}, D_1 = \begin{bmatrix} 1 \\ 2 \end{bmatrix}, D_2 = \begin{bmatrix} 1.2 \\ 0.1 \end{bmatrix}, D_3 = \begin{bmatrix} 0.2 \\ 0.2 \end{bmatrix},$$
$$F_1 = \begin{bmatrix} 0.1 \\ 0.2 \end{bmatrix}, F_2 = F_3 = \begin{bmatrix} 1 \\ 1 \end{bmatrix}, W_1 = \begin{bmatrix} 1.1 \\ 1.2 \end{bmatrix}, W_2 = \begin{bmatrix} 0.2 \\ 0.11 \end{bmatrix}, W_3 = \begin{bmatrix} 0.11 \\ 0.2 \end{bmatrix},$$
$$G_1 = \begin{bmatrix} 1.11 \\ 0.2 \end{bmatrix}, G_2 = \begin{bmatrix} 1.10 \\ 1.20 \end{bmatrix}, G_3 = \begin{bmatrix} 1.20 \\ 0.20 \end{bmatrix}.$$

Solving the conditions (5.18) and (5.19), the parameters P_i and Q_i, the observer gain matrices K_{1i} and K_{2i} in Theorem 5.1 are given. For the simulation, the switched signal is shown in Fig. 5.1, a white noise input signal $d(t)$ is described in Fig. 5.2, and a piecewise constant fault is considered, which is defined as

$$f_a(t) = \begin{cases} 0, & others \\ 100, & 20s \le t \le 80s. \end{cases}$$

The FE result is shown in Fig. 5.3. One can get that the FE $\hat{f}_a(t)$ accurately converges to the real fault $f_a(t)$.

For the simulation, a piecewise slop fault is considered, which is defined as

$$f_s(t) = \begin{cases} 10, & others \\ t, & 20s \le t \le 40s, 60s \le t \le 80s. \end{cases}$$

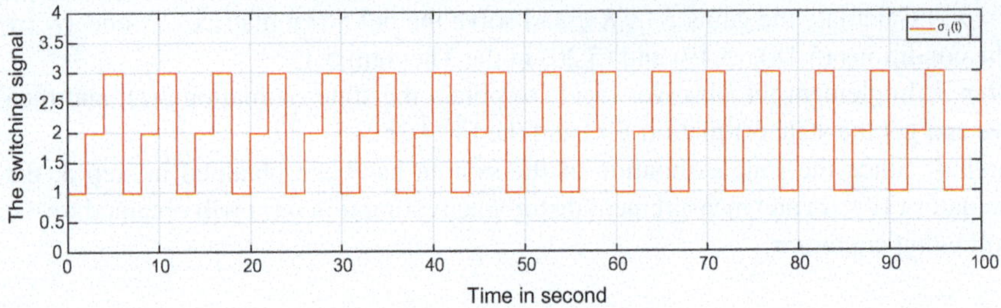

Fig. 5.1 The switching signal $\sigma_i(t)$

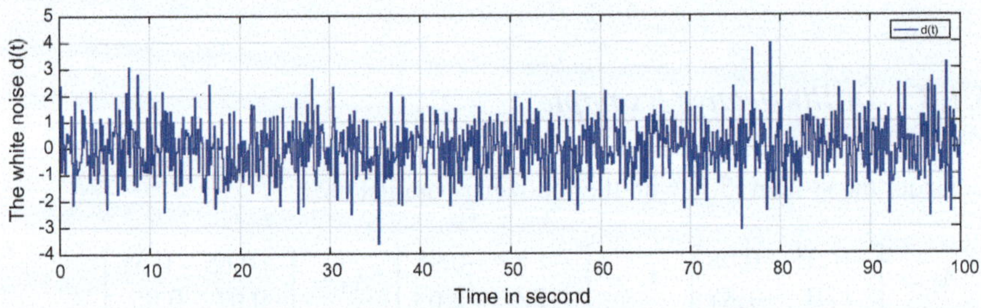

Fig. 5.2 The white noise $d(t)$

The FE result is shown in Fig. 5.4. One can get that the FE $\hat{f}_s(t)$ accurately converges to the real fault $f_s(t)$.

For the simulation, a time-varying fault is considered, which is defined as

$$\omega(t) = \begin{cases} sin(t) + 1, & 30s \leq t \leq 60s \\ cos(t), & others. \end{cases}$$

The FE result is shown in Fig. 5.5. One can get that the FE $\hat{\omega}(t)$ accurately converges to the real fault $\omega(t)$.

5.1.5 Summary

In this section, an adaptive FE observer with actuator and sensor faults and measurement disturbance is designed. An algorithm suitable for various faults is proposed, the observer can accurately estimate the system state, actuator fault, and sensor fault online. Finally, the effectiveness of the algorithm is illustrated with an example. Later, the fault-tolerant control of the system will be considered to enrich this section.

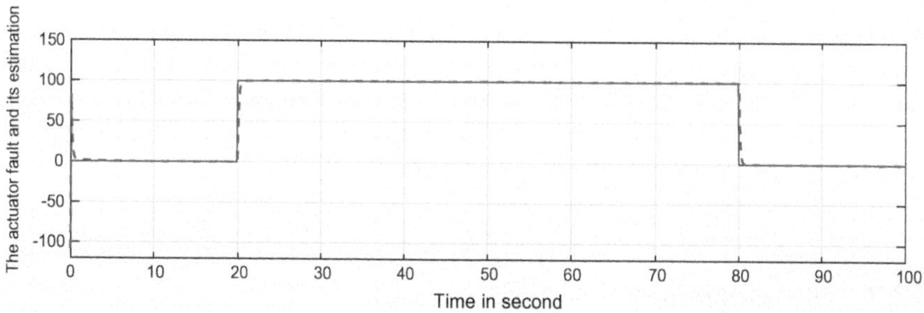

Fig. 5.3 The actuator fault $f_a(t)$ and its estimation $\hat{f}_a(t)$

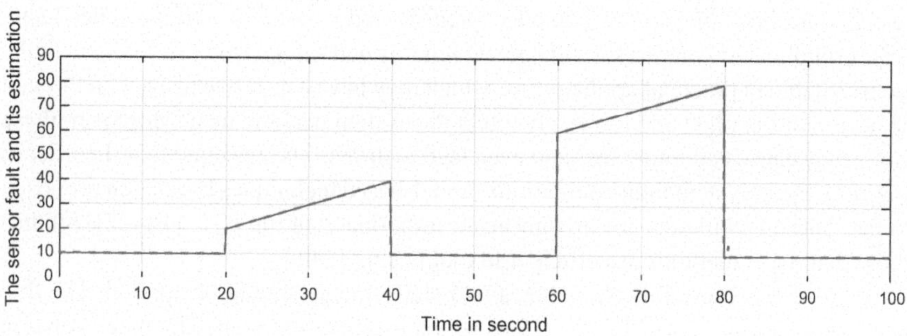

Fig. 5.4 The sensor fault $f_s(t)$ and its estimation $\hat{f}_s(t)$

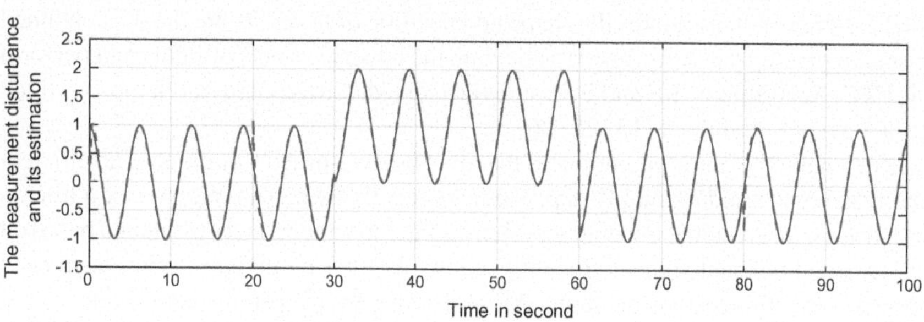

Fig. 5.5 The measure disturbance $\omega(t)$ and its estimation $\hat{\omega}(t)$

5.2 Asynchronous FEA with Actuator and Sensor Faults

In order to extend the FTC technique of the switched system, asynchronous FE and tolerant control is investigated in the present study. To this end, continuous-time switched systems in the presence of actuator faults, sensor faults, and external disturbances are considered. Based on equivalent transformations, a new augmented system is constructed, where faults of actuators and sensors are considered as a part of the state. Then, a desired asynchronous state observer is proposed for the augmented switched system. In order to address this issue, sufficient conditions are obtained to design the asynchronous state observer in the framework of the ADT technique. Then, an asynchronous observer-based fault-tolerant technique is described to ensure the stability of the closed-loop system. Finally, two simulation examples are applied to verify the effectiveness of the proposed methodology.

5.2.1 Introduction

Considering the fast industrial development in the past few decades, engineering systems are becoming more complex, precise, and expensive. On the other hand, these complicated systems should operate reliably and safely. However, system faults are inevitable in practical applications, which may have a negative impact on the safe operation of the plant and adversely affect the system performance. Moreover, these faults may cause economic loss and even lead to disastrous consequences. In order to diagnose these system faults, numerous model-based techniques have been proposed so far. These techniques are of significant importance in the FTC [18–22]. Studies show that these methods are efficient in diagnosing faults.

In order to achieve this objective, a two-step strategy should be applied. The first step, which is called the FDI step, detects and localizes the fault. The main function of the FDI is judging the system operating status and discovering the fault location. Promising results have been reported in this regard [23, 24]. The next step is called the FE step, which estimates the shape and position of faults. Then, the data obtained from the FE step can be utilized to alleviate the adverse effects of faults and perform the FTC. Accordingly, FE and FTC steps are considered as hot research topics, which attract many researchers [13, 25, 26].

Reviewing the literature indicates that the observer-based technique is an efficient and feasible method in the FE step. Studies show that when an observer or a filter is constructed as a residual generator, system faults can be accurately estimated based on the obtained residual signals. Recently, a few observer-based approaches have been proposed to estimate possible faults. More specifically, [27] applied the sliding-mode observer-based method and showed that actuator and sensor faults can be possibly reconstructed. Moreover, [28] studied the FTC problem and found that sensor faults can be successfully estimated and tolerated with Ito-type stochastic disturbances. [29] proposed UIO-based FE approaches and studied the FE observer and the fault-tolerant controller. In order to reduce the estimation cost, [30] achieved the reduced-order observer to reconstruct system faults. Then they reported the observer-based FTC approaches accordingly. Furthermore, [10] proposed an approach for designing an adjustable dynamic observer to achieve a balance between the accuracy and computational expenses of the estimation. Based on the output information, [31] applied the proportional-integral observer to reconstruct the actuator fault. Moreover, [12] proposed the proportional-integral observer and reported the observer-based FTC technique accordingly.

The switched system consists of a set of subsystems and a switching law that indicates the active subsystem at the switching instant. The switching concept allows the modeling of a broad class of systems, including the robotic control system, chemical reaction process, automotive industry, and switched power converters. In fact, these physical processes can be modeled as the switched system. Therefore, studying the switched system has attracted many researchers so that significant improvements have been obtained in the past few decades in this area [32–36]. Moreover, several interesting and innovative approaches have been proposed so far to address different aspects of the fault diagnosis and FTC. For example, [37] applied the reduced-order observer under the arbitrary switching signal and proposed actuators and sensors for the FE and FTC in switched systems. Reference [38] acquired the robust fault-tolerant stabilization problem for time delayed uncertain switched discrete-time systems. They applied the delta operator approach to impose actuator faults and input saturation to studied systems. Moreover, [39] investigated the adaptive fuzzy finite-time FTC issue for switched nonlinear large-scale systems with actuator and sensor faults. They combined the common Lyapunov function approach and the recursive design method. Furthermore, [40, 41] applied the data-based projection method and studied the sensor and actuator FE and compensation for hybrid switched systems. Recently, [42] investigated a proportional-integral observer-based FE and FTC for switched fuzzy stochastic systems based on the piecewise Lyapunov function and ADT methods [43].

It should be indicated that the main assumption in all of the aforementioned studies is that the filter observer is synchronous with subsystems. However, it is not possible for the system to identify the observer or the filter in practical applications. It is inferred that asynchronous switching between a system and its observer or filter definitely exists for these cases [44]. It is worth noting that the fault diagnosis of coupling asynchronous and FTC for switched systems is becoming a popular topic in this research field. Moreover, the asynchronous FD has attracted many scholars in the past few years [45]. [46] designed an asynchronous FD filter for continuous-time switched delayed systems. They proposed a novel switching strategy in accordance with the state and switching delays. Moreover, [47] designed an asynchronous FD observer and FTCler. In the aforementioned researches, the piecewise Lyapunov function and ADT approach were effectively employed. [48] comprehensively discussed the synchronously adaptive observer-based FDI technique and the corresponding applications in the boost-buck power converter. However, reviewing the literature indicates that designing the asynchronous FE and FTCler for the switched system have not been studied in detail.

The present study intends to propose FE and FTC for the continuous-time switched system with actuator and sensor faults. The main contributions can be summarized as follows:

- External disturbances, actuator faults and sensor faults, which are more general in practical applications, are considered.
- A novel asynchronous FE observer is designed to estimate the system state, actuator, and sensor faults.

- Utilizing the estimated system state and fault information, an observer-based fault-tolerant controller is designed, which can ensure that the closed-loop system is asymptotically stable with H_∞ performance.

5.2.2 *The Preliminaries*

In the present study, it is intended to investigate the following continuous-time switched system:

$$\begin{cases} \dot{x}(t) = A_{\sigma(t)}x(t) + B_{\sigma(t)}u(t) + D_{\sigma(t)}d(t) \\ y(t) = C_{\sigma(t)}x(t) \end{cases} \tag{5.28}$$

where $x(t) \in \mathbb{R}^n$, $u(t) \in \mathbb{R}^{n_u}$, and $y(t) \in \mathbb{R}^m$ are the system's state, control input, and the measurement output, respectively. Moreover, $d(t) \in \mathbb{R}^{n_d}$ denotes the external disturbance, which is assumed to be norm bounded. On the other hand, $\sigma(t)$ is the switching signal, which is assumed to be time-dependent. The switching sequence is defined by $t_0 < t_1 < \cdots < t_k < \cdots$, where t_k is the switching instant. When $\sigma(t) = i$, the ith subsystem is activated and $i \in \mathscr{N} = \{1, 2, \ldots, N\}$. In this case, the system (5.28) can be simplified in the form below:

$$\begin{cases} \dot{x}(t) = A_i x(t) + B_i u(t) + D_i d(t) \\ y(t) = C_i x(t) \end{cases} \tag{5.29}$$

It should be indicated that the parameter matrices in the abovementioned system are assumed to be constant real matrices with appropriate dimensions.

It is assumed that the system (5.29) encounters both actuator and sensor faults. Then, (5.29) can be described as the following:

$$\begin{cases} \dot{x}(t) = A_i x(t) + B_i(u(t) + f_a(t)) + D_i d(t) \\ y(t) = C_i x(t) + f_s(t) \end{cases} \tag{5.30}$$

where $f_a(t) \in \mathbb{R}^{n_u}$ and $f_s(t) \in \mathbb{R}^m$ denote the additive actuator fault and the sensor fault in the system, respectively. It is worth noting that the following assumptions are made in the present study:

Assumption 7 The pair of (A_i, B_i) is controllable, while the pair of (A_i, C_i) is observable.

Assumption 8 The derivative of the actuator fault is norm bounded. In other words, $\dot{f}_a(t) \in L_2[0, \infty)$.

Remark 5.2 According to the study of [49], the conditions in assumption 2 imply that the actuator fault is energy-bounded. Therefore, assumption 2 is reasonable in practice.

According to the investigation performed by [50], the classic Luenberger observer may amplify the effect of the sensor fault, which has been completely solved by constructing an augmented state observer instead of the class observer. In order to obtain the estimations of the actuator and sensor faults simultaneously, the following matrices are defined in the form below:

$$J_i = \begin{bmatrix} J_1 \\ J_{2i} \\ J_3 \end{bmatrix} = \begin{bmatrix} I_n & 0 & 0 \\ C_i & I_m & 0 \\ 0 & 0 & I_{n_u} \end{bmatrix},$$
$$H_i = \begin{bmatrix} H_{1i} & H_2 & H_3 \end{bmatrix} = \begin{bmatrix} I_n & 0 & 0 \\ -C_i & I_m & 0 \\ 0 & 0 & I_{n_u} \end{bmatrix}. \tag{5.31}$$

Then, the dynamic (5.30), is rewritten as the following:

$$\begin{cases} J_1\dot{\bar{x}}(t) = \bar{A}_i\bar{x}(t) + B_i u(t) + D_i d(t) \\ y(t) = J_{2i}\bar{x}(t) \end{cases} \tag{5.32}$$

where $\bar{x}(t) = \begin{bmatrix} x^T(t) & f_s^T(t) & f_a^T(t) \end{bmatrix}^T$, $\bar{A}_i = \begin{bmatrix} A_i & 0 & B_i \end{bmatrix}$.

It can be proved that $H_i J_i = I_{n+m+n_u}$.

$$H_i J_i = H_{1i} J_1 + H_2 J_{2i} + H_3 J_3 = I_{n+m+n_u} \tag{5.33}$$

Therefore, Eq. (5.32), can be rewritten as the following:

$$\begin{cases} \dot{\bar{x}}(t) = H_{1i}\bar{A}_i\bar{x}(t) + H_{1i}B_i u(t) + H_{1i}D_i d(t) \\ \qquad +H_2 J_{2i}\dot{\bar{x}}(t) + H_3 J_3\dot{\bar{x}}(t) \\ y(t) = J_{2i}\bar{x}(t) \end{cases} \tag{5.34}$$

For Eq. (5.34), the following observer can be constructed:

$$\begin{cases} \dot{\hat{\bar{x}}}(t) = H_{1\sigma'(t)}\bar{A}_{\sigma'(t)}\hat{\bar{x}}(t) + H_{1\sigma'(t)}B_{\sigma'(t)}u(t) \\ \qquad + H_2 J_{2\sigma'(t)}\dot{\bar{x}}(t) + L_{\sigma'(t)}(y(t) - \hat{y}(t)) \\ \hat{y}(t) = J_{2\sigma'(t)}\hat{\bar{x}}(t) \end{cases} \tag{5.35}$$

where $\hat{\bar{x}}(t) \in \mathbb{R}^{n+m+n_u}$ is the state estimation of $\bar{x}(t)$ and $\hat{y}(t) \in \mathbb{R}^m$ is the observer's measurement output. Moreover, $L_{\sigma'(t)}$ and $\sigma'(t)$ denote the gain of the observer to be designed and the switching signal of the observer (5.35), respectively. Figure 5.6 shows the system and the corresponding observer.

It is assumed that an inevitable delay exists between the observer and the activated subsystems of (5.34). Figure 5.7 shows the correlation between $\sigma(t)$ and $\sigma'(t)$. Moreover, t_k and $t_k + \Delta_k$, $k = 0, 1, \ldots$ represent the switching instant of $\sigma(t)$ and $\sigma'(t)$, respectively. It should be indicated that Δ_k denotes the switching delay, and it is assumed that the maximal switching delay Δ_{max} is a priori. In other words, the period

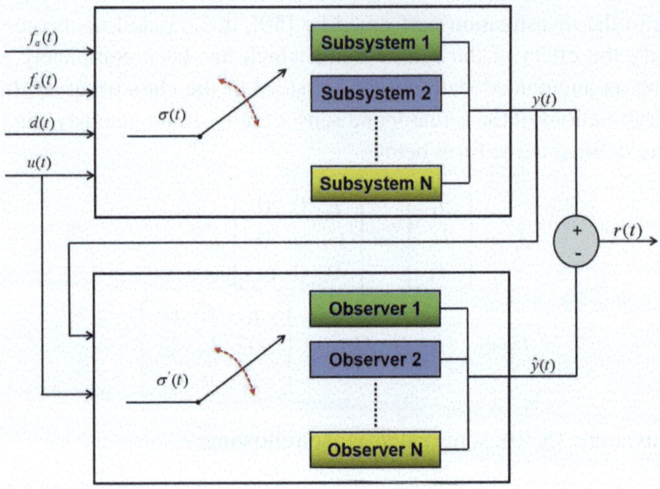

Fig. 5.6 The switched system and the corresponding asynchronously observer

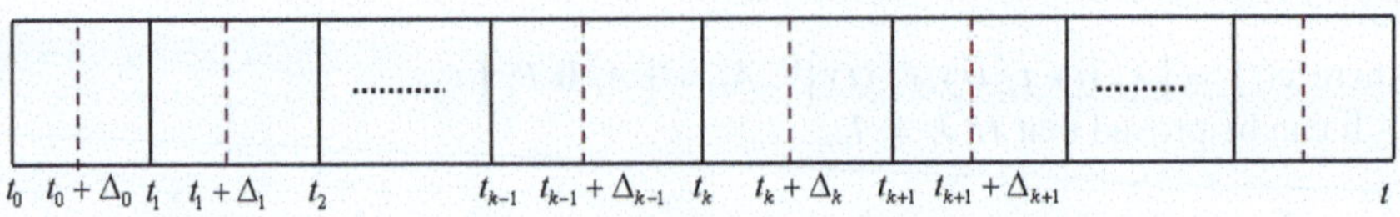

Fig. 5.7 The diagram of asynchronous switching

of $\Omega = [t_0, t_1) \cup [t_1 + \Delta_1, t_2) \cup \cdots \cup [t_k + \Delta_k, t_{k+1}) \cup \cdots$ indicates the matched time, while the period of $\Omega^{'} = [t_1, t_1 + \Delta_1) \cup \cdots \cup [t_k, t_k + \Delta_k) \cup \cdots$ implies the mismatched time.

Defining function $e(t)$ as $e(t) = \bar{x}(t) - \hat{\bar{x}}(t)$, then the derivation of $e(t)$ can be expressed as

$$\begin{aligned}\dot{e}(t) &= (H_{1i}\bar{A}_i - L_i J_{2i})e(t) + H_{1i}D_i d(t) + H_3 J_3 \dot{\bar{x}}(t)\\ &= (H_{1i}\bar{A}_i - L_i J_{2i})e(t) + \bar{D}_i \bar{d}(t), \qquad t \in \Omega\end{aligned} \tag{5.36}$$

Moreover,

$$\begin{aligned}\dot{e}(t) &= (H_{1j}\bar{A}_j - L_i J_{2j})e(t) + H_{1j}D_j d(t) + H_3 J_3 \dot{\bar{x}}(t)\\ &= (H_{1j}\bar{A}_j - L_i J_{2j})e(t) + \bar{D}_j \bar{d}(t), \qquad t \in \Omega^{'}\end{aligned} \tag{5.37}$$

where

$$\bar{D}_i = \begin{bmatrix} D_i & 0 \\ -C_i D_i & 0 \\ 0 & I_{n_u} \end{bmatrix}, \quad \bar{d}(t) = \begin{bmatrix} d(t) \\ \dot{f}_a(t) \end{bmatrix}. \tag{5.38}$$

Definition 5.2 ([45]) The switched system (5.36) (or (5.37)) is considered as a stable system with H_∞ performance γ, when the following conditions are satisfied:

- With $\bar{d}(t) = 0$, the system (5.36) (or (5.37)) is asymptotically stable.
- Under zero initial condition $e(t) = 0$, for any nonzero $\bar{d}(t) \in [0, \infty)$, $\zeta^* > 0$, and $\gamma > 0$, the following condition occurs:

$$\int_{t_0}^{\infty} e^{-\zeta^*(t-t_0)} e^T(t)e(t)dt \leq \gamma^2 \int_{t_0}^{\infty} \bar{d}^T(t)\bar{d}(t)dt \tag{5.39}$$

Therefore, the main purpose of the present study can be summarized as follows:

- Designing an asynchronous observer (5.35) so that the error system (5.36) or (5.37) is asymptotically stable with a prescribed H_∞ performance index.
- Designing a fault-tolerant controller so that the close-loop systems is asymptotically stable with a prescribed H_∞ performance index.

It is observed that an unknown term $H_2 J_{2\sigma'(t)} \dot{\bar{x}}(t)$ exists in Eq. (5.35). In order to improve the estimation accuracy of the observer, an auxiliary variable is defined as the following:

$$\eta(t) = \hat{\bar{x}}(t) - H_2 J_{2\sigma'(t)} \bar{x}(t) \tag{5.40}$$

In this case, an accuracy state estimation observer for Eq. (5.34), can be obtained as the following:

$$\begin{cases} \dot{\eta}(t) = \tilde{A}_{\sigma'(t)} \eta(t) + \tilde{B}_{\sigma'(t)} u(t) + \tilde{L}_{\sigma'(t)} \hat{y}(t) \\ \hat{\bar{x}}(t) = \eta(t) + \tilde{C}_{\sigma'(t)} \hat{y}(t) \end{cases} \tag{5.41}$$

where $\tilde{A}_i$, $\tilde{B}_i$, $\tilde{C}_i$, and $\tilde{L}_i$ are the observer's gains that should be determined later.

Remark 5.3 When an augmented vector $\bar{x}(t)$ is constructed, the actuator and sensor fault signals are all involved in the vector. When the observer (5.41), is successfully solved, then the state, actuator fault, and the sensor fault of (5.28) can be obtained simultaneously.

5.2.3 *The Observer Design*

Theorem 5.2 *For given constants $\alpha > 0$, $\beta > 0$, $\mu_1 > 1$, $\mu_2 > 1$, and $\gamma > 0$, if there exist positive definite matrices $P_i > 0$ and $P_{ij} > 0$, and matrices Q_i and Q_{ij}, where $i \neq j, i, j \in \mathcal{N}$, then*

$$P_j \leq \mu_1 P_{ij}, \quad P_{ij} \leq \mu_2 P_i \tag{5.42}$$

$$\begin{bmatrix} \Xi_i & P_i \bar{D}_i \\ * & -\gamma^2 I \end{bmatrix} < 0 \tag{5.43}$$

$$\begin{bmatrix} \Xi_{ij} & P_{ij} \bar{D}_j \\ * & -\gamma^2 I \end{bmatrix} < 0 \tag{5.44}$$

where $\Xi_i = P_i H_{1i} \bar{A}_i + \bar{A}_i^T H_{1i}^T P_i - Q_i J_{2i} - J_{2i}^T Q_i^T + \alpha P_i + I$, $\Xi_{ij} = P_{ij} H_{1j} \bar{A}_j + \bar{A}_j^T H_{1j}^T P_{ij} - Q_{ij} J_{2j} - J_{2j}^T Q_{ij}^T - \beta P_{ij} + I$, *then error systems (5.36) and (5.37) are asymptotically stable with* H_∞ *performance index* γ *for any switching signal with the ADT satisfying*

$$\tau_a > \tau_a^* = \frac{\ln(\mu_1 \mu_2)}{\zeta^*}, \quad \frac{T^-(t_0, t)}{T^+(t_0, t)} \geq \frac{\beta + \zeta^*}{\alpha - \zeta^*}, \quad 0 < \zeta^* < \alpha.$$

where $T^-(t_0, t)$ *and* $T^+(t_0, t)$ *denote the total matched and mismatched period during* $[t_0, t)$, *respectively. Meanwhile, the observer parameters in (5.41) are described as the following:*

$$\begin{aligned} \tilde{A}_i &= H_{1i} \bar{A}_i - L_i J_{2i}, \quad \tilde{B}_i = H_{1i} B_i, \quad \tilde{C}_i = H_2 J_{2i}, \\ \tilde{L}_i &= L_i + (H_{1i} \bar{A}_i - L_i J_{2i}) H_2, \quad L_i = P_i^{-1} Q_i. \end{aligned} \tag{5.45}$$

Proof The stability for the error systems (5.36) and (5.37) are initially established. When $t \in \Omega$, the following piecewise Lyapunov function is considered:

$$V_i(t) = e^T(t) P_i e(t) \tag{5.46}$$

Then along the trajectory of the error system (5.36), the following expression holds:

$$\begin{aligned} &\dot{V}_i(t) + \alpha V_i(t) \\ &= \dot{e}^T(t) P_i e(t) + e^T(t) P_i \dot{e}(t) + \alpha e^T(t) P_i e(t) \\ &= e^T(t)(P_i(H_{1i} \bar{A}_i - L_i J_{2i}) + (H_{1i} \bar{A}_i - L_i J_{2i})^T P_i + \alpha P_i) e(t) \end{aligned} \tag{5.47}$$

Defining $Q_i = P_i L_i$, then Eq. (5.47) is transformed into the following form:

$$\dot{V}_i(t) + \alpha V_i(t) = e^T(t)(P_i H_{1i} \bar{A}_i + P_i \bar{A}_i^T H_{1i}^T + \alpha P_i - Q_i J_{2i} - J_{2i}^T Q_i^T) e(t) \tag{5.48}$$

When Eq. (5.43) is applied, it yields the following expression:

$$\dot{V}_i(t) + \alpha V_i(t) < 0 \tag{5.49}$$

Based on the foregoing equations, it is concluded that $V_i(t)$ is satisfied:
in the period of $t_0 \leq t \leq t_1$,

$$V_i(t) < V_i(t_0) \tag{5.50}$$

in the period of $t_{k-1} + \Delta_{k-1} \leq t < t_k$,

$$V_i(t) < V_i(t_{k-1} + \Delta_{k-1})e^{-\alpha(t-t_{k-1}-\Delta_{k-1})} \tag{5.51}$$

When $t \in \Omega'$, error system can be defined in accordance with Eq. (5.37). In this case, the following piecewise Lyapunov function is applied:

$$V_{ij}(t) = e^T(t)P_{ij}e(t) \tag{5.52}$$

Along the trajectory of system (5.37), this expression can be rewritten in the form below

$$\begin{aligned} &\dot{V}_{ij}(t) - \beta V_{ij}(t) \\ &= \dot{e}^T(t)P_{ij}e(t) + e^T(t)P_{ij}\dot{e}(t) - \beta e^T(t)P_{ij}e(t) \\ &= e^T(t)(P_{ij}(H_1\bar{A}_j - L_iJ_{2j}) + (H_1\bar{A}_j - L_iJ_{2j})^T P_{ij} - \beta P_{ij})e(t) \end{aligned} \tag{5.53}$$

Defining $Q_{ij} = P_{ij}L_i$, expression (5.53), can be transformed into the following form:

$$\begin{aligned} &\dot{V}_{ij}(t) + \alpha V_{ij}(t) \\ &= e^T(t)(P_{ij}H_{1j}\bar{A}_j + P_{ij}\bar{A}_j^T H_{1j}^T - Q_{ij}J_{2j} - J_{2j}^T Q_{ij}^T - \beta P_{ij})e(t) \end{aligned} \tag{5.54}$$

When Eq. (5.44), is applied, the expression above can be rewritten as the following:

$$\dot{V}_{ij}(t) - \beta V_{ij}(t) < 0 \tag{5.55}$$

Therefore, it is concluded that $V_{ij}(t)$ is satisfied in the unmatched period.

$$V_{ij}(t) < V_{ij}(t_k)e^{\beta(t-t_k)}, \quad t \in [t_k, t_k + \Delta_k), k = 1, 2, \ldots \tag{5.56}$$

Consider the following piecewise Lyapunov functional candidate, previously defined in Eqs. (5.36) and (5.37):

$$V(t) = \begin{cases} e^T(t)P_ie(t), & t \in \Omega \\ e^T(t)P_{ij}e(t), & t \in \Omega' \end{cases} \tag{5.57}$$

When $t \in \Omega'$ with constraints in accordance with Eqs. (5.42) and (5.55), it can be proved that

$$V(t) = V_{ij}(t) < V_{\sigma'(t_{k-1}+\Delta_{k-1})\sigma(t_k)}(t) \leq V_{\sigma'(t_{k-1}+\Delta_{k-1})\sigma(t_k)}(t_k)e^{\beta(t-t_k)} \tag{5.58}$$

According to the similar process in [45], the following expression can be written

$$V(t) \leq \mu_1^{-1} V_{\sigma(t_0)}(t_0) e^{N_0 ln(\mu_1 \mu_2)} e^{-(\zeta^* - \frac{ln(\mu_1 \mu_2)}{\tau_a})(t-t_0)} \tag{5.59}$$

Therefore, it is found that when the ADT condition in Eq. (5.45), is satisfied, then $V(t) \to 0$ as $t \to \infty$. Under these circumstances, the condition of H_∞ performance is considered. For zero initial condition (i.e., $e(t) = 0$) and $d(t) \in L_2[0, \infty)$, consider the Lyapunov function similar to that in the investigation performed by (5.46) when $t \in \Omega$. It is found that:

$$\begin{aligned}
&\dot{V}_i(t) + \alpha V_i(t) \\
&= \dot{e}^T(t) P_i e(t) + e^T(t) P_i \dot{e}(t) + \alpha e^T(t) P_i e(t) \\
&= e^T(t)(P_i H_{1i} \bar{A}_i + P_i \bar{A}_i^T H_{1i}^T + \alpha P_i - Q_i J_{2i} \\
&\quad - J_{2i}^T Q_i^T + I) e(t) + 2e^T(t) P_i \bar{D}_i \bar{d}(t) - e^T(t) e(t) \\
&\quad - \gamma^2 \bar{d}^T(t) \bar{d}(t) + \gamma^2 \bar{d}^T(t) \bar{d}(t) \\
&= \begin{bmatrix} e^T(t) & \bar{d}^T(t) \end{bmatrix} \begin{bmatrix} \Xi_i & \bar{P}_i D_i \\ * & -\gamma^2 I \end{bmatrix} \begin{bmatrix} e(t) \\ \bar{d}(t) \end{bmatrix} + \Gamma(t)
\end{aligned} \tag{5.60}$$

where $\Gamma(t) = -e^T(t)e(t) + \gamma^2 \bar{d}^T(t) \bar{d}(t)$. If the condition (5.43) holds, then the following expression can be obtained:

$$\dot{V}_i(t) + \alpha V_i(t) \leq \Gamma(t) \tag{5.61}$$

That is:

$$\dot{V}_i(t) \leq -\alpha V_i(t) + \Gamma(t) \tag{5.62}$$

Based on Eq. (5.62), it is found that
in the period of $t_0 \leq t \leq t_1$, one has

$$V_i(t) \leq e^{-\alpha(t-t_0)} V_i(t_0) + \int_{t_0}^{t} e^{-\alpha(t-s)} \Gamma(s) ds \tag{5.63}$$

in the period of $t_{k-1} + \Delta_{k-1} \leq t < t_k$, one has

$$V_i(t) \leq e^{-\alpha(t-t_{k-1}-\Delta_{k-1})} V_i(t_{k-1} + \Delta_{k-1}) + \int_{t_{k-1}+\Delta_{k-1}}^{t} e^{-\alpha(t-s)} \Gamma(s) ds \tag{5.64}$$

when $t \in [t_k, t_k + \Delta_k)$, $k = 1, 2, \ldots$, the Lyapunov function is considered similar to that in Eq. (2.76). According to the similar proof process for the case of $t \in \Omega$, it is found that

$$\dot{V}_{ij}(t) - \beta V_{ij}(t) = \begin{bmatrix} e^T(t) & \bar{d}^T(t) \end{bmatrix} \begin{bmatrix} \Xi_{ij} & \bar{P}_{ij} D_j \\ * & -\gamma^2 I \end{bmatrix} \begin{bmatrix} e(t) \\ \bar{d}(t) \end{bmatrix} + \Gamma(t) \tag{5.65}$$

Then, it is concluded that

$$\dot{V}_{ij}(t) \le \beta V_{ij}(t) + \Gamma(t) \tag{5.66}$$

Based on Eq. (5.66), it is found that

$$\dot{V}_{ij}(t) \le e^{\beta(t-t_k)} V_{ij}(t_k) + \int_{t_k}^{t} e^{\beta(t-s)} \Gamma(s) ds, t_k \le t < t_k + \Delta_k, k = 1, 2, \ldots \tag{5.67}$$

Considering Eqs. (5.63), (5.64), and (5.67) and similar to the proof of the H_∞ design technique established in [45], it can be said that

$$\int_{t_0}^{\infty} e^{-\zeta^*(t-t_0)} e^T(t)e(t)dt \le \gamma^2 \int_{t_0}^{\infty} \bar{d}^T(t)\bar{d}(t)dt \tag{5.68}$$

According to Definition 5.2, the error system (5.36) or (5.37) is asymptotically stable with H_∞ performance, for the disturbance attenuation level γ. Therefore, observer gains in Eq. (5.41), are given. Considering the definition of the auxiliary variable $\eta(t)$ in (5.40), it is observed that

$$\begin{aligned}
\dot{\eta}(t) &= \dot{\hat{\bar{x}}}(t) - H_2 J_{2\sigma'(t)} \dot{\bar{x}}(t) \\
&= H_{1\sigma'(t)} \bar{A}_{\sigma'(t)} \hat{\bar{x}}(t) + J_1 B_{\sigma'(t)} u(t) + H_2 J_{2\sigma'(t)} \dot{\bar{x}}(t) \\
&\quad + L_{\sigma'(t)}(y(t) - \hat{y}(t)) - H_2 J_{2\sigma'(t)} \dot{\bar{x}}(t) \\
&= H_{1\sigma'(t)} \bar{A}_{\sigma'(t)} \hat{\bar{x}}(t) + J_1 B_{\sigma'(t)} u(t) + L_{\sigma'(t)}(y(t) - J_{2\sigma'(t)} \hat{\bar{x}}(t)) \\
&= (H_{1\sigma'(t)} \bar{A}_{\sigma'(t)} - L_{\sigma'(t)} J_{2\sigma'(t)}) \hat{\bar{x}}(t) + J_1 B_{\sigma'(t)} u(t) + L_{\sigma'(t)} y(t) \\
&= (H_{1\sigma'(t)} \bar{A}_{\sigma'(t)} - L_{\sigma'(t)} J_{2\sigma'(t)}) \eta(t) + J_1 B_{\sigma'(t)} u(t) \\
&\quad + (J_1 \bar{A}_{\sigma'(t)} - L_{\sigma'(t)} J_{2\sigma'(t)}) H_{2\sigma'(t)} J_{2\sigma'(t)} \bar{x}(t) + L_{\sigma'(t)} y(t) \\
&= (H_{1\sigma'(t)} \bar{A}_{\sigma'(t)} - L_{\sigma'(t)} J_{2\sigma'(t)}) \eta(t) + J_1 B_{\sigma'(t)} u(t) \\
&\quad + (L_{\sigma'(t)} + (J_1 \bar{A}_{\sigma'(t)} - L_{\sigma'(t)} J_{2\sigma'(t)}) H_{2\sigma'(t)}) y(t)
\end{aligned} \tag{5.69}$$

Compared with the form in (5.41), we have

$$\tilde{A}_i = J_1 \bar{A}_i - L_i J_{2i}, \tag{5.70}$$

$$\tilde{B}_i = J_1 B_i, \tag{5.71}$$

$$\tilde{C}_i = H_{2i} J_{2i}, \tag{5.72}$$

$$\tilde{L}_i = L_i + (J_1 \bar{A}_i - L_i J_{2i}) H_{2i}. \tag{5.73}$$

which indicates the ideal observer parameter design in (5.45). Then the proof is completed.

5.2.4 Designing the Fault-Tolerant Controller

In this section, it is intended to design an asynchronous fault-tolerant controller to compensate the adverse influence of faults, which can verify that the closed-loop system is asymptotically stable with H_∞ performance. To this end, the controller is selected as follows:

$$u(t) = K_{\sigma'(t)} J_1 \hat{\bar{x}}(t) - J_3 \hat{\bar{x}}(t) \tag{5.74}$$

where K_i is the controller gain to be designed. Substituting Eq. (5.74) into Eq. (5.28), the following expression can be obtained:

$$\begin{aligned}
\dot{x}(t) &= A_{\sigma(t)}x(t) + B_{\sigma(t)}(u(t) + f_a(t)) + D_{\sigma(t)}d(t) \\
&= A_{\sigma(t)}x(t) + B_{\sigma(t)}K_{\sigma'(t)}J_1\hat{\bar{x}}(t) - B_{\sigma(t)}J_3\hat{\bar{x}}(t) \\
&\quad + D_{\sigma(t)}d(t) + B_{\sigma(t)}f_a(t) \\
&= A_{\sigma(t)}x(t) + B_{\sigma(t)}K_{\sigma'(t)}J_1\bar{x}(t) + D_{\sigma(t)}d(t) \\
&\quad - B_{\sigma(t)}K_{\sigma'(t)}J_1\bar{x}(t) + B_{\sigma(t)}K_{\sigma'(t)}J_1\hat{\bar{x}}(t) \\
&\quad + B_{\sigma(t)}f_a(t) + B_{\sigma(t)}J_3\bar{x}(t) - B_{\sigma(t)}J_3\bar{x}(t) - B_{\sigma(t)}J_3\hat{\bar{x}}(t) \\
&= (A_{\sigma(t)} + B_{\sigma(t)}K_{\sigma'(t)})x(t) + D_{\sigma(t)}d(t) - B_{\sigma(t)}(K_{\sigma'(t)}J_1 - J_3)e(t) \\
&= (A_{\sigma(t)} + B_{\sigma(t)}K_{\sigma'(t)})x(t) - B_{\sigma(t)}d_1(t) + B_{\sigma(t)}d_2(t) + \tilde{D}_{\sigma(t)}\bar{d}(t)
\end{aligned} \tag{5.75}$$

where $d_1(t) = K_{\sigma'(t)}J_1e(t)$, $d_2(t) = J_3e(t)$, $\tilde{D}_{\sigma(t)} = \begin{bmatrix} D_{\sigma(t)} & 0 \end{bmatrix}$. Then

$$\dot{x}(t) = (A_i + B_iK_i)x(t) - B_id_1(t) + B_id_2(t) + \tilde{D}_i\bar{d}(t), \quad t \in \Omega \tag{5.76}$$

Moreover

$$\dot{x}(t) = (A_j + B_jK_i)x(t) - B_jd_1(t) + B_jd_2(t) + \tilde{D}_j\bar{d}(t), \quad t \in \Omega' \tag{5.77}$$

Theorem 5.3 *For given scalars $\alpha_1 > 0$, $\beta_1 > 0$, $\varepsilon_1 > 0$, $\varepsilon_2 > 0$, $\gamma_1 > 0$, $\bar{\mu}_1 > 1$, $\bar{\mu}_2 > 1$, if positive definite matrices $R_i > 0$, $R_{ij} > 0$ and matrices W_i exist:*

$$R_{ij} \le \bar{\mu}_1 R_j, \quad R_i \le \bar{\mu}_2 R_{ij} \tag{5.78}$$

$$\begin{bmatrix}
\Lambda_i R_i & -B_i & B_i & \tilde{D}_i \\
* \; -I & 0 & 0 & 0 \\
* \;\; * & -\varepsilon_1 I & 0 & 0 \\
* \;\; * & * & -\varepsilon_2 I & 0 \\
* \;\; * & * & * & -\gamma_2^2 I
\end{bmatrix} < 0 \tag{5.79}$$

$$\begin{bmatrix} \Lambda_{ij} & R_j & -B_j & B_j & -\tilde{D}_j \\ * & -I & 0 & 0 & 0 \\ * & * & -\varepsilon_2 I & 0 & 0 \\ * & * & * & \varepsilon_2 I & 0 \\ * & * & * & * & -\gamma_2^2 I \end{bmatrix} < 0 \tag{5.80}$$

where $\Lambda_i = A_i R_i + R_i A_i^T + \alpha_1 R_i + B_i W_i + W_i^T B_i^T$, $\Lambda_{ij} = A_j R_j + R_j A_j^T - \beta_1 R_j + B_j W_{ij} + W_{ij}^T B_i^T$. *Then, the closed system (5.76) (or (5.77)) is asymptotically stable with* H_∞ *performance index* γ_1 *for any switching signal with satisfying the ADT*

$$\tau_a > \tau_a^* = \frac{\ln(\mu_1 \mu_2)}{\zeta^*}, \quad \frac{T^-(t_0, t)}{T^+(t_0, t)} \geq \frac{\beta + \zeta^*}{\alpha - \zeta^*}, \quad 0 < \zeta^* < \alpha.$$

where $T^-(t_0, t)$ *and* $T^+(t_0, t)$ *denote the total matched and mismatched period during* $[t_0, t)$*, respectively. Meanwhile, the observer parameters in (5.41), can be described in the form below*

$$K_i = W_i R_i^{-1} \tag{5.81}$$

Proof When $t \in \Omega$, the following Lyapunov function is selected:

$$V_i(t) = x^T(t) X_i x(t) \tag{5.82}$$

Then the following expression can be written along the trajectory of system (5.76):

$$\begin{aligned} &\dot{V}_i(t) + \alpha V_i(t) \\ &= \dot{x}^T(t) X_i x(t) + x^T(t) X_i \dot{x}(t) + \alpha V_i(t) \\ &= x^T(t)(X_i(A_i + B_i K_i) + (A_i + B_i K_i)^T X_i) x(t) - 2x^T(t) X_i B_i d_1(t) \\ &\quad + 2x^T(t) X_i B_i d_2(t) + 2x^T(t) X_i \tilde{D}_j \bar{d}(t) + \alpha x^T(t) X_i x(t) \\ &= x^T(t)(X_i(A_i + B_i K_i) + (A_i + B_i K_i)^T X_i) x(t) - 2x^T(t) X_i B_i d_1(t) \\ &\quad + 2x^T(t) X_i B_i d_2(t) + 2x^T(t) X_i \tilde{D}_i \bar{d}(t) + x^T(t) x(t) - x^T(t) x(t) \\ &\quad + \varepsilon_1 d_1^T(t) d_1(t) - \varepsilon_1 d_1^T(t) d_1(t) + \varepsilon_2 d_2^T(t) d_2(t) - \varepsilon_2 d_2^T(t) d_2(t) \\ &\quad + \gamma_2 \bar{d}^T(t) \bar{d}(t) - \gamma_2 \bar{d}^T(t) \bar{d}(t) + \alpha x^T(t) X_i x(t) \\ &= \xi^T(t) \Theta_{1i} \xi(t) + \Gamma_1(t) \end{aligned} \tag{5.83}$$

where $\xi(t) = \begin{bmatrix} x^T(t) & d_2^T(t) & d_2^T(t) & \bar{d}^T(t) \end{bmatrix}^T$ and $\Gamma_1(t) = -x^T(t)x(t) + \varepsilon_1 d_1^T(t) d_1(t) + \varepsilon_2 d_2^T(t) d_2(t) + \gamma_2^2 \bar{d}^T(t) \bar{d}(t)$.

$$\Theta_{1i} = \begin{bmatrix} \tilde{\Lambda}_i & -X_i B_i & X_i B_i & -X_i \tilde{D}_i \\ * & -\varepsilon_2 I & 0 & 0 \\ * & * & \varepsilon_2 I & 0 \\ * & * & * & -\gamma_2^2 I \end{bmatrix}, \tag{5.84}$$

$$\tilde{\Lambda}_i = X_i A_i + A_i^T X_i + \alpha X_i + X_i B_i K_i + K_i^T B_i^T X_i + I. \tag{5.85}$$

Defining $R_i = X_i^{-1}$ and pre- and post-multiplying $diag\{R_i, I, I, I\}$ to Θ_{1i}, it is found that

$$\Theta_{2i} = \begin{bmatrix} \bar{\Lambda}_i & -B_i & B_i & -\tilde{D}_i \\ * & -\varepsilon_2 I & 0 & 0 \\ * & * & \varepsilon_2 I & 0 \\ * & * & * & -\gamma_2^2 I \end{bmatrix}. \tag{5.86}$$

with $\bar{\Lambda}_i = A_i R_i + R_i A_i^T + \alpha R_i + B_i W_i + W_i^T B_i^T + R_i R_i$. Based on Eq. (5.79), it is found that $\Theta_{2i} < 0$. Then

$$\dot{V}_i(t) < -\alpha V_i(t) + \Gamma_1(t) \tag{5.87}$$

From (5.62), it is concluded that: in the period of $t_0 \le t \le t_1$, one has

$$V_i(t) \le e^{-\alpha(t-t_0)} V_i(t_0) + \int_{t_0}^{t} e^{-\alpha(t-s)} \Gamma_1(s) ds \tag{5.88}$$

in the period of $t_{k-1} + \Delta_{k-1} \le t < t_k$, one has

$$V_i(t) \le e^{-\alpha(t-t_{k-1}-\Delta_{k-1})} V_i(t_{k-1} + \Delta_{k-1}) + \int_{t_{k-1}+\Delta_{k-1}}^{t} e^{-\alpha(t-s)} \Gamma_1(s) ds \tag{5.89}$$

When $t \in \Omega'$, the Lyapunov function is selected in the form below

$$V_{ij}(t) = x^T(t) X_{ij} x(t) \tag{5.90}$$

Then along the trajectory of the system (5.77), and according to the similar proof process for $t \in \Omega$, it is found that

$$\dot{V}_{ij}(t) - \beta V_{ij}(t) = \xi^T(t) \Theta_{1ij} \xi(t) + \Gamma_1(t) \tag{5.91}$$

where

$$\Theta_{1ij} = \begin{bmatrix} \tilde{\Lambda}_{ij} & -X_j B_j & X_j B_j & -X_j \tilde{D}_j \\ * & -\varepsilon_2 I & 0 & 0 \\ * & * & \varepsilon_2 I & 0 \\ * & * & * & -\gamma_2^2 I \end{bmatrix}, \tag{5.92}$$

$$\tilde{\Lambda}_{ij} = X_j A_j + A_j^T X_j - \beta X_j + X_j B_j K_i + K_i^T B_j^T X_j + I. \tag{5.93}$$

Similar to the foregoing section, defining $R_j = X_j^{-1}$ and pre- and post-multiplying $diag\{R_j, I, I, I\}$ to Θ_{1ij}, it results in the form below

$$\Theta_{2ij} = \begin{bmatrix} \bar{\Lambda}_{ij} & -B_j & B_j & -\tilde{D}_j \\ * & -\varepsilon_2 I & 0 & 0 \\ * & * & \varepsilon_2 I & 0 \\ * & * & * & -\gamma_2^2 I \end{bmatrix}. \tag{5.94}$$

with $\bar{\Lambda}_{ij} = A_j R_j + R_j A_j^T - \beta R_j + B_j W_{ij} + W_{ij}^T B_i^T + R_j R_j$. Based on Eq. (5.80), it is found that $\Theta_{2ij} < 0$. Then

$$\dot{V}_i(t) < \beta V_i(t) + \Gamma_1(t) \tag{5.95}$$

Considering Eq. (5.95), it is found that

$$\dot{V}_{ij}(t) \le e^{\beta(t-t_k)} V_{ij}(t_k) + \int_{t_k}^{t} e^{\beta(t-s)} \Gamma_1(s) ds, \; t_k \le t < t_k + \Delta_k, k = 1, 2, \ldots \tag{5.96}$$

Following by Eqs. (5.88), (5.89) and (5.96) and according to the similar proof process in [45], the following expression can be obtained:

$$\begin{aligned} & \int_{t_0}^{\infty} e^{-\zeta^*(t-t_0)} x^T(t) x(t) dt \\ \le & \int_{t_0}^{\infty} (\varepsilon_1 d_1^T(t) d_1(t) + \varepsilon_2 d_2^T(t) d_2(t) + \gamma_2^2 \bar{d}^T(t) \bar{d}(t)) dt \\ \le & \int_{t_0}^{\infty} (\varepsilon_1 e^T(t) (K_i J_1)^T (K_i J_1) e(t) + \varepsilon_2 e^T(t) J_3^T J_3 e(t) + \gamma_2^2 \bar{d}^T(t) \bar{d}(t)) dt \\ \le & \int_{t_0}^{\infty} (\varepsilon_1 \lambda_{max}(K_i^T K_i) e^T(t) J_1^T J_1 e(t) + \varepsilon_2 e^T(t) J_3^T J_3 e(t) + \gamma_2^2 \bar{d}^T(t) \bar{d}(t)) dt \\ \le & \int_{t_0}^{\infty} (\varepsilon_1 \lambda_{max}(K_i^T K_i) \gamma_2^2 \bar{d}^T(t) \bar{d}(t) + \varepsilon_2 \gamma_2^2 \bar{d}^T(t) \bar{d}(t) + \gamma_2^2 \bar{d}^T(t) \bar{d}(t)) dt \\ = & \gamma_1^2 \int_{t_0}^{\infty} (\bar{d}^T(t) \bar{d}(t)) dt \end{aligned} \tag{5.97}$$

where $\gamma_1 = \gamma_2 \sqrt{\varepsilon_1 \lambda_{max}(K_i^T K_i) + \varepsilon_2 + 1}$. Moreover, the constraint in accordance with Eq. (5.79) (or (5.80)), guarantees $\dot{V}_i(t) + \alpha V_i(t) < 0$ (or $\dot{V}_{ij}(t) - \beta V_{ij}(t) < 0$) when $\bar{d}(t) = 0$. Therefore, it is concluded that the closed-loop system (5.76) (or (5.77)), has asymptotical stability with H_∞ performance γ_1. (□)

5.2.5 Simulation Example

In order to evaluate the performance of the proposed framework, it is applied on two test cases.

Example 1 Consider the system in accordance with Eq. (5.30), with the following parameters:

$$A_1 = \begin{bmatrix} -0.4 & 0.1 \\ 0 & -0.2 \end{bmatrix}, \; A_2 = \begin{bmatrix} -0.4 & -2.1 \\ 1.5 & -2 \end{bmatrix}, \; A_3 = \begin{bmatrix} -2.25 & 0.4 \\ 0 & -5.9 \end{bmatrix},$$
$$B_1 = \begin{bmatrix} 0.2 \\ 0.1 \end{bmatrix}, \; B_2 = \begin{bmatrix} 0.1 \\ 0.2 \end{bmatrix}, \; B_3 = \begin{bmatrix} 0.3 \\ 0.4 \end{bmatrix},$$
$$C_1 = \begin{bmatrix} 1 \; 3 \end{bmatrix}, \; C_2 = \begin{bmatrix} 1 \; 2 \end{bmatrix}, \; C_3 = \begin{bmatrix} 1 \; 1 \end{bmatrix},$$
$$D_1 = \begin{bmatrix} 0.2 \\ -0.1 \end{bmatrix}, \; D_2 = \begin{bmatrix} -0.2 \\ 0.1 \end{bmatrix}, \; D_3 = \begin{bmatrix} 0.2 \\ 0.4 \end{bmatrix}.$$

When known parameters are set as $\mu_1 = \mu_2 = 1.2$ and $\alpha = 1.2$, the corresponding observer gains can be solved as follows:

$$L_1 = \begin{bmatrix} -2.1437 \\ -3.4937 \\ 1.0563 \\ -0.0000 \end{bmatrix}; \; L_2 = \begin{bmatrix} 15.6137 \\ 127.9779 \\ -27.2840 \\ 0.3185 \end{bmatrix}; \; L_3 = \begin{bmatrix} -9.3132 \\ -44.8452 \\ 5.2884 \\ 6.5548 \end{bmatrix}.$$

Moreover, the disturbance is assumed to be a band-limited white noise. The simulation time is set to 200s with the same corresponding platform. It is worth noting that subsystems 1, 2, and 3 are activated when $\sigma(t) = 1$, $\sigma(t) = 2$, and $\sigma(t) = 3$, respectively. Figure 5.9 shows the switching signals of $\sigma(t)$ and $\sigma'(t)$. In this test, actuator and sensor faults are assumed in the form below:

$$f_{1a}(t) = \begin{cases} 0, & 0 \le t < 35 \\ 2sin(0.1\pi t) + 2, & 35 \le t < 45 \\ 4, & 45 \le t < 120 \\ 2cos(0.1\pi t) + 2, & 120 \le t < 130 \\ 0, & 130 \le t \end{cases}$$

$$f_{1s}(t) \begin{cases} 0, & 0 \le t < 35 \\ 2, & 35 \le t < 120 \\ 0, & 120 \le t \end{cases}$$

Figures 5.10 and 5.11 describe the actuator and sensor fault estimation results and the corresponding fault estimation error. Solid line shows the fault, while dotted line is the estimation for the actuator fault and the corresponding FE error. In this test, parameters are set as $\alpha_1 = 0.7$, $\beta_1 = 1.05$, $\varepsilon_1 = 0.05$, $\varepsilon_2 = 0.05$, $\gamma_1 = 0.5$, $\bar{\mu}_1 = 1.05$, and $\bar{\mu}_2 = 1.2$. Solving the problem with constraints presented in theorem 2, the controller gains can be obtained as

$$K_1 = \begin{bmatrix} 13.5645 & -5.0943 \end{bmatrix}; K_2 = \begin{bmatrix} 8.2348 & 2.9745 \end{bmatrix}; K_3 = \begin{bmatrix} -0.3245 & -1.2345 \end{bmatrix}.$$

Figure 5.12 shows the output responses of $y(t)$ without and with accommodation. Obtained simulation results show that the designed controller has an efficient fault-tolerant performance.

Example 2 In the second example, a boost-buck converter circuit [51], is considered to further evaluate the performance of the proposed techniques. Figure 5.8 illustrates the configuration of the boost-buck converter. In the present study, the state vector is defined as $x(t) = \begin{bmatrix} i_{in}^T & i_{out}^T & v_c^T \end{bmatrix}^T$, while the model matrices are described as

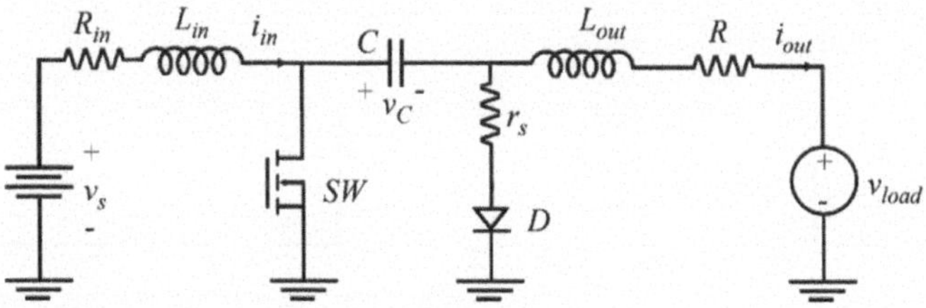

Fig. 5.8 The boost-buck converter

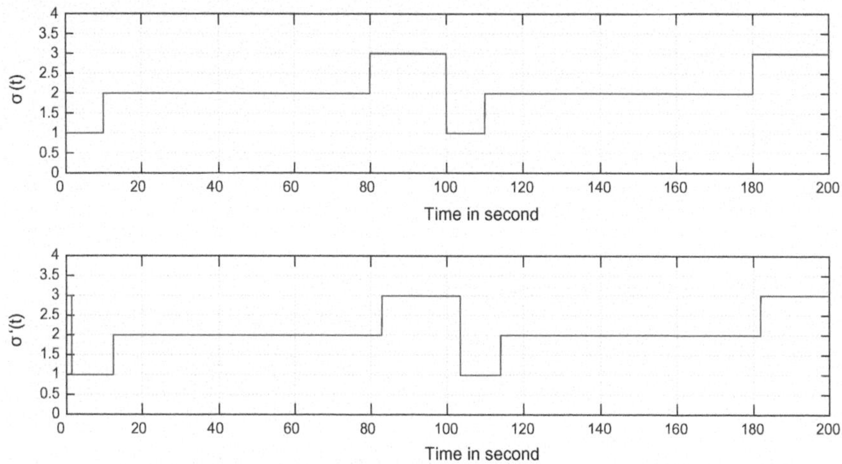

Fig. 5.9 The switching signal $\sigma(t)$ and $\sigma'(t)$

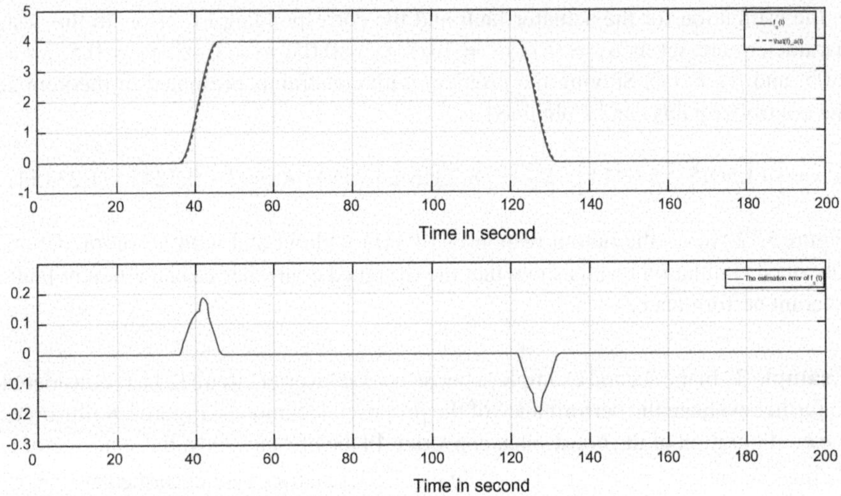

Fig. 5.10 The fault estimation of $f_{1a}(t)$ and its estimation error

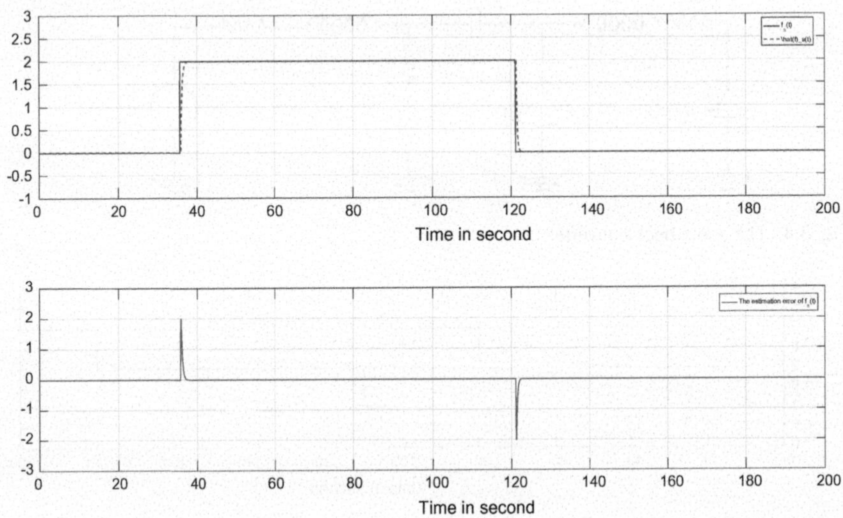

Fig. 5.11 The fault estimation of $f_{1s}(t)$ and its estimation error

$$A_1 = \begin{bmatrix} -\frac{R_{in}+r_s}{L_{in}} & \frac{r_s}{L_{in}} & -\frac{1}{L_{in}} \\ \frac{r_s}{L_{out}} & -\frac{R+r_s}{L_{out}} & 0 \\ -\frac{1}{C} & 0 & 0 \end{bmatrix}, B_1 = \begin{bmatrix} \frac{1}{L_{in}} \\ 0 \\ 0 \end{bmatrix};$$

$$A_2 = \begin{bmatrix} -\frac{R_{in}}{L_{in}} & 0 & 0 \\ 0 & -\frac{R}{L_{out}} & -\frac{1}{L_{out}} \\ 0 & -\frac{1}{C} & 0 \end{bmatrix}, B_2 = \begin{bmatrix} 0 \\ -\frac{1}{L_{out}} \\ 0 \end{bmatrix}.$$

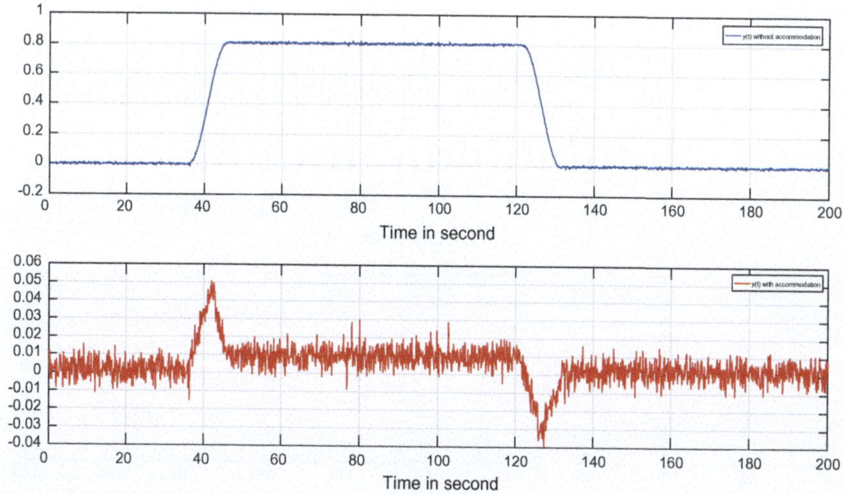

Fig. 5.12 The output $y(t)$ without and with accommodation

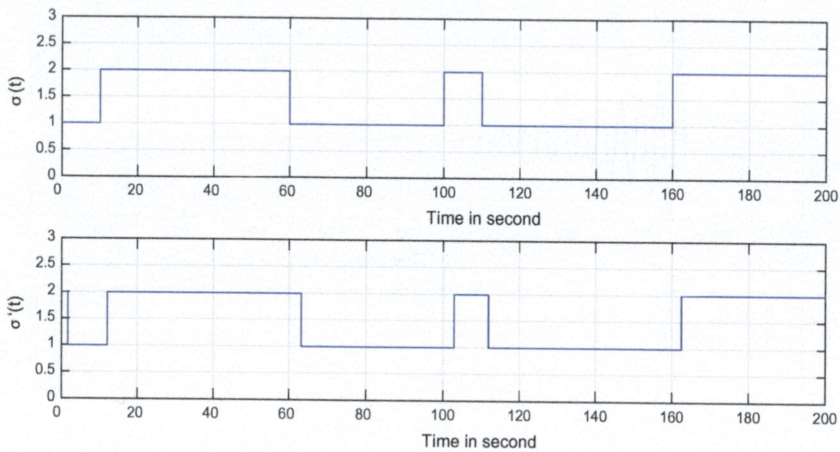

Fig. 5.13 The switching signal $\sigma(t)$ and $\sigma'(t)$

The values of the parameters are set as $R_{in} = 25\Omega$, $L_{in} = 20mH$, $C = 15\mu F$, $r_s = 10\Omega$, $L_{out} = 30mH$, $R = 25\Omega$, and $v_s = 18V$. When $\mu_1 = \mu_2 = 1.2$ and $\alpha = 0.9$, and the ADT condition in Eq. (2.71), is applied, the observer gains can be obtained by solving the constraints presented in Eqs. (5.42)–(5.44), as the following:

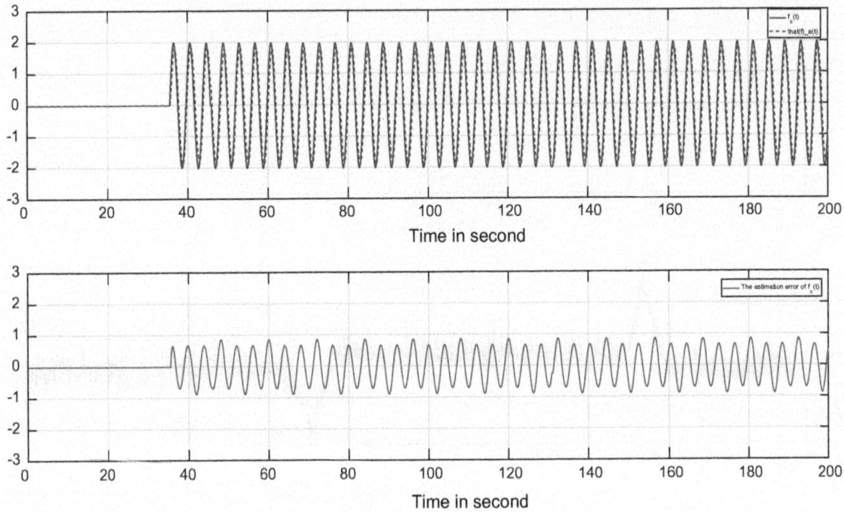

Fig. 5.14 The fault estimation of $f_{2a}(t)$ and its estimation error

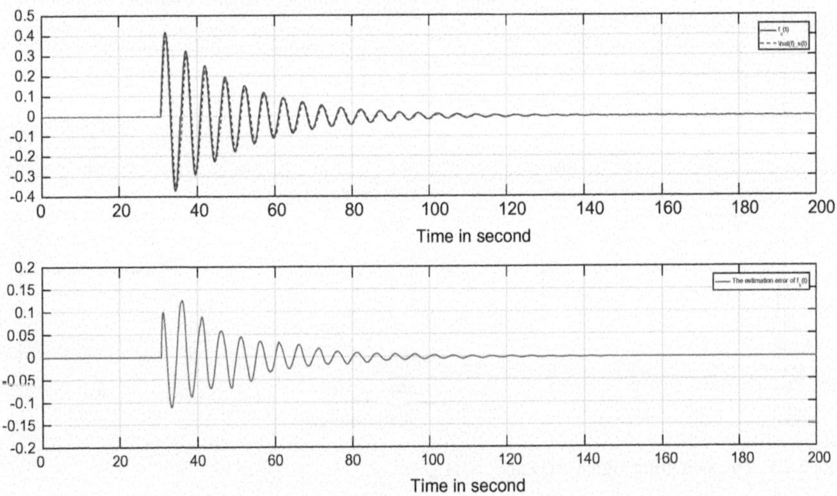

Fig. 5.15 The fault estimation of $f_{2s}(t)$ and its estimation error

$$L_1 = \begin{bmatrix} 0.1548 \\ -4.3368 \\ 0.4367 \\ 5.8756 \\ -26.9923 \end{bmatrix}; L_2 = \begin{bmatrix} -8.1436 \\ 0.3546 \\ 0.7641 \\ -2.6783 \\ -18.8239 \end{bmatrix}.$$

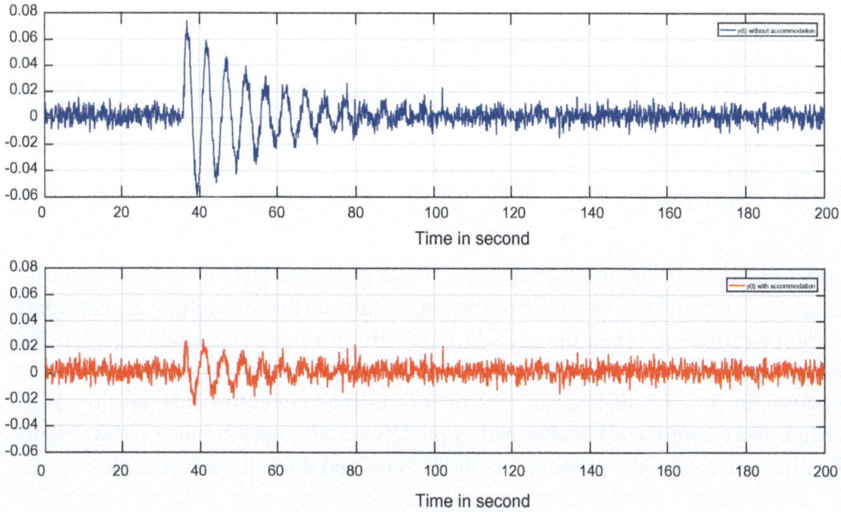

Fig. 5.16 The output $y(t)$ without and with accommodation

It should be indicated that for both cases, the simulation is performed for 200s in the same platform. Subsystems 1 and 2 are activated when $\sigma(t) = 1$, and $\sigma(t) = 2$, respectively. Moreover, Fig. 5.13, represents the switching signals of $\sigma(t)$ and $\sigma'(t)$. In this case, the actuator and senor faults are defined as

$$f_{2a}(t) = \begin{cases} 2(1 - e^{-0.2t})cos(0.5\pi t), & t \geq 35 \\ 0, & 0 \leq t < 35 \end{cases}$$

$$f_{2s}(t) = \begin{cases} 2e^{-0.05t}sin(0.4\pi t), & t \geq 30 \\ 0, & 0 \leq t < 30 \end{cases}$$

Figures 5.14 and 5.15 illustrate the simulation results obtained from the FE results and the corresponding FE error. According to the simulation results in Figs. 5.13 and 5.14, it is concluded that the proposed approach can achieve the objects of FE successfully.

When parameters are set to $\alpha_1 = 0.7$, $\beta_1 = 1.05$, $\epsilon_1 = 0.05$, $\epsilon_2 = 0.05$, $\gamma_1 = 0.5$, $\bar{\mu}_1 = 1.05$ and $\bar{\mu}_2 = 1.2$, the controller gains can be obtained as the following:

$$K_1 = \begin{bmatrix} -0.2343 \ 16.2365 \ 3.8344 \end{bmatrix};$$
$$K_2 = \begin{bmatrix} 0.4523 \ -8.4235 \ 0.0013 \end{bmatrix}.$$

Figure 5.16 shows the output responses of $y(t)$ with and without the accommodation. The simulation results show that the designed controller has an excellent fault-tolerant performance.

5.2.6 Summary

In the present section, the asynchronous observer-based FTC problem is addressed for a class of switched systems. Both actuator fault and the sensor fault are considered in the investigation. Then the asynchronous FE observer is designed to reconstruct the system state, actuator, and sensor faults. According to the estimation information, an observer-based fault-tolerant controller is designed to ensure that the closed-loop system is asymptotically stable with a predefined H_∞ performance index. Finally, the effectiveness of the proposed methods is proved through two examples.

References

1. Jiang, B., Zhang, K., Shi, P.: Integrated fault estimation and accommodation design for discrete-time Takagi-Sugeno fuzzy systems with actuator faults. IEEE Trans. Fuzzy Syst. **19**(2), 291–304 (2011)
2. Du, D.: Fault detection for discrete-time linear systems based on descriptor observer approach. Appl. Math. Comput. **293**(15), 575–585 (2017)
3. Zhang, M., Ding, S.X., Zhou, D.: A new scheme of fault detection for linear discrete time-varying systems. IEEE Trans. Autom. Control **61**(9), 2597–2602 (2016)
4. Ding, S.X., Li, L., Krger, M.: Application of randomized algorithms to assessment and design od observer-based fault detection systems. Automatica **107**, 175–182 (2019)
5. Xiong, J., Chang, X., Yi, X.: Design of robust nonfragile fault detection filter for uncertain dynamic systems with quantization. Appl. Math. Comput. **338**, 774–788 (2018)
6. Xie, X., Yue, D., Hu, S.: Fault estimation observer design of discrete-time nonlinear systems via a joint real-time scheduling law. IEEE Trans. Syst. ManCybern. Syst. **47**(7), 1451–1463 (2017)
7. Yan, X., Edwards, C.: Robust decentralized actuator fault detection and estimation for large-scale systems using a sliding mode observer. Int. J. Control **81**(4), 591–606 (2007)
8. Li, H., Gao, Y., Shi, P., Lam, H.K.: "Observer-based fault detection for nonlinear systems with sensor fault and limited communication capacity. IEEE Trans. Autom. Control **61**(9), 2745–2751 (2016)
9. Zhang, K., Jiang, B., Cocquempot, V.: Robust fault estimation observer design with finite-time convergence specification. IET Control Theory Appl. **11**(1), 1–9 (2017)
10. Zhang, H., Han, J., Wang, Y.: Sensor fault estimation of switched fuzzy systems with unknown input. IEEE Trans. Fuzzy Syst. **26**(3), 1114–1124 (2018)
11. Clemente, C.R., Lozano, E.: A multi-criteria computer package for power transformer fault detection and diagnosis. Appl. Math. Comput. **319**, 153–164 (2018)
12. Zhang, K., Jiang, B., Shi, P.: A new approach to observer-based fault-tolerant controller design fot Takagi-Sugeno fuzzy systems with state delay. Cicuits, Syst. Signal Process. **28**(5), 679–697 (2009)
13. Du, D., Cocquempot, V., Jiang, B.: Robust fault estimation observer design for switched systems with unknown input. Appl. Math. Comput. **348**, 70–83 (2019)

14. Li, J., Pan, K., Su, Q.: Sensor fault detection and estimation for switched power electronics systems based on silding mode observer. Appl. Math. Comput. **535**(15), 282–294 (2019)
15. Daafouz, J., Riedinger, P., Lung, C.: Stability synthesis for switched systems: a switched Lyapunov function approach. In: Proceedings of the IFAC World Congress, pp. 1883–1887 (2002)
16. Gahinet, P., Apkarian, P.: An linear matrix inequality approach to H_∞ control. Int. J. Robust Nonlinear Control **4**, 422–448 (1994)
17. Zhou, L., Ho, D.W.C., Zhai, G.: Stability analysis of switched linear singular systems. Automatica **49**, 1481–1487 (2013)
18. Ding, S.X.: Model-Based Fault Diagnosis Techniques: Design Schemes, Algorithms and Tools. Springer, London, U.K. (2013)
19. Chen, J., Patton, R.J.: Robust Model-Based Fault Diagnosis for Dynamic Systems. Springer, New York, NY, USA (2012)
20. Tang, L., Zhao, J.: Switched threshold-based fault detection for switched nonlinear systems with its application to chuaFs circuit system. IEEE Trans. Circuits Syst. I: Regul. Pap. **66**(2), 733–741 (2019)
21. Xie, X., Yue, D., Park, J.H.: Observer-based fault estimation for discrete-time nonlinear systems and its application: a weighted switching approach. IEEE Trans. Circuits Syst. I: Regul. Pap. **66**(11), 4377–4387 (2019)
22. Zhang, K., Jiang, B., Shi, P., et al.: Observer-Based Fault Estimation Techniques. Springer, Switzerland (2018)
23. Frank, P.M.: Survey of robust residual generation and evaluation methods in observer-based fault detection systems. J. Process Control **7**(6), 403–424 (1997)
24. Kumar, R.S., Manimozhi, M., Enosh, M.T.: A survey of fault detection and isolation in wind turbine drives. In: 2013 International Conference on Power, Energy and Control, Sri Rangalatchum Dindigul, India, 6–8 Feb (2013)
25. Liu, C., Jiang, B., Patton, R.J., Zhang, K.: Hierarchical-structure-based fault estimation and fault-tolerant control for multiagent systems. IEEE Trans. Control Netw. Syst. **6**(2), 586–597 (2019)
26. Zhao, D., Karimi, H.R., Sakthivel, R., et al.: Non-fragile fault-tolerant control for nonlinear Markovian jump systems with intermittent actuator fault. Nonlinear Anal. Hybrid Syst. **32**, 337–350 (2019)
27. Yin, S., Gao, H., Qiu, J., Kaynak, O.: Descriptor reduced-order sliding mode observers design for switched systems with sensor and actuator faults. Automatica **76**, 282–292 (2017)
28. Liu, M., Cao, X., Shi, P.: Fault estimation and tolerant control for fuzzy stochastic systems. IEEE Trans. Fuzzy Syst. **21**(2), 221–229 (2013)
29. Lan, J., Patton, R.J.: A new strategy for integration of fault estimation within fault-tolerant control. Automatica **6**, 48–59 (2016)
30. Du, D., Jiang, B.: Actuator fault estimation and accommodation for switched systems with time delay: discrete-time case. ISA Trans. **62**, 137–144 (2016)
31. Youssef, T., Chadli, M., Karimi, H.R., et al.: Actuator and sensor faults estimation based on proportional integral observer for T-S fuzzy model. J. Frankl. Inst. **354**(6), 2524–2542 (2017)
32. Hespanha, J.P., Morse, A.S.: Stability of switched systems with average dwell time. IEEE Conf. Decis Control **3**(3), 2655–2660 (2002)
33. Qi, W., Zong, G., Karimi, H.R.: Finite-time observer-based sliding mode control for quantized semi-Markov switching systems with application. IEEE Trans. Ind. Inform. **16**(2), 1259–1271 (2020)
34. Meng, X., Zhai, D., Fu, Z., Xie X.: Adaptive FTC for a class of switched nonlinear systems with unknown control directions. In: Applied Mathematics and Computation, vol. 370, in press (2020)
35. Zhang, L., Ning, Z., Shi, Y.: Analysis and synthesis for a class of stochastic switching systems against delayed mode switching: a framework of integrating mode weights. Automatica **99**, 99–111 (2019)

36. Lee, T., Tan, Y., Mareels, I.: Analyzing the stability of switched systems using common zeroing-output systems. IEEE Trans. Autom. Control **62**(10), 5138–5153 (2017)
37. Han, J., Liu, X., Wei, X., Hu, X., Zhang, H.: Reduced-order observer based fault estimation and fault-tolerant control for switched stochastic systems with actuator and sensor faults. ISA Trans. **88**, 91–101 (2019)
38. Sakthivel, R., Ramya, L.S., Ma, Y.K., et al.: Stabilization of uncertain switched discrete-time systems against actuator faults and input saturation. Nonlinear Anal. : Hybrid Syst. **35** (2020)
39. Zhang, J., Li, S., Xiang, Z.: Adaptive fuzzy finite-time fault-tolerant control for switched nonlinear large-scale systems with actuator and sensor faults. In: Journal of the Franklin Institute in press (2020)
40. Yahia, S., Bedoui, S., Abderrahim, K.: Sensor fault estimation and compensation for hybrid switched system. In: 2016 20th International Conference on System Theory, Control and Computing, 13–15 Oct (2016)
41. Yahia, S., Bedoui, S., Abderrahim, K.: Actuator fault estimation and compensation for hybrid switched system. In:2017 14th International Multi-Conference on Systems, Signals & Devices, 28–31 March (2017)
42. Han, J., Zhang, H., Wang, Y., Zhang, K.: Fault estimation and fault-tolerant control for switched fuzzy stochastic systems. IEEE Trans. Fuzzy Syst. **26**(5), 2993–3003 (2018)
43. Chen, L., Zhao, Y., Fu, S., et al.: Fault estimation observer design for descriptor switched systems with actuator and sensor failures. IEEE Trans. Circuits Syst. I: Regul. Pap. **66**(2), 810–819 (2019)
44. Tan, Y., Fei, S., Liu, J., Zhang, D.: Asynchronous adaptive event-triggered tracking control for multi-agent systems with stochastic actuator faults. Appl. Math. Comput. **355**, 482–496 (2019)
45. Du, D., Jiang, B., Shi, P., Karimi, H.R.: Fault detection for continues-time switched systems under asynchronous switching. Int. J. Robust Nonlinear Control **24**(11), 1694–1706 (2014)
46. Zhao, X., Zhao, J.: Asynchronous fault detection for continuous-time switched delay systems. J. Frankl. Inst. **352**(12), 5915–5935 (2015)
47. Eddoukali, Y., Benzaouia, A., Ouladsine, M.: Integrated fault detection and control design for continuous-time switched systems under asynchronous switching. ISA Trans. **84**, 12–19 (2019)
48. Raza, M.T., Khan, A.Q., Mustafa, G., et al.: Design of fault detection and isolation filter for switched control systems under asynchronous switching. IEEE Trans. Control Syst. Technol. **24**(1), 13–23 (2016)
49. Huang, S., Yang, G.: FTCler design for T-S fuzzy systems with time-varying delay and actuator faults: a k-step fault-estimation approach. IEEE Trans. Fuzzy Syst. **22**(6), 1526–1540 (2014)
50. Gao, Z., Wang, H.: Descriptor observer approaches for multivariable systems with measurement noises and application in fault detection and diagnosis. Syst. Control Lett. **55**(4), 304–313 (2006)
51. Tanwani, A., Dominguez-Garcia, A.D., Liberzon, D.: An inversion-based approach to fault detection and isolation in switching electrical networks. IEEE Trans. Control Syst. Technol. **19**(5), 1059–1074 (2011)

Chapter 6
Adaptive Observer-Based FEA for Discrete-Time Switched Systems

6.1 FD and Estimation

In this section, a novel FD and FE method is proposed for discrete-time switched systems. In the proposed design, an FD observer with online learning capability is utilized to generate a residual, which is obtained by comparing the system output with respect to the observer output. A fault is successfully detected if the generated residual exceeds a priori chosen threshold. Subsequently, the fault magnitude is estimated online by using a suitable parameter update law.

6.1.1 Introduction

In complex systems, dependability is as important as performances. Once a fault occurred in the system, the system behavior may be drastically changed, ranging from performance degradation to instability. FTC is an effective method in order to reach the system objectives, or if this turns to be impossible, to achievable new objectives to avoid catastrophic behaviors. In recent years, a great amount of effort has been devoted to this significant issue, and a number of achievements have been obtained, which can be found in several excellent papers, see for example [1–5] and the references therein.

As for FTC, which includes two main approaches: passive and active. While passive fault tolerance considers systems faults as a special kind of uncertainties [6], active fault tolerance is based on FDI and accommodation technique [7]. Though FDI can give information whether there exist faults occurring [8], the magnitude of the fault cannot be precisely provided. Therefore, the next step is to search an efficient approach to estimate the magnitude of the fault, which is called fault estimate or fault reconstruction. Finally, using the estimated fault information, a fault-tolerant controller can be designed to compensate the effect of the fault.

D. Du et al., *Observer-Based Fault Diagnosis and Fault-Tolerant Control for Switched Systems*, Studies in Systems, Decision and Control 323,
https://doi.org/10.1007/978-981-15-9073-3_6

From the above discussion, it can be seen that FD and FE plays a very important role in active FTC. Therefore, the study of FE has become a hot research topic owing to its importance in active FTC. In this section, a novel FD and estimation method is proposed for discrete-time switched systems. In the proposed design, an FD observer with online learning capability is utilized to generate a residual which is obtained by comparing the system output with respect to the observer output. A fault is successfully detected if the generated residual exceeds an a priori chosen threshold. Subsequently, the fault magnitude is estimated online by using a suitable parameter update law.

6.1.2 Model Description

Consider the following discrete-time switched linear system:

$$
\begin{aligned}
x(t+1) = & \sum_{i=1}^{N} \delta_i(t) A_i x(t) + \sum_{i=1}^{N} \delta_i(t) B_i u(t) + \sum_{i=1}^{N} \delta_i(t) D_i \omega(t) \\
& + \varpi(\theta(t), y(t))
\end{aligned} \tag{6.1}
$$

$$
y(t) = \sum_{i=1}^{N} \delta_i(t) C_i x(t) \tag{6.2}
$$

where $x(t) \in \mathbb{R}^n$ is the state vector, $u(t) \in \mathbb{R}^m$ is the control input vector, and $y(t) \in \mathbb{R}^q$ is the measurable output vector. The vector $\delta_i(t)$ is called switching signal, which specifies which subsystem will be activated at the discrete time t, where

$$
\delta_i(t) : Z^+ = \{0, 1, 2 \cdots\} \rightarrow \{0, 1\}, \quad \sum_{i=1}^{N} \delta_i(t) = 1, \quad \forall t \in Z^+.
$$

A_i, B_i, C_i, and D_i, are constant matrices with appropriate dimensions, and $\varpi(\theta(t), y(t))$ represents the fault signal. In addition,

$$
\varpi(\theta(t), y(t)) = \eta(t - t_0)\theta(t) f(t) \tag{6.3}
$$

where the function $\eta(t - t_0)$ is given by

$$
\eta(t - t_0) = \begin{cases} 0, & t \leq t_0 \\ 1, & t > t_0 \end{cases}
$$

where t_0 is the time of fault occurring. That is, $f(t)$ is zero prior to the failure time $t \leq t_0$ and is $\theta(t) f(t)$ after the failure occurs $t > t_0$. $\theta(t) \in \mathbb{R}^{n \times p}$ is the unknown fault magnitude, and the fault basis function $f(t) \in \mathbb{R}^p$ is considered to be known.

In the fault-free case, we have $\theta(t) = \theta_H$, where θ_H is a known scalar (subscript H stands for healthy case).

Remark 6.1 Similar to [9], the fault basis function could be a sinusoidal function, linear function, polynomial function, etc. In this section, sigmoid functions are not considered as fault basis functions.

6.1.3 FD Observer Design

To detect the fault, the following observer is constructed:

$$\hat{x}(t+1) = \sum_{i=1}^{N} \delta_i(t) A_i \hat{x}(t) + \sum_{i=1}^{N} \delta_i(t) B_i u(t) + \hat{\varpi}(\theta_H(t), y(t)) + \sum_{i=1}^{N} \delta_i(t) K_i (y(t) - \hat{y}(t)) \tag{6.4}$$

$$\hat{y}(t) = \sum_{i=1}^{N} \delta_i(t) C_i \hat{x}(t) \tag{6.5}$$

where $\hat{x}(t) \in \mathbb{R}^n$ is the state estimate. $\hat{y}(t) \in \mathbb{R}^q$ is the measured output, $K_i \in \mathbb{R}^{n\times q}$ is the observer gain, $\hat{\varpi}(\theta_H(t), y(t)) = \theta_H(t) f(t)$.

We define

$$e(t) = x(t) - \hat{x}(t), \quad e_y(t) = y(t) - \hat{y}(t) \tag{6.6}$$

Then the observation error and output error equations are given by

$$e(t+1) = \sum_{i=1}^{N} \delta_i(t) \overline{A}_i \hat{x}(t) + \sum_{i=1}^{N} \delta_i(t) D_i \omega(t) + (\theta(t) - \theta_H) f(t) \tag{6.7}$$

$$e_y(t) = \sum_{i=1}^{N} \delta_i(t) C_i e(t) \tag{6.8}$$

where $\overline{A}_i = A_i - K_i C_i$.

The following lemma is inspired from [10], but the analytical equations are developed exclusively for the discrete-time switched system defined in (6.1)–(6.2).

Lemma 6.1 *Define the following residual signal*

$$e_y(t) = y(t) - \hat{y}(t) \tag{6.9}$$

the above equation can be written as

$$e_y(t) = \sum_{i=1}^{N} \delta_i(t) C_i \left[\left(sI - \sum_{i=1}^{N} \delta_i(t) \overline{A}_i \right)^{-1} \times \left(\sum_{i=1}^{N} \delta_i(t) D_i \omega(t) \right) \right] \quad (6.10)$$

Therefore, the disturbance $\omega(t)$ in the aforementioned equation could be separated from the FD residual if either of the following conditions are satisfied:

$$\sum_{i=1}^{N} \delta_i(t) C_i \left(\sum_{i=1}^{N} \delta_i(t) D_i \omega(t) \right) = 0, \; \sum_{i=1}^{N} \delta_i(t) C_i \left(\sum_{i=1}^{N} \delta_i(t) \overline{A}_i \right) = 0 \quad (6.11)$$

$$\sum_{i=1}^{N} \delta_i(t) C_i \left(\sum_{i=1}^{N} \delta_i(t) D_i \omega(t) \right) = 0, \; \left(\sum_{i=1}^{N} \delta_i(t) D_i \omega(t) \right) \left(\sum_{i=1}^{N} \delta_i(t) \overline{A}_i \right) = 0 \quad (6.12)$$

Remark 6.2 From the conditions given in (6.11) and (6.12), one can get that the residual signal $e_y(t)$ is independent of the system uncertainty. This means that once a fault occurs, it can be detected immediately.

The following theorem can be directly deduced by Lemma 6.1:

Theorem 6.1 *Consider the system described by (6.1)–(6.2), and its observer described by (6.4)–(6.5). If either of the condition in (6.11) or (6.12) is satisfied, the observer is asymptotically convergent when no fault occurs ($\theta(t) = \theta_H$), i.e., $lim_{t\to 0} e(t) = 0$ and $lim_{t\to 0} e_y(t) = 0$.*

According to Theorem 6.1, after the convergence of the observer, the FD can be carried out as follows:

$$e_y(t) = 0 \quad no\ fault\ occurs \quad (6.13)$$

$$e_y(t) > 0 \quad fault\ has\ occured \quad (6.14)$$

6.1.4 FE Algorithm Design

To estimate the fault after the alarm (6.13)–(6.14), has been generated, the following observer is constructed:

$$\hat{x}(t+1) = \sum_{i=1}^{N} \delta_i(t) A_i \hat{x}(t) + \sum_{i=1}^{N} \delta_i(t) B_i u(t) + \varpi(\hat{\theta}(t), y(t)) + \sum_{i=1}^{N} \delta_i(t) K_i (y(t) - \hat{y}(t)) \quad (6.15)$$

$$\hat{y}(t) = \sum_{i=1}^{N} \delta_i(t) C_i \hat{x}(t) \tag{6.16}$$

where $\hat{x}(t) \in \mathbb{R}^n$ is the state estimate. $\hat{y}(t) \in \mathbb{R}^q$ is the measured output, $K_i \in \mathbb{R}^{n\times q}$ is the observer gain, $\hat{\varpi}(\hat{\theta}(t), y(t)) = \hat{\theta}(t) f(t)$, where $\hat{\theta}(t)$ is an estimate of $\theta(t)$. The value of $\hat{\theta}(t)$ is set to $\theta_H(t)$ until a fault is detected. It is assumed that after a fault occurs, $\theta(t) \neq \theta_H(t)$.

We define

$$e(t) = x(t) - \hat{x}(t), \quad e_y(t) = y(t) - \hat{y}(t), \quad \tilde{\theta}(t) = \theta(t) - \hat{\theta}(t) \tag{6.17}$$

then it can be obtained that

$$e(t+1) = \sum_{i=1}^{N} \delta_i(t) \overline{A}_i e(t) + \sum_{i=1}^{N} \delta_i(t) D_i \omega(t) + \tilde{\theta}(t) f(t) \tag{6.18}$$

$$e_y(t) = \sum_{i=1}^{N} \delta_i(t) C_i e(t) \tag{6.19}$$

Theorem 6.2 *Under the condition in (6.11) or (6.12), the observer described by (6.15)–(6.16), and following FE algorithm:*

$$\hat{\theta}(t+1) = \hat{\theta}(t) + \alpha f(t) e_y^T(t) \Theta - \gamma \| I - \alpha f(t) f^T(t) \| \hat{\theta}(t) \tag{6.20}$$

can realize the output residual $e_y(t)$ and the parameter estimation error $\tilde{\theta}(t)$ are uniformly ultimately bounded, where $\alpha > 0$ is the learning rate and $\gamma > 0$ is a designed constant. Θ is a known matrix and satisfies $\|\Theta\| \leq \rho,\ \rho > 0$.

Proof Consider the following Lyapunov function candidate:

$$V(t) = e^T(t) e(t) + \frac{1}{\alpha} tr\{\tilde{\theta}(t) \tilde{\theta}^T(t)\} \tag{6.21}$$

The first difference of the aforementioned equation is given by

$$\begin{aligned}
\Delta V(t) &= e^T(t+1) e(t+1) - e^T(t) e(t) + \frac{1}{\alpha} tr\{\tilde{\theta}(t+1) \tilde{\theta}^T(t+1) - \tilde{\theta}(t) \tilde{\theta}^T(t)\} \\
&= \left[\sum_{i=1}^{N} \delta_i(t) \overline{A}_i e(t) + \sum_{i=1}^{N} \delta_i(t) D_i \omega(t) + \tilde{\theta}(t) f(t) \right]^T \\
&\quad \times \left[\sum_{i=1}^{N} \delta_i(t) \overline{A}_i e(t) + \sum_{i=1}^{N} \delta_i(t) D_i \omega(t) + \tilde{\theta}(t) f(t) \right] - e^T(t) e(t) \\
&\quad + \frac{1}{\alpha} tr \left\{ \left[\tilde{\theta}(t) - \alpha f(t) e_y^T(t) \Theta + \gamma \| I - \alpha f(t) f^T(t) \| \hat{\theta}(t) \right]^T \right.
\end{aligned}$$

$$
\begin{aligned}
&\times\left[\bar{\theta}(t)-\alpha f(t)e_y^T(t)\Theta+\gamma\|I-\alpha f(t)f^T(t)\|\hat{\theta}(t)\right]-\tilde{\theta}(t)\tilde{\theta}^T(t)\Big\}\\
=&\left(\sum_{i=1}^{N}\delta_i(t)\overline{A}_ie(t)\right)^T\left(\sum_{i=1}^{N}\delta_i(t)\overline{A}_ie(t)\right)+\\
&2\left(\sum_{i=1}^{N}\delta_i(t)\overline{A}_ie(t)\right)^T\left(\sum_{i=1}^{N}\delta_i(t)D_i\omega(t)\right)\\
&+2\left(\sum_{i=1}^{N}\delta_i(t)\overline{A}_ie(t)\right)^T\left(\tilde{\theta}(t)f(t)\right)\\
&+\left(\sum_{i=1}^{N}\delta_i(t)D_i\omega(t)\right)^T\left(\sum_{i=1}^{N}\delta_i(t)D_i\omega(t)\right)\\
&+2\left(\sum_{i=1}^{N}\delta_i(t)D_i\omega(t)\right)^T\left(\tilde{\theta}(t)f(t)\right)\\
&+\left(\tilde{\theta}(t)f(t)\right)^T\left(\tilde{\theta}(t)f(t)\right)-e^T(t)e(t)\\
&+\frac{1}{\alpha}tr\left\{\left[\left(I-\gamma\|I-\alpha f(t)f^T(t)\|I\right)\tilde{\theta}(t)\right.\right.\\
&\left.-\alpha f(t)e_y^T(t)\Theta+\gamma\|I-\alpha f(t)f^T(t)\|\theta(t)\right]^T\\
&\times\left[\left(I-\gamma\|I-\alpha f(t)f^T(t)\|I\right)\tilde{\theta}(t)\right.\\
&\left.\left.-\alpha f(t)e_y^T(t)\Theta+\gamma\|I-\alpha f(t)f^T(t)\|\theta(t)\right]-\tilde{\theta}(t)\tilde{\theta}^T(t)\right\}\\
=&\left(\sum_{i=1}^{N}\delta_i(t)\overline{A}_ie(t)\right)^T\left(\sum_{i=1}^{N}\delta_i(t)\overline{A}_ie(t)\right)\\
&+2\left(\sum_{i=1}^{N}\delta_i(t)\overline{A}_ie(t)\right)^T\left(\sum_{i=1}^{N}\delta_i(t)D_i\omega(t)\right)\\
&+2\left(\sum_{i=1}^{N}\delta_i(t)\overline{A}_ie(t)\right)^T\left(\tilde{\theta}(t)f(t)\right)\\
&+\left(\sum_{i=1}^{N}\delta_i(t)D_i\omega(t)\right)^T\left(\sum_{i=1}^{N}\delta_i(t)D_i\omega(t)\right)\\
&+2\left(\sum_{i=1}^{N}\delta_i(t)D_i\omega(t)\right)^T\left(\tilde{\theta}(t)f(t)\right)+\left(\tilde{\theta}(t)f(t)\right)^T\left(\tilde{\theta}(t)f(t)\right)-e^T(t)e(t)\\
&+\frac{1}{\alpha}tr\left\{\left[(I-\gamma\|I-\alpha f(t)f^T(t)\|I)\tilde{\theta}^T(t)\tilde{\theta}(t)(I-\gamma\|I-\alpha f(t)f^T(t)\|I)\right.\right.\\
&-2\alpha\Theta^Te_y(t)f^T(t)(I-\gamma\|I-\alpha f(t)f^T(t)\|I)\tilde{\theta}(t)\\
&+2\gamma\|I-\alpha f(t)f^T(t)\|\theta(t)(I-\gamma\|I-\alpha f(t)f^T(t)\|I)\tilde{\theta}^T(t)\theta(t)\\
&+\alpha^2\Theta^Te_y(t)f^T(t)f(t)e_y^T(t)\Theta-2\alpha\gamma\Theta^Te_y(t)f^T(t)\|I-\alpha f(t)f^T(t)\|\theta(t)\\
&\left.\left.+\gamma^2\|I-\alpha f(t)f^T(t)\|^2\theta^T(t)\theta(t)\right]-\tilde{\theta}(t)\tilde{\theta}^T(t)\right\}
\end{aligned}
$$

$$
\begin{aligned}
&\leq \left(\sum_{i=1}^{N} \delta_i(t)\overline{A}_i e(t)\right)^T \left(\sum_{i=1}^{N} \delta_i(t)\overline{A}_i e(t)\right) \\
&+2\left(\sum_{i=1}^{N} \delta_i(t)\overline{A}_i e(t)\right)^T \left(\sum_{i=1}^{N} \delta_i(t)D_i\omega(t)\right) \\
&+2\left(\sum_{i=1}^{N} \delta_i(t)\overline{A}_i e(t)\right)^T \left(\tilde{\theta}(t)f(t)\right) \\
&+\left(\sum_{i=1}^{N} \delta_i(t)D_i\omega(t)\right)^T \left(\sum_{i=1}^{N} \delta_i(t)D_i\omega(t)\right) \\
&+2\left(\sum_{i=1}^{N} \delta_i(t)D_i\omega(t)\right)^T \left(\tilde{\theta}(t)f(t)\right) + \left(\tilde{\theta}(t)f(t)\right)^T \left(\tilde{\theta}(t)f(t)\right) - e^T(t)e(t) \\
&+\frac{1}{\alpha}tr\left\{\left[3(I-\gamma\|I-\alpha f(t)f^T(t)\|I)\tilde{\theta}^T(t)\tilde{\theta}(t)(I-\gamma\|I-\alpha f(t)f^T(t)\|I)\right.\right. \\
&+3\alpha^2\Theta^T e_y(t)f^T(t)f(t)e_y^T(t)\Theta + 3\gamma^2\|I-\alpha f(t)f^T(t)\|^2\theta^T(t)\theta(t)\Big] \\
&\left.-\tilde{\theta}(t)\tilde{\theta}^T(t)\right\} \\
&\leq \left(\sum_{i=1}^{N} \delta_i(t)\overline{A}_i e(t)\right)^T \left(\sum_{i=1}^{N} \delta_i(t)\overline{A}_i e(t)\right) \\
&+3\left(\sum_{i=1}^{N} \delta_i(t)D_i\omega(t)\right)^T \left(\sum_{i=1}^{N} \delta_i(t)D_i\omega(t)\right) \\
&+3\left(\tilde{\theta}(t)f(t)\right)^T \left(\tilde{\theta}(t)f(t)\right) - e^T(t)e(t) \\
&+\frac{1}{\alpha}tr\left\{2\tilde{\theta}^T(t)\tilde{\theta}(t) - 6\gamma\|I-\alpha f(t)f^T(t)\|\tilde{\theta}^T(t)\tilde{\theta}(t)\right. \\
&+3\gamma^2\|I-\alpha f(t)f^T(t)\|^2\tilde{\theta}^T(t)\tilde{\theta}(t) \\
&+3\alpha^2\Theta^T e_y(t)f^T(t)f(t)e_y^T(t)\Theta \\
&\left.+3\gamma^2\|I-\alpha f(t)f^T(t)\|^2\theta^T(t)\theta(t)\right\}
\end{aligned} \tag{6.22}
$$

Apply Cauchy-Schwarz inequality and using results of [11], we have

$$
\begin{aligned}
&3\left(\sum_{i=1}^{N} \delta_i(t)D_i\omega(t)\right)^T \left(\sum_{i=1}^{N} \delta_i(t)D_i\omega(t)\right) \\
&+3\frac{\gamma^2}{\alpha}\|I-\alpha f(t)f^T(t)\|^2\theta^T(t)\theta(t) \leq \beta_M
\end{aligned} \tag{6.23}
$$

where $\beta_M > 0$, then we have

$$
\begin{aligned}
\Delta V(t) \leq & \left(\sum_{i=1}^{N} \delta_i(t)\overline{A}_i e(t)\right)^T \left(\sum_{i=1}^{N} \delta_i(t)\overline{A}_i e(t)\right) \\
& +3\left(\tilde{\theta}(t) f(t)\right)^T \left(\tilde{\theta}(t) f(t)\right) - e^T(t)e(t) \\
& +\frac{1}{\alpha} tr\left\{2\tilde{\theta}^T(t)\tilde{\theta}(t) - 6\gamma\|I - \alpha f(t) f^T(t)\|\tilde{\theta}^T(t)\tilde{\theta}(t)\right. \\
& +3\gamma^2\|I - \alpha f(t) f^T(t)\|^2\tilde{\theta}^T(t)\tilde{\theta}(t) \\
& \left.+3\alpha^2\Theta^T e_y(t) f^T(t) f(t) e_y^T(t)\Theta\right\} + \beta_M \\
\leq & \left(\sum_{i=1}^{N} \delta_i(t)\overline{A}_i e(t)\right)^T \left(\sum_{i=1}^{N} \delta_i(t)\overline{A}_i e(t)\right) \\
& +3\left(\tilde{\theta}(t) f(t)\right)^T \left(\tilde{\theta}(t) f(t)\right) - e^T(t)e(t) \\
& +\frac{1}{\alpha} tr\left\{2\tilde{\theta}^T(t)\tilde{\theta}(t) - 6\gamma\|I - \alpha f(t) f^T(t)\|\tilde{\theta}^T(t)\tilde{\theta}(t)\right. \\
& +3\gamma^2\|I - \alpha f(t) f^T(t)\|^2\tilde{\theta}^T(t)\tilde{\theta}(t) \\
& \left.+3\alpha^2\Theta^T \left(\sum_{i=1}^{N} \delta_i(t) C_i e(t)\right) f^T(t) f(t) \left(\sum_{i=1}^{N} \delta_i(t) C_i e(t)\right)^T \Theta\right\} \\
& +\beta_M
\end{aligned} \tag{6.24}
$$

It is known that $\|f(t)\| \leq f_{max}$, and let $\|C_i\| \leq C_{max}$ and $\|\overline{A}_i\| \leq \overline{A}_{max}$ for $i \in \psi$. Applying the Frobenius norm to (6.24), we have

$$
\begin{aligned}
\Delta V(t) \leq & -(1 - 3\overline{A}_{max}^2 - 3\alpha f_{max}^2 C_{max}^2 \rho_{max}^2)\|e(t)\|^2 - \left(\frac{6\gamma}{\alpha}\|I - \alpha f(t) f^T(t)\|\right. \\
& \left. -3 f_{max}^2 - \frac{2}{\alpha} - \frac{3\gamma^2}{\alpha}\|I - \alpha f(t) f^T(t)\|^2)\right)\tilde{\theta}(t)\|^2 + \beta_M
\end{aligned} \tag{6.25}
$$

If the following conditions are provided

$$
\overline{A}_{max} \geq \sqrt{\frac{1 - 3\alpha f_{max}^2 C_{max}^2 \rho_{max}^2}{3}} \tag{6.26}
$$

$$
\tilde{\theta}(t)\| \geq \sqrt{\frac{\beta_M}{\frac{6\gamma}{\alpha}\|I - \alpha f(t) f^T(t)\| - 3 f_{max}^2 - \frac{2}{\alpha} - \frac{3\gamma^2}{\alpha}\|I - \alpha f(t) f^T(t)\|^2)}} \tag{6.27}
$$

Then we have $\Delta V(t) < 0$.

Thus, we could show that the state residual $\|e(t)\|$ and the parameter estimation error $\|\tilde{\theta}(t)\|$ are UUB. Additionally, it is straight forward to show that the output residual $\|e_y(t)\|$ is UUB since the state residual $\|e(t)\|$ is bounded. Therefore, the aforementioned analytical result guarantees that the estimation of the fault magnitude by the OLAD is stable. This completes the proof. (□)

6.2 Actuator FE in Finite-Frequency

This section deals with the problem of actuator FE observer design for discrete-time switched systems with finite-frequency. The frequencies of the fault and the unknown input disturbance are assumed to be in a finite range. A finite-frequency FE observer is proposed to estimate the actuator fault in the entire frequency domain. Based on the generalized KYP lemma and the switched Lyapunov function method, the efficient conditions are obtained to achieve the FE observer design. Finally, an example is given to illustrate the proposed technique.

6.2.1 Introduction

For the past few decades, a great amount of attention has been paid to propose efficient fault diagnosis and FTC techniques due to the safety, reliability, and high performance demands of various complex dynamics systems [12, 13]. Fault diagnosis techniques are generally composed of two steps: the first step is to detect if a fault has occurred, which is called FD [14]; then the following step is to localize the fault and to estimate its magnitude, which is called fault identification. FE is often a difficult task and the literature is not so abundant as for FD. In recent years, several approaches have been developed by using observer-based methods and solutions were given under LMIs form [15–18].

Switched systems are a class of hybrid system defined by a collection of subsystems together with a switching signal which specifies how the system switches from one subsystem to another one [19, 20]. Due to the importance of switched systems in various fields, such as mechanical systems, chemical process control, electrical power systems, and traffic control systems, etc. [21, 22], fault diagnosis and FTC for switched system have been extensively investigated in the past few years. For example, FD filter design for continuous-time and discrete-time switched system was separately considered in [23] and [24]. Based on the descriptor observer method, sensor FE issue was studied in [25], while the discrete-time case was developed in [26]. In [27], actuator FE for continuous-time switched systems with state delay was considered by using an adaptive fault diagnosis observer approach. Recently, based on a reduced-order observer and the switched Lyapunov function method, actuator FE for discrete-time switched systems was proposed in [28]. An observer-based fault-tolerant control method for a class of switched nonlinear systems with unfixed or fixed dwell-time periods was investigated in [29].

It is worth noting that the works cited previously are all considered in the full frequency domain. However, for some real practical plants, the fault and the unknown disturbance frequency ranges are known in advance. Under such cases, the techniques developed for full frequency design cause great conservatism because of the overdesign. Therefore, it is necessary and more practical to design the fault diagnosis observer for switched system in finite-frequency domain. Some results with finite-frequency domain have been shown in recent studies. For instance, [30] was concerned with the robust H_∞ filter design for discrete-time switched system in finite-frequency domain. The issue of FD filter design in finite-frequency specification was addressed in [31], for discrete-time switched systems. The FD filter design with H_- and H_∞ performance indexes in finite-frequency domain was investigated in [32], for discrete-time T-S fuzzy systems. In [33], an H_∞ filter with pole-placement constraint was proposed to achieve FE for discrete-time T-S fuzzy systems.

In this work, the problem of actuator FE observer design for discrete-time switched systems is considered. We consider that the switched system is affected by finite-frequency disturbance and actuator fault. Based on the generalized KYP lemma and the switched Lyapunov function method, an actuator FE observer is designed such that the error system is asymptotically stable with an H_∞ performance index, and the results are given in LMIs form.

The main contributions of this section can be summarized as follows:

- a finite-frequency robust FE observer design for discrete-time switched systems is proposed, which can give an accurate estimation of the actuator fault;
- the FE observer performances, i.e., robustness against the disturbance and the fault, are guaranteed by the design procedure;
- the entire frequency range is divided into three ranges, which greatly reduces the computational burden.

6.2.2 Problem Description and Preliminaries

Consider the following discrete-time switched system:

$$\begin{cases} x(t+1) = A_{\sigma(t)}x(t) + B_{\sigma(t)}(u(t) + f(t)) + D_{1\sigma(t)}d(t) \\ y(t) = C_{\sigma(t)}x(t) + D_{2\sigma(t)}d(t) \end{cases} \tag{6.28}$$

where $x(t) \in \mathbb{R}^{n_x}$ is the state vector, $u(t) \in \mathbb{R}^{n_u}$ is the known input vector, $y(t) \in \mathbb{R}^{n_y}$ is the measured output vector; $f(t) \in \mathbb{R}^{n_f}$ and $d(t) \in \mathbb{R}^{n_d}$ are, respectively, the actuator fault signal and the unknown input disturbance, which belong to $l_2[0, \infty)$ and whose frequency ranges are finite and known in advance. Here, we define that $\vartheta_d \in \Theta_d$ and $\vartheta_f \in \Theta_f$, respectively, represent the frequency of the disturbance and the fault. Θ_d and Θ_f are divided into low-frequency (LF), middle-frequency (MF), and high-frequency (HF) three ranges and shown in the Table 6.1

Table 6.1 Three ranges of the disturbance and the fault frequencies

Frequency	LF	MF	HF
Θ_d	$\|\vartheta_d\| \leq \vartheta_{dl}$	$\vartheta_{dm1} \leq \vartheta_d \leq \vartheta_{dm2}$	$\|\vartheta_d\| \geq \vartheta_{dh}$
Θ_f	$\|\vartheta_f\| \leq \vartheta_{fl}$	$\vartheta_{fm1} \leq \vartheta_f \leq \vartheta_{fm2}$	$\|\vartheta_f\| \geq \vartheta_{fh}$

where ϑ_{dl}, ϑ_{dm1}, ϑ_{dm2}, ϑ_{dh}, ϑ_{fl}, ϑ_{fm1}, ϑ_{fm2}, and ϑ_{fh} are all assumed to be known. The piecewise function $\sigma(t) : [0, \infty) \to \aleph = \{1, 2, \ldots, N\}$ is the switching signal, which specifies which subsystem is activated at the instant t. When $\sigma(t) = i$, it means that the i-th subsystem is activated. The switching signal is assumed to be *a priori* unknown, but its instantaneous value is available in real time. A_i, B_i, C_i, D_{1i}, and D_{2i} are constant real matrices of appropriate dimensions.

In order to estimate the actuator fault, the following FE observer is proposed:

$$\begin{cases} \hat{x}(t+1) = A_{\sigma(t)}\hat{x}(t) + B_{\sigma(t)}(u(t) + \hat{f}(t)) - L_{\sigma(t)}(\hat{y}(t) - y(t)) \\ \hat{y}(t) = C_{\sigma(t)}\hat{x}(t) \\ \hat{f}(t+1) = \hat{f}(t) - F_{\sigma(t)}(\hat{y}(t) - y(t)) \end{cases} \tag{6.29}$$

where $\hat{x}(t) \in \mathbb{R}^{n_x}$ is the observer state, $\hat{y}(t) \in \mathbb{R}^{n_y}$ is the observer output, $\hat{f}(t) \in \mathbb{R}^{n_f}$ is the estimation of the actuator fault $f(t)$. The matrices $L_{\sigma(t)}$ and $F_{\sigma(t)}$ are the observer parameters to be determined later.

Define

$$e_x(t) = \hat{x}(t) - x(t), \quad e_f(t) = \hat{f}(t) - f(t), \quad \Delta f(t) = f(t+1) - f(t).$$

Then from the systems (6.28) and (6.29), one can get the following augmented system:

$$\begin{cases} \bar{e}(t+1) = \tilde{A}_{\sigma(t)}\bar{e}(t) + \tilde{D}_{\sigma(t)}d(t) + \bar{I}\Delta f(t) \\ e_f(t) = \bar{I}\bar{e}(t) \end{cases} \tag{6.30}$$

where

$$\tilde{A}_i = \bar{A}_i - \bar{L}_i\bar{C}_i, \ \tilde{D}_i = \bar{L}_i D_{2i} - \bar{D}_{1i}, \ \bar{I} = \begin{bmatrix} 0 & I \end{bmatrix}, \ \bar{C}_i = \begin{bmatrix} C_i & 0 \end{bmatrix},$$
$$\bar{e}(t) = \begin{bmatrix} e_x(t) \\ e_f(t) \end{bmatrix}, \ \bar{A}_i = \begin{bmatrix} A_i & B_i \\ 0 & I \end{bmatrix}, \ \bar{L}_i = \begin{bmatrix} L_i \\ F_i \end{bmatrix}, \ \bar{D}_{1i} = \begin{bmatrix} D_{1i} \\ 0 \end{bmatrix}.$$

In this section, we mainly focus on designing a finite-frequency robust FE observer for the discrete-time switched systems, such that the FE error is robust to the disturbance and the fault. Therefore, the problem of this section can be formulated as follows:

- For $d(t) = 0$ and $\Delta f(t) = 0$, system (6.30) is asymptotically stable;

- For $\Delta f(t) = 0$, the H_∞ performance from $d(t)$ to $e_f(t)$ is less than a given positive scalar γ_d;
- For $d(t) = 0$, the H_∞ performance from $\Delta f(t)$ to $e_f(t)$ is less than a given positive scalar γ_f.

The following lemmas are recalled:

Lemma 6.2 (Generalized KYP Lemma [30]) *Consider the following discrete-time switched system Σ_e:*

$$\begin{cases} x(t+1) = \mathscr{A}_{\sigma(t)}(\lambda)x(t) + \mathscr{B}_{\sigma(t)}(\lambda)d(t) \\ e(t) = \mathscr{C}_{\sigma(t)}(\lambda)x(t) + \mathscr{D}_{\sigma(t)}(\lambda)d(t) \end{cases} \tag{6.31}$$

For a given symmetric matrix

$$\Pi = \begin{bmatrix} I & 0 \\ 0 & -\gamma^2 I \end{bmatrix} \tag{6.32}$$

the following statements are equivalent:
(i): The finite-frequency inequality

$$\rho_{max}(G^i_{de}(e^{j\vartheta_d})) < \gamma, \tag{6.33}$$

where $\rho_{max}(G^i_{de}(e^{j\vartheta_d}))$ is the maximum singular value of the transfer function matrix $G^i_{de}(e^{j\vartheta_d})$ for the i-th subsystem from the noise input $d(t)$ to error output $e(t)$;
(ii): There exist Hermitian matrix functions $P_i(\lambda)$, $Q_i(\lambda)$ satisfying $Q_i(\lambda) > 0$, and

$$\begin{bmatrix} \mathscr{A}_i(\lambda) & \mathscr{B}_i(\lambda) \\ I & 0 \end{bmatrix}^T \Xi_i(\lambda) \begin{bmatrix} \mathscr{A}_i(\lambda) & \mathscr{B}_i(\lambda) \\ I & 0 \end{bmatrix} + \begin{bmatrix} \mathscr{C}_i(\lambda) & \mathscr{D}_i(\lambda) \\ 0 & I \end{bmatrix}^T \Pi \begin{bmatrix} \mathscr{C}_i(\lambda) & \mathscr{D}_i(\lambda) \\ 0 & I \end{bmatrix} < 0 \tag{6.34}$$

where

$$\Xi_i(\lambda) = \begin{bmatrix} -P_i(\lambda) & Q_i(\lambda) \\ Q_i(\lambda) & P_i(\lambda) - 2cos(\vartheta_{dl})Q_i(\lambda) \end{bmatrix} \tag{6.35}$$

for low-frequency range $|\vartheta_d| \le \vartheta_{dl}$,

$$\Xi_i(\lambda) = \begin{bmatrix} -P_i(\lambda) & e^{j\vartheta_{dc}}Q_i(\lambda) \\ e^{-j\vartheta_{dc}}Q_i(\lambda) & P_i(\lambda) - 2cos(\vartheta_{dw})Q_i(\lambda) \end{bmatrix}, \\ \vartheta_{dc} = \frac{\vartheta_{dm1} + \vartheta_{dm2}}{2}, \quad \vartheta_{dw} = \frac{\vartheta_{dm1} - \vartheta_{dm2}}{2}, \tag{6.36}$$

for middle-frequency range $\vartheta_{dm1} < \vartheta_d < \vartheta_{dm2}$, and

$$\Xi_i(\lambda) = \begin{bmatrix} -P_i(\lambda) & -Q_i(\lambda) \\ -Q_i(\lambda) & P_i(\lambda) + 2cos(\vartheta_{dh})Q_i(\lambda) \end{bmatrix} \tag{6.37}$$

for high-frequency range $|\vartheta_d| \geq \vartheta_{dh}$.

Remark 6.3 Because $\rho_{max}(G^i_{de}(e^{j\vartheta_d}))$ is the maximum singular value of the transfer function matrix $G^i_{de}(e^{j\vartheta_d})$ for the i-th subsystem from the noise input $d(t)$ to the error output $e(t)$, one can get that (6.33) is equivalent to minimize the error $e(t)$ over the energy bounded noise $d(t)$, i.e.,

$$\min \sup_{d(t)\neq 0} \frac{\|e(t)\|_2}{\|d(t)\|_2} < \gamma$$

Remark 6.4 In [30], the matrices of system Σ_e are defined in the following form:

$$(A_i(\lambda), B_i(\lambda), C_i(\lambda), D_i(\lambda)) = \sum_{l=1}^{s} \lambda_l (A_{i,l}, B_{i,l}, C_{i,l}, D_{i,l}), \quad \sum_{l=1}^{s} \lambda_l = 1$$

If we set $A_{i,1} = A_{i,2} = \cdots = A_{i,s} = A_i$, $B_{i,1} = B_{i,2} = \cdots = B_{i,s} = B_i$, $C_{i,1} = C_{i,2} = \cdots = C_{i,s} = C_i$, $D_{i,1} = D_{i,2} = \cdots = D_{i,s} = D_i$, then the system Σ_e becomes a common linear switched system. Therefore, the result can be directly used by such system.

Lemma 6.3 (Projection Lemma [34]) *Given a symmetric matrix Σ and two matrices Γ, Λ of appropriate column dimensions, the problem*

$$\Sigma + \Gamma^T \Theta \Lambda + \Lambda^T \Theta^T \Gamma < 0 \tag{6.38}$$

is solvable with respect to the matrix Θ of compatible dimension if and only if the following inequalities hold:

$$\mathcal{N}_\Gamma^T \Sigma \mathcal{N}_\Gamma < 0, \quad \mathcal{N}_\Lambda^T \Sigma \mathcal{N}_\Lambda < 0, \tag{6.39}$$

where $\mathcal{N}_\Gamma$ and $\mathcal{N}_\Lambda$ are respectively any basis of the nullspace of Γ and Λ.

Lemma 6.4 ([35]) *For discrete-time switched systems (6.28) with $f(t) = 0$ and $d(t) = 0$, the following statements are equivalent:*

(i) There exists a Lyapunov function of the form $V(t, x(t)) = x^T(t)P_{\sigma(t)}x(t)$, whose difference is negative definite, proving asymptotic stability of (1);
(ii) There exist N symmetric matrices $P_1, \ldots, P_N$, satisfying

$$\begin{bmatrix} P_i & A_i^T P_j \\ P_j A_i & P_j \end{bmatrix} > 0 \tag{6.40}$$

6.2.3 Main Results

In this section, finite-frequency robust FE observer will be designed for system (6.28), which guarantees that the error system (6.30), is asymptotically stable with H_∞ performance.

Theorem 6.3 *For given positive scalars γ_d and γ_f, the system (6.30) is asymptotically stable with H_∞ performances index γ_d and γ_f, if there exist positive definite matrices $\bar{P}_{1i}$, $\bar{P}_{1j}$, Hermitian matrices $\bar{P}_{2i}$, $\bar{P}_{3i}$, $\bar{Q}_{1i} > 0$, $\bar{Q}_{2i} > 0$, and matrices $\bar{S}_i$, $\bar{Y}_i$, for all $i, j \in \{1, 2, \ldots, N\}$, which satisfy*

$$\begin{bmatrix} \bar{P}_{1j} - \bar{S}_i - \bar{S}_i^T & \bar{S}_i \bar{A}_i - \bar{Y}_i \bar{C}_i \\ * & -\bar{P}_{1i} \end{bmatrix} < 0 \tag{6.41}$$

$$\begin{bmatrix} -\bar{P}_{2i} & \bar{Q}_{1i} - \bar{S}_i^T & 0 & 0 \\ * & \Phi_{li} & \bar{Y}_i D_{2i} - \bar{S}_i \bar{D}_{1i} & \bar{I}^T \\ * & * & -\gamma_d^2 I & 0 \\ * & * & * & -I \end{bmatrix} < 0 \tag{6.42}$$

for the low-frequency disturbance $|\vartheta_d| \leq \vartheta_{dl}$;

$$\begin{bmatrix} -\bar{P}_{2i} & e^{j\vartheta_{dc}} \bar{Q}_{1i} - \bar{S}_i^T & 0 & 0 \\ * & \Phi_{mi} & \bar{Y}_i D_{2i} - \bar{S}_i \bar{D}_{1i} & \bar{I}^T \\ * & * & -\gamma_d^2 I & 0 \\ * & * & * & -I \end{bmatrix} < 0 \tag{6.43}$$

for the middle-frequency disturbance $\vartheta_{dm1} \leq \vartheta_d \leq \vartheta_{dm2}$;

$$\begin{bmatrix} -\bar{P}_{2i} & -\bar{Q}_{1i} - \bar{S}_i^T & 0 & 0 \\ * & \Phi_{hi} & \bar{Y}_i D_{2i} - \bar{S}_i \bar{D}_{1i} & \bar{I}^T \\ * & * & -\gamma_d^2 I & 0 \\ * & * & * & -I \end{bmatrix} < 0 \tag{6.44}$$

for the high-frequency disturbance $|\vartheta_d| \geq \vartheta_{dh}$;
and

$$\begin{bmatrix} -\bar{P}_{3i} & \bar{Q}_{2i} - \bar{S}_i^T & 0 & 0 \\ * & \Psi_{li} & \bar{Y}_i D_{2i} - \bar{S}_i \bar{D}_{1i} & \bar{I}^T \\ * & * & -\gamma_f^2 I & 0 \\ * & * & * & -I \end{bmatrix} < 0 \tag{6.45}$$

for the low-frequency fault $|\vartheta_f| \leq \vartheta_{fl}$;

$$\begin{bmatrix} -\bar{P}_{3i} & e^{j\vartheta_{fc}}\bar{Q}_{2i} - \bar{S}_i^T & 0 & 0 \\ * & \Psi_{mi} & \bar{Y}_i D_{2i} - \bar{S}_i \bar{D}_{1i} & \bar{I}^T \\ * & * & -\gamma_f^2 I & 0 \\ * & * & * & -I \end{bmatrix} < 0 \tag{6.46}$$

for the middle-frequency fault $\vartheta_{fm1} \le \vartheta_f \le \vartheta_{fm2}$;

$$\begin{bmatrix} -\bar{P}_{3i} & -\bar{Q}_{2i} - \bar{S}_i^T & 0 & 0 \\ * & \Psi_{hi} & \bar{Y}_i D_{2i} - \bar{S}_i \bar{D}_{1i} & \bar{I}^T \\ * & * & -\gamma_f^2 I & 0 \\ * & * & * & -I \end{bmatrix} < 0 \tag{6.47}$$

for the high-frequency fault $|\vartheta_f| \ge \vartheta_{fh}$;
where

$$\begin{aligned}
\Phi_{li} &= \bar{S}_i\bar{A}_i - \bar{Y}_i\bar{C}_i + (\bar{S}_i\bar{A}_i - \bar{Y}_i\bar{C}_i)^T + \bar{P}_{2i} - 2cos(\vartheta_{dl})\bar{Q}_{1i};\\
\Phi_{mi} &= \bar{S}_i\bar{A}_i - \bar{Y}_i\bar{C}_i + (\bar{S}_i\bar{A}_i - \bar{Y}_i\bar{C}_i)^T + \bar{P}_{2i} - 2cos(\vartheta_{dw})\bar{Q}_{1i};\\
\Phi_{hi} &= \bar{S}_i\bar{A}_i - \bar{Y}_i\bar{C}_i + (\bar{S}_i\bar{A}_i - \bar{Y}_i\bar{C}_i)^T + \bar{P}_{2i} + 2cos(\vartheta_{dh})\bar{Q}_{1i};\\
\vartheta_{dc} &= \frac{\vartheta_{dm1} + \vartheta_{dm2}}{2}, \vartheta_{dw} = \frac{\vartheta_{dm1} - \vartheta_{dm2}}{2};\\
\Psi_{li} &= \bar{S}_i\bar{A}_i - \bar{Y}_i\bar{C}_i + (\bar{S}_i\bar{A}_i - \bar{Y}_i\bar{C}_i)^T + \bar{P}_{3i} - 2cos(\vartheta_{fl})\bar{Q}_{2i};\\
\Psi_{mi} &= \bar{S}_i\bar{A}_i - \bar{Y}_i\bar{C}_i + (\bar{S}_i\bar{A}_i - \bar{Y}_i\bar{C}_i)^T + \bar{P}_{3i} - 2cos(\vartheta_{fw})\bar{Q}_{2i};\\
\Psi_{hi} &= \bar{S}_i\bar{A}_i - \bar{Y}_i\bar{C}_i + (\bar{S}_i\bar{A}_i - \bar{Y}_i\bar{C}_i)^T + \bar{P}_{3i} + 2cos(\vartheta_{fh})\bar{Q}_{2i};\\
\vartheta_{fc} &= \frac{\vartheta_{fm1} + \vartheta_{fm2}}{2}, \vartheta_{fw} = \frac{\vartheta_{fm1} - \vartheta_{fm2}}{2}.
\end{aligned}$$

Furthermore, if the above conditions are satisfied, the FE observer parameters in (6.29), are given by $\bar{L}_i = \bar{S}_i^{-1}\bar{Y}_i$.

Proof From $\bar{L}_i = \bar{S}_i^{-1}\bar{Y}_i$, one can get that $\bar{Y}_i = \bar{S}_i\bar{L}_i$. Substituting it into (6.41), implies

$$\begin{bmatrix} \bar{P}_{1j} - \bar{S}_i - \bar{S}_i^T & \bar{S}_i(\bar{A}_i - \bar{L}_i\bar{C}_i) \\ * & -\bar{P}_{1i} \end{bmatrix} < 0 \tag{6.48}$$

which is equivalent to

$$\begin{bmatrix} \bar{P}_{1j} & 0 \\ 0 & -\bar{P}_{1i} \end{bmatrix} + \begin{bmatrix} \bar{S}_i \\ 0 \end{bmatrix} \begin{bmatrix} -I & \bar{A}_i - \bar{L}_i\bar{C}_i \end{bmatrix} + \begin{bmatrix} -I \\ \bar{A}_i^T - \bar{C}_i^T\bar{L}_i^T \end{bmatrix} \begin{bmatrix} \bar{S}_i^T & 0 \end{bmatrix} < 0 \tag{6.49}$$

By the projection Lemma, if we define

$$\Sigma = \begin{bmatrix} \bar{P}_{1j} & 0 \\ 0 & -\bar{P}_{1i} \end{bmatrix}, \Lambda = \begin{bmatrix} -I & \bar{A}_i - \bar{L}_i\bar{C}_i \end{bmatrix}, \Theta = I, \Gamma = \begin{bmatrix} \bar{S}_i^T & 0 \end{bmatrix}, \tag{6.50}$$

and select

$$\mathscr{N}_\Lambda = \begin{bmatrix} \bar{A}_i - \bar{L}_i\bar{C}_i \\ I \end{bmatrix}, \tag{6.51}$$

then one can get from (6.49) that

$$\begin{aligned} &\mathscr{N}_\Lambda^T \Sigma \mathscr{N}_\Lambda \\ &= \begin{bmatrix} (\bar{A}_i - \bar{L}_i\bar{C}_i)^T & I \end{bmatrix} \begin{bmatrix} \bar{P}_{1j} & 0 \\ 0 & -\bar{P}_{1i} \end{bmatrix} \begin{bmatrix} \bar{A}_i - \bar{L}_i\bar{C}_i \\ I \end{bmatrix} \\ &= (\bar{A}_i - \bar{L}_i\bar{C}_i)^T \bar{P}_{1j}(\bar{A}_i - \bar{L}_i\bar{C}_i) - \bar{P}_{1i} < 0 \end{aligned} \tag{6.52}$$

For the system (6.30) with $\omega(t) = 0$, one has

$$\bar{e}(t+1) = \tilde{A}_{\sigma(t)}\bar{e}(t) \tag{6.53}$$

For system (6.53), we construct a switched Lyapunov function as

$$V(t, \bar{e}(t)) = \bar{e}^T(t)\bar{P}_{1\sigma(t)}\bar{e}(t) \tag{6.54}$$

Then by Lemma 6.3 and the condition (6.52), one can get that the system (6.53), is asymptotically stable.

In the following, we will consider the finite-frequency H_∞ performance. By Schur complement, (6.42) can be expressed as

$$\begin{bmatrix} -\bar{P}_{2i} & \bar{Q}_{1i} - \bar{S}_i^T & 0 \\ * & \Pi_i & \bar{S}_i(\bar{L}_i D_{2i} - \bar{D}_{1i}) \\ * & * & -\gamma_d^2 I \end{bmatrix} < 0 \tag{6.55}$$

where $\Pi_i = \bar{S}_i(\bar{A}_i - \bar{L}_i\bar{C}_i) + (\bar{A}_i - \bar{L}_i\bar{C}_i)^T\bar{S}_i^T + \bar{P}_{2i} - 2cos(\vartheta_{dl})\bar{Q}_{1i} + \bar{I}^T\bar{I}$. Then, (6.55) can be rewritten as

$$\begin{aligned} &\begin{bmatrix} -\bar{P}_{2i} & \bar{Q}_{1i} & 0 \\ * & \bar{P}_{2i} - 2cos(\vartheta_{dl})\bar{Q}_{1i} + \bar{I}^T\bar{I} & 0 \\ * & * & -\gamma_d^2 I \end{bmatrix} \\ &+ \begin{bmatrix} 0 \\ \bar{S}_i \\ 0 \end{bmatrix} \begin{bmatrix} -I & \bar{A}_i - \bar{L}_i\bar{C}_i & \bar{L}_i D_{2i} - \bar{D}_{1i} \end{bmatrix} \\ &+ \begin{bmatrix} -I \\ (\bar{A}_i - \bar{L}_i\bar{C}_i)^T \\ (\bar{L}_i D_{2i} - \bar{D}_{1i})^T \end{bmatrix} \begin{bmatrix} 0 & \bar{S}_i^T & 0 \end{bmatrix} < 0 \end{aligned} \tag{6.56}$$

By the projection Lemma, if we define

$$\Sigma = \begin{bmatrix} \bar{P}_{2i} & \bar{Q}_{1i} & 0 \\ \bar{Q}_{1i} & \bar{P}_{2i} - 2cos(\vartheta_{dl})\bar{Q}_{1i} + \bar{I}^T\bar{I} & 0 \\ 0 & 0 & -\gamma_d^2 I \end{bmatrix}, \quad \Theta = I,$$
$$\Lambda = \begin{bmatrix} -I \ \bar{A}_i - \bar{L}_i\bar{C}_i \ \bar{L}_i D_{2i} - \bar{D}_{1i} \end{bmatrix}, \quad \Gamma = \begin{bmatrix} 0 \ \bar{S}_i^T \ 0 \end{bmatrix}, \tag{6.57}$$

and select

$$\mathcal{N}_\Lambda = \begin{bmatrix} \bar{A}_i - \bar{L}_i\bar{C}_i & \bar{L}_i D_{2i} - \bar{D}_{1i} \\ I & 0 \\ 0 & I \end{bmatrix}, \tag{6.58}$$

then one can get from (6.56) that

$$\begin{aligned} &\mathcal{N}_\Lambda^T \Sigma \mathcal{N}_\Lambda \\ &= \begin{bmatrix} \bar{A}_i - \bar{L}_i\bar{C}_i & \bar{L}_i D_{2i} - \bar{D}_{1i} \\ I & 0 \\ 0 & I \end{bmatrix}^T \begin{bmatrix} -\bar{P}_{2i} & \bar{Q}_{1i} & 0 \\ \bar{Q}_{1i} & \bar{P}_{2i} - 2cos(\vartheta_{dl})\bar{Q}_{1i} + \bar{I}^T\bar{I} & 0 \\ 0 & 0 & -\gamma_d^2 I \end{bmatrix} \\ &\times \begin{bmatrix} \bar{A}_i - \bar{L}_i\bar{C}_i & \bar{L}_i D_{2i} - \bar{D}_{1i} \\ I & 0 \\ 0 & I \end{bmatrix} = \begin{bmatrix} \Omega_{11} & \Omega_{12} \\ \Omega_{12}^T & \Omega_{22} \end{bmatrix} < 0 \end{aligned} \tag{6.59}$$

where

$$\begin{aligned} \Omega_{11} &= -(\bar{A}_i - \bar{L}_i\bar{C}_i)^T \bar{P}_{2i}(\bar{A}_i - \bar{L}_i\bar{C}_i) + \bar{Q}_{1i}(\bar{A}_i - \bar{L}_i\bar{C}_i) + (\bar{A}_i - \bar{L}_i\bar{C}_i)^T \bar{Q}_{1i} \\ &\quad -2cos(\vartheta_{dl})\bar{Q}_{1i} + \bar{P}_{2i} + \bar{I}^T\bar{I}; \\ \Omega_{12} &= -(\bar{A}_i - \bar{L}_i\bar{C}_i)^T \bar{P}_{2i}(\bar{L}_i D_{2i} - \bar{D}_{1i}) + \bar{Q}_{1i}(\bar{L}_i D_{2i} - \bar{D}_{1i}); \\ \Omega_{22} &= -(\bar{L}_i D_{2i} - \bar{D}_{1i})^T \bar{P}_{2i}(\bar{L}_i D_{2i} - \bar{D}_{1i}) - \gamma_d^2 I. \end{aligned}$$

Therefore, if (6.42) holds, then we can deduce that (6.59) holds, which further implies that (6.34) holds. By Lemma 1, one can obtain that the system (6.30), satisfies the finite-frequency H_∞ performance index γ_d.

The proofs of the conditions in (6.43) and (6.44), can be deduced by setting

$$\Sigma = \begin{bmatrix} -\bar{P}_{2i} & e^{\mathrm{j}\vartheta_{dc}}\bar{Q}_{1i} & 0 \\ e^{-\mathrm{j}\vartheta_{dc}}\bar{Q}_{1i} & \bar{P}_{2i} - 2cos(\vartheta_{dm})\bar{Q}_{1i} + \bar{I}^T\bar{I} & 0 \\ 0 & 0 & -\gamma_d^2 I \end{bmatrix} \tag{6.60}$$

and

$$\Sigma = \begin{bmatrix} -\bar{P}_{2i} & -\bar{Q}_{1i} & 0 \\ -\bar{Q}_{1i} & \bar{P}_{2i} + 2cos(\vartheta_{dh})\bar{Q}_{1i} + \bar{I}^T\bar{I} & 0 \\ 0 & 0 & -\gamma_d^2 I \end{bmatrix} \tag{6.61}$$

By following the same proof lines, (6.45), (6.46), and (6.47) can be proven. Then the theorem is concluded. (□)

Remark 6.5 In this section, switched Lyapunov function is utilized to guarantee the asymptotical stability of the discrete-time switched systems. Compared with common Lyapunov function, switched Lyapunov function is less conservative and easy to design. However, it also causes difficult to obtain an efficient and necessary conclusion.

6.2.4 An Illustrative Example

Consider the discrete-time switched system (6.28), which consists of two subsystems, and the following parameters:

$$
\begin{aligned}
A_1 &= \begin{bmatrix} -1.2 & 0.08 \\ 0.1 & -0.03 \end{bmatrix}, B_1 = \begin{bmatrix} 0.1 \\ 0.1 \end{bmatrix}, D_{11} = \begin{bmatrix} 0.5 \\ 0.5 \end{bmatrix}, \\
C_1 &= \begin{bmatrix} 1 & 1 \end{bmatrix}, D_{21} = 0.05; \\
A_2 &= \begin{bmatrix} 0.085 & -0.02 \\ 0 & 1.3 \end{bmatrix}, B_2 = \begin{bmatrix} 0.1 \\ 0.1 \end{bmatrix}, D_{12} = \begin{bmatrix} 0.5 \\ 0.5 \end{bmatrix}, \\
C_2 &= \begin{bmatrix} 1 & 1 \end{bmatrix}, D_{22} = 0.5.
\end{aligned}
$$

Solving the conditions in Theorem 6.3 with the H_∞ performance index $\gamma_d = 1.0$ and $\gamma_f = 0.8$, one can get a set of solution as follows:

$$
P_{11} = \begin{bmatrix} 1.5917 & -0.2779 & -0.0445 \\ -0.2779 & 0.6844 & -0.0078 \\ -0.0445 & -0.0078 & 1.7395 \end{bmatrix}, P_{12} = \begin{bmatrix} 0.8046 & -0.1818 & -0.0324 \\ -0.1818 & 1.2718 & -0.0072 \\ -0.0324 & -0.0072 & 1.9106 \end{bmatrix},
$$

$$
P_{21} = \begin{bmatrix} 0.0248 & -0.2366 & 0.0132 \\ -0.2366 & 0.7566 & 0.0380 \\ 0.0132 & 0.0380 & 2.4505 \end{bmatrix}, P_{22} = \begin{bmatrix} 1.2098 & -0.6422 & -0.0049 \\ -0.6422 & 0.7708 & 0.0087 \\ -0.0049 & 0.0087 & 2.1800 \end{bmatrix},
$$

$$
P_{31} = \begin{bmatrix} 0.0135 & -0.2571 & 0.0129 \\ -0.2571 & 0.7197 & 0.0376 \\ 0.0129 & 0.0376 & 2.4505 \end{bmatrix}, P_{32} = \begin{bmatrix} 0.5869 & -0.3720 & -0.0179 \\ -0.3720 & 0.1103 & 0.0071 \\ -0.0179 & 0.0071 & 1.6895 \end{bmatrix},
$$

$$
Q_{11} = \begin{bmatrix} -0.8146 & -0.0079 & 0.0203 \\ -0.0079 & -0.3841 & 0.0324 \\ 0.0203 & 0.0324 & 0.0436 \end{bmatrix}, Q_{12} = \begin{bmatrix} -0.1303 & -0.4476 & 0.0261 \\ -0.4476 & -0.0456 & 0.0104 \\ 0.0261 & 0.0104 & -0.0509 \end{bmatrix},
$$

$$
Q_{21} = \begin{bmatrix} -0.8146 & -0.0079 & 0.0203 \\ -0.0079 & -0.3841 & 0.0324 \\ 0.0203 & 0.0324 & 0.0436 \end{bmatrix}, Q_{22} = \begin{bmatrix} -0.1303 & -0.4476 & 0.0261 \\ -0.4476 & 0.0456 & 0.0104 \\ 0.0261 & 0.0104 & -0.0509 \end{bmatrix},
$$

$$S_1 = \begin{bmatrix} 0.8113 & -0.1410 & -0.0069 \\ -0.1410 & 1.1968 & 0.0080 \\ -0.0069 & 0.0080 & 1.9041 \end{bmatrix}, S_2 = \begin{bmatrix} 1.3951 & -0.2051 & -0.0166 \\ -0.2051 & 0.8527 & -0.0049 \\ -0.0166 & -0.0049 & 1.7298 \end{bmatrix},$$

$$Y_1 = \begin{bmatrix} -0.2252 \\ 0.2203 \\ -0.0051 \end{bmatrix}, Y_2 = \begin{bmatrix} 0.1358 \\ 0.6996 \\ -0.0036 \end{bmatrix}.$$

The actuator FE observer parameters are calculated as

$$L_1 = \begin{bmatrix} -0.2508 \\ 0.1546 \\ -0.0043 \end{bmatrix}, \quad L_2 = \begin{bmatrix} 0.2260 \\ 0.8748 \\ 0.0026 \end{bmatrix}.$$

To illustrate the efficiency of the FE, the disturbance is simulated as a white noise in high-frequency domain, i.e., $\vartheta_{dh} = \frac{5\pi}{6}$, as shown in Fig. 6.1. The actuator fault is assumed to be in a low-frequency domain, i.e., $\vartheta_{fl} = \frac{\pi}{6}$, and two cases are considered. Firstly, a piecewise constant fault case is considered

$$f_1(t) = \begin{cases} 0, & 0s \le t < 20s; \\ 10, & 20s \le t < 300s; \\ -10, & 300s \le t \le 600s. \end{cases}$$

The simulation result is shown in Fig. 6.2, one can get that the FE observer gives an accurate estimation of the actuator fault. Then, the actuator fault is considered as a time-varying fault case, and it is set as

$$f_2(t) = \begin{cases} 0, & 0s \le t < 20s; \\ 0.5sin(\frac{\pi}{10}t) + 1, & 20s \le t < 300s; \\ -0.5sin(\frac{\pi}{10}t) - 1, & 300s \le t \le 600s. \end{cases}$$

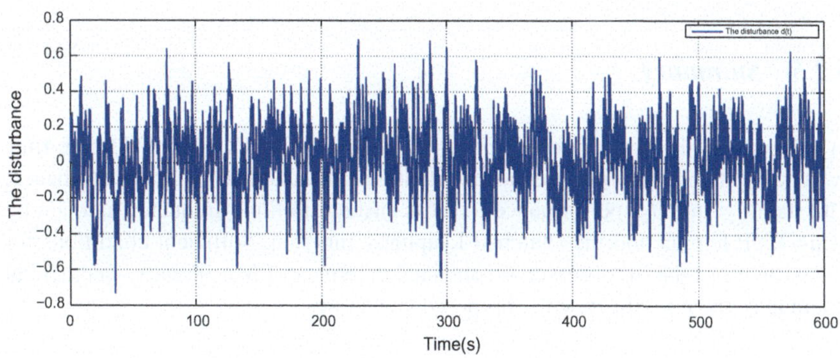

Fig. 6.1 The output disturbance $d(t)$

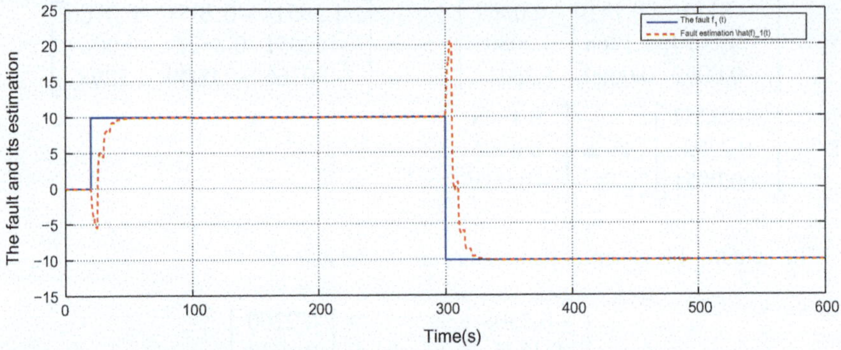

Fig. 6.2 Actuator fault $f(t)$ (solid line) and its estimation $\hat{f}(t)$ (dashed line)

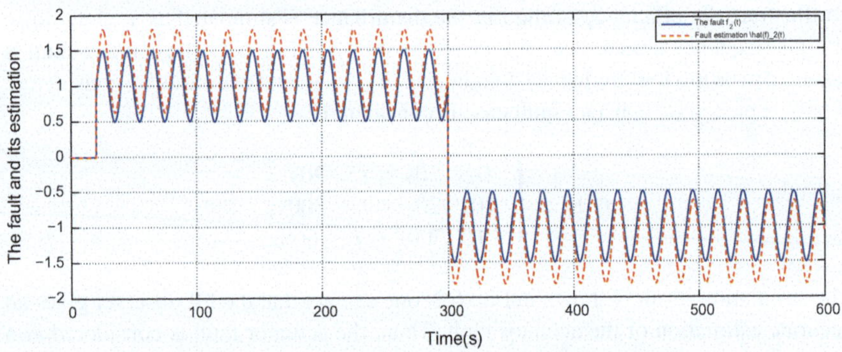

Fig. 6.3 Actuator fault $f_2(t)$ (solid line) and its estimation $\hat{f}_2(t)$ (dashed line)

The simulation result is shown in Fig. 6.3, one can get that the FE observer can also give a good FE performance.

6.2.5 Summary

In this section, the problem of FE observer design is investigated for discrete-time switched systems with finite-frequency actuator fault and unknown input disturbance. The stability and H_∞ performance index is ensured in finite-frequency domain by using KYP lemma and the switched Lyapunov function. Sufficient conditions for the existence of the FE observer are obtained in terms of LMIs. Finally, a numerical example is given to illustrate the proposed technique.

References

1. Basseville, M.: Detecting changes in signals and systems-a survey. Automatica **24**(3), 309–326 (1988)
2. Patton, R.J.: Robustness in model-based fault diagnosis: the 1995 situation. Annu. Rev. Control **21**, 103–123 (1997)
3. Bodson, M., Groszkiewicz, J.: Multivariable adaptive algorithms for reconfigurable flight control. IEEE Trans. Control Syst. Technol. **5**(2), 217–229 (1997)
4. Zuo, Z., Daniel, W.C.H., Wang, Y.: FTC for singular systems with actuator saturation and nonlinear perturbation. Automatica **46**(3), 569–576 (2010)
5. Gao, Z., Antsaklis, P.J.: Reconfigurable control system design via perfect model following. Int. J. Control **56**(4), 783–798 (1992)
6. Yang, G., Ye, D.: Reliable H_∞ control of linear systems with adaptive mechanism. IEEE Trans. Autom. Control **55**(1), 242–247 (2010)
7. Zhang, K., Jiang, B., Shi, P.: Fast fault estimation and accommodation for dynamical systems. IET Control Theory Appl. **3**(2), 189–199 (2009)
8. Zhong, M., Li, S., Zhao, Y.: Robust H_∞ fault detection for uncertain LDTV systems using Krein space approach. Int. J. Innov. Comput. Inf. Control **9**(4), 1637–1649 (2013)
9. Thumati, B.T., Jagannathan, S.: A model based fault detection and prediction scheme for nonlinear multivariable discrete-time systems with asymptotic stability guarantees. IEEE Trans. Neural Netw. **21**(3), 404–423 (2010)
10. Zheng, Y., Fang, H., Wang, H.: Takagi-Sugeno fuzzy-model-based fault detection for networked control systems with Markov delays. IEEE Trans. Syst. Man Cybern. **36**(4), 924–929 (2006)
11. Jagannathan, S.: Neural Network Control of Nonlinear Discrete-Time Systems. CRC, New York, NY, USA (2006)
12. Fekih, A.: Fault diagnosis and FTC design for aerospace systems: a bibliographical review. In: 2014 American Control Conference, June 4–6, Portland, American, pp. 1286–1291 (2014)
13. Song, Y., Wang, B.: Survey on reliability of power electronic systems. IEEE Trans. Power Electron. **28**(1), 591–604 (2013)
14. Yang, Y., Ding, S.X., Li, L.: On observer-based fault detection for nonlinear systems. Syst. Control Lett. **82**, 18–25 (2015)
15. Xie, X., Yue, D., Hu, S.: Fault estimation observer design of discrete-time nonlinear systems via a joint real-time scheduling law. IEEE Trans. Syst. Man Cybern. Syst. **47**(7), 1451–1463 (2017)
16. Gao, Z., Cecati, C., Ding, S.: A survey of fault diagnosis and fault-tolerant techniques-part I: fault diagnosis with model-based and signal-based approaches. IEEE Trans. Ind. Electron. **62**(6), 3757–3767 (2015)
17. Wang, X., Tan, C., Zhou, D.: A novel sliding mode observer for state and fault estimation in systems not satisfying matching and minimum phase conditions. Automatica **79**, 290–295 (2017)
18. Li, X., Lu, D., Zeng, G., Liu, J., Zhang, W.: Integrated fault estimation and non-fragile fault-tolerant control design for uncertain Takagi-Sugeno fuzzy systems with actuator fault and sensor fault. IET Control Theory Appl. **11**(10), 1542–1553 (2017)
19. Perdon, A.M., Zattoni, E., Conte, G.: Model matching with strong stability in switched linear systems. Syst. Control Lett. **97**, 98–107 (2016)
20. Zhao, X., Yin, S., Li, H., Niu, B.: Switching stabilization for a class of slowly switched systems. IEEE Trans. Autom. Control **60**(1), 221–226 (2015)
21. Zhao, X., Yin, Y., Niu, B., Zheng, X.: Stabilization for a class of switched nonlinear systems with novel average dwell time switching by T-S fuzzy modeling. IEEE Trans. Cybern. **46**(8), 1952–1957 (2016)
22. Yang, H., Jiang, B., Cocquempot, V., Chen, M.: Spacecraft formation stabilization and fault tolerance: a state-varying switched system approach. Syst. Control Lett. **62**(9), 715–722 (2013)
23. Wang, D., Wang, W., Shi, P.: Robust fault detection for switched linear systems with state delays. IEEE Trans. Syst. Man Cybern. B: Cybern. **39**(3), 800–805 (2009)

24. Wang, D., Shi, P., Wang, W.: Robust fault detection for continuous-time switched delay systems: an linear matrix inequality approach. IET Control Theory Appl. **4**(1), 100–108 (2010)
25. Du, D., Jiang, B., Shi, P.: Sensor fault estimation and compensation for time-delay switched systems. Int. J. Syst. Sci. **43**(4), 629–640 (2012)
26. Du, D., Jiang, B., Shi, P.: Sensor fault estimation and accommodation for discrete-time switched linear systems. IET Control Theory Appl. **8**(11), 960–967 (2014)
27. Du, D., Jiang, B., Shi, P.: Active fault-tolerant control for switched systems with time delay. Int. J. Adapt. Control Signal Process. **25**(5), 466–480 (2011)
28. Du, D., Jiang, B.: Actuator fault estimation and accommodation for switched systems with time delay: discrete-time case. ISA Trans. **62**, 137–144 (2016)
29. Yang, H., Jiang, B., Staroswiecki, M.: Observer-based fault-tolerant control for a class of switched nonlinear systems. IET Control Theory Appl. **1**(5), 523–1532 (2007)
30. Ding, D., Yang, G.: Finite frequency H_∞ filtering for uncertain discrete-time switched linear systems. Progress Nat. Sci. **19**, 1625–1633 (2009)
31. Zhao, X., Li, J., Ye, D.: Fault detection for switched systems with finite-frequency specifications. Nonlinear Dyn. **70**(1), 409–420 (2012)
32. Yang, H., Xia, Y., Liu, B.: Fault detection for T-S fuzzy discrete systems in finite-frequency. IEEE Trans. Syst. Man Cybern. Part B-Cybern. **41**(4), 911–920 (2011)
33. Zhang, K., Jiang, B., Shi, P., Xu, J.: Analysis and design of robust H_∞ fault estimation observer with finite-frequency specifications for discrete-time fuzzy systems. IEEE Trans. Fuzzy Syst. **45**(7), 1225–1235 (2015)
34. Gahinet, P., Apkarian, P.: A linear matrix inequality approach to H_∞ control. Int. J. Robust Nonlinear Control **4**(4), 421–448 (1994)
35. Daafouz, J., Riedinger, P., Iung, C.: Stability analysis and control synthesis for switched systems: a switched Lyapunov function approach. IEEE Trans. Autom. Control **47**(11), 1883–1887 (2002)

Chapter 7
UIO-Based Fault Diagnosis for Continuous-Time Switched Systems

7.1 UIO-Based FD for Continuous-Time Switched Systems

In this section, the problem of FD is investigated for continuous-time switched systems. A UIO is designed as an FD filter to generate the residual signal. To make the residual signal sensitive to the fault and robust to the unknown disturbances, we proposed H_-/H_∞ designed method by employing the switched Lyapunov function and the ADT techniques. Furthermore, sufficient conditions for the existence of such an FD filter are exploited to guarantee that the error systems under ADT condition are asymptotically stable with an H_∞ / H_- performance index. Finally, a practical example is provided and simulation results are conducted to demonstrate the effectiveness of the proposed approach.

7.1.1 Introduction

With the rapid development of industrial technology, the faults may have serious effect on system performance or even cause disastrous damage [1–6]. As an effective way of coping with faults, fault diagnosis have been widely studied in the past decades. FD technology includes three parts: FD, fault isolation, and FE [7]. FD, the first step of FD technique, is to detect if a fault has occurred as soon as possible. Thus, it can be seen that the FD is the most important step, only after the fault has been determined can the isolation and the estimation of the fault be further completed, as well as the basis for the final control decision can be provided. Therefore, FD technology have received extensive attention in academic circles.

Among various FD methods, observer-based FD approach is regarded as an efficient and popular technology, the basic idea is to use state observer or filter to generate a residual signal and based on this to construct a residual evaluation function to com-

D. Du et al., *Observer-Based Fault Diagnosis and Fault-Tolerant Control for Switched Systems*, Studies in Systems, Decision and Control 323,
https://doi.org/10.1007/978-981-15-9073-3_7

pare with a predefined threshold. When the residual evaluation function has a value larger than the threshold, then the fault occurs [8, 9]. Until now, various observer-based approaches have been proposed. For examples, sliding-mode observer [10], adaptive observer [11], PIO [8], and UIO [12], etc. Among them, PIO and adaptive observer are suitable to actuator faults, sliding-mode observer requires that upper bound of the disturbances is known, while the UIO is an effective state observer with good robustness to the unknown disturbance signal and sensitivity to the fault signal, which can detect not only the sensor fault but also the actuator fault and only need to decouple the partial disturbance or attenuating the disturbance that cannot be decoupled completely.

Switched system belongs to a type of hybrid system, which includes several subsystems (continuous or discrete) and a rule that orchestrates the switching among them. Many real engineering systems can be modeled as such systems, such as circuit system, aerospace systems [13], and chemical systems [14], etc. ADT technology, one of the constrained switching signals, requires that the average time of each subsystem running in a limited time period is larger than a constant [15], which is more general and less conservative than classical dwell time. In [12], FD based on UIO for switched discrete-time systems was studied. The problem of actuator FD for discrete-time switched systems with output disturbance was investigated in [16]. In [17], The authors proposed an FD filter for discrete-time switched systems with interval time-varying delays.

From the proposed results for switched systems, most of the researches on the FD are mainly considered for the discrete-time systems, which motivates us to develop the effective FD filters for continuous-time switched systems. Presently, there have some technologies proposed to deal with the above problems. For examples, the FD filter design problems for continuous-time switched system were studied in [18–20]. However, both of the papers mainly considered the robust performance and the fault sensitivity was not given. Moreover, [19] and [18], used the traditional observer that cannot guarantee robustness to the disturbances and sensitivity to the faults very well. In [21], robust control and FD for continuous-time switched systems subject to a dwell-time constraint were studied, while the authors followed the rule of classical dwell-time switching signal to detect faults, which is more conservative and less general than ADT. Therefore, motivated by the problems existing in the obtained achievements, we study an effective UIO-based FD filter for continuous-time switched systems, and followed the ADT switching rule, the H_∞ and H_- performance indices are guaranteed by the design procedure.

The main contributions of this paper can be summarized as follows:

- under ADT switching signal, UIO-based FD filter is designed for continuous-time switched systems in full frequency domain with the same term of the actuator and sensor faults;
- the H_∞ and H_- performance indices are guaranteed by the design procedure to describe the disturbance robustness performance and fault sensitivity performance of the FD filter;

- a practical example on a liquid level control system is provided to illustrate the application of the proposed techniques.

7.1.2 Model Description and Preliminaries

Consider the following continuous-time switched system:

$$\begin{cases} \dot{x}(t) = \sum_{i=1}^{N} \sigma_i(t)(A_i x(t) + B_i u(t) + D_{1i} d(t) + F_{1i} f(t)) \\ y(t) = \sum_{i=1}^{N} \sigma_i(t)(C_i x(t) + D_{2i} d(t) + F_{2i} f(t)) \end{cases} \tag{7.1}$$

where $x(t) \in \mathbb{R}^n, u(t) \in \mathbb{R}^u, y(t) \in \mathbb{R}^m, d(t) \in \mathbb{R}^d$ and $f(t) \in \mathbb{R}^l$ are the state vector, the control input vector, the measured output vector, the unknown input vector, and the fault vector, respectively. A_i, B_i, C_i, D_{1i}, D_{2i}, F_{1i}, and F_{2i} are known constant real matrices with appropriate dimensions.

The piecewise constant function

$$\sigma_i(t) : [0, \infty) \to \{0, 1\}(i \in \mathbb{N})$$

is the switching signal satisfying $\sum_{i=1}^{N} \sigma_i(t) = 1$, which specifies which subsystem is activated at the switching instant.

Definition 7.1 ([22]) For any switching signal $\sigma_i(t)$ and any $t_2 > t_1 > 0$, let $N_{\sigma_i(t_1,t_2)}$ denote the number of switchings $\sigma_i(t)$ on an interval (t_1, t_2). If

$$N_{\sigma_i(t)}(t_1, t_2) \leq N_0 + \frac{t_1 - t_2}{\tau_a}, \tag{7.2}$$

holds for given $N_0 \geq 0$ and $\tau_a > 0$, then the constant τ_a is called the ADT and N_0 is the chattering bound.

Remark 7.1 This is a kind of slowly switching signals, which is less conservative than classical DT switching signal. For simplicity, we chose the chattering bound N_0 is zero in this section.

To detect the fault for switched system (7.1), the following UIO is employed to generate the residual signal:

$$\begin{cases} \dot{z}(t) = \sum_{i=1}^{N} \sigma_i(t)(N_i z(t) + G_i u(t) + L_i y(t)) \\ \hat{x}(t) = z(t) - \sum_{i=1}^{N} \sigma_i(t) E_i y(t) \\ \hat{y}(t) = \sum_{i=1}^{N} \sigma_i(t) C_i \hat{x}(t) \end{cases} \tag{7.3}$$

where $z(t) \in \mathbb{R}^n$ is the state of the observer, $\hat{x}(t) \in \mathbb{R}^n$ represents the state estimation of $x(t)$ and $\hat{y}(t) \in \mathbb{R}^m$ is the filter's output vector. Matrices N_i, G_i, E_i and L_i are the gains of UIO to be designed later.

Remark 7.2 In this section, an UIO-based FD approach for continuous-time switched systems is provided by employing the switched Lyapunov function and the ADT techniques, which is less conservative and more general. In addition, the UIO in the paper can achieve effective FE while performing the FD.

Define

$$e(t) = x(t) - \hat{x}(t), \quad r(t) = y(t) - \hat{y}(t), \tag{7.4}$$

then one can get

$$\begin{cases} e(t) = \sum_{i=1}^{N} \sigma_i(t)(T_i x(t) + E_i D_{2i} d(t) + E_i F_{2i} f(t)) - z(t) \\ r(t) = \sum_{i=1}^{N} \sigma_i(t)(C_i e(t) + D_{2i} d(t) + F_{2i} f(t)) \end{cases} \tag{7.5}$$

where $T_i = I_n + E_i C_i$.

From (7.1), (7.3), and (7.5), the following error system can be derived:

$$\begin{cases} \dot{e}(t) = \sum_{i=1}^{N} \sigma_i(t)((T_i A_i - L_i C_i)e(t) + (L_i C_i E_i - T_i A_i E_i)y(t) + (T_i B_i - G_i)u(t) \\ \qquad +(T_i A_i - N_i - L_i C_i)z(t) + (T_i D_{1i} - L_i D_{2i})d(t) \\ \qquad +(T_i F_{1i} - L_i F_{2i})f(t) + E_i D_{2i} d(t) + E_i F_{2i} f(t)) \\ r(t) = \sum_{i=1}^{N} \sigma_i(t)(C_i e(t) + D_{2i} d(t) + F_{2i} f(t)) \end{cases} \tag{7.6}$$

If the following relationships hold:

$$\begin{cases} (L_i C_i - T_i A_i)E_i = 0, \\ T_i A_i - N_i - L_i C_i = 0, \\ T_i B_i - G_i = 0, \\ E_i(D_{2i}, F_{2i}) = 0. \end{cases} \tag{7.7}$$

Then the state error estimation (7.6), reduces to

$$\begin{cases} \dot{e}(t) = \sum_{i=1}^{N} \sigma_i(t)(N_i e(t) + \bar{D}_i d(t) + \bar{F}_i f(t)) \\ r(t) = \sum_{i=1}^{N} \sigma_i(t)(C_i e(t) + D_{2i} d(t) + F_{2i} f(t)) \end{cases} \tag{7.8}$$

where $\bar{D}_i = T_i D_{1i} - L_i D_{2i}$, $\bar{F}_i = T_i F_{1i} - L_i F_{2i}$.

A desirable FD observer should generate a residual signal that is robust against disturbances $d(t)$ and sensitive to the faults $f(t)$. Therefore, the main objective of this paper is to design UIO (7.3), such that

- The error system (7.8) is stable;
- Under zero initial condition, the generated residual $r(t)$ is robust against unknown disturbance $d(t)$, i.e.

$$\int_0^{\infty} r^T(t) r(t) dt \le \gamma^2 \int_0^{\infty} d^T(t) d(t) dt; \tag{7.9}$$

- The generated residual $r(t)$ is sensitive to the fault $f(t)$, i.e., under zero initial condition, the following H_- fault sensitivity condition holds

$$\int_0^{\infty} r^T(t)r(t)dt \geq \beta^2 \int_0^{\infty} f^T(t)f(t)dt; \tag{7.10}$$

where γ and β are both the positive scalar.

The following lemmas are given

Lemma 7.1 ([23]) *If there exist function $\mathscr{V}$ and $\mathscr{U}$ satisfying*

$$\dot{\mathscr{V}} \leq -\zeta\mathscr{V} + \kappa\mathscr{U}$$

then

$$\mathscr{V} \leq e^{-\zeta(t-t_0)}\mathscr{V}(t_0) + \kappa \int_0^{t-t_0} e^{-\zeta\tau}\mathscr{U}(t-\tau)d\tau.$$

Lemma 7.2 ([24]) *Consider the continuous-time switched system $\dot{x}(t) = f_{\sigma_i(t)}(x(t))$, and let $\alpha > 0$, $\mu > 1$ be given constants. Suppose that there exist $\mathscr{C}^1$ function $V_{\sigma_i(t)} : \mathbb{R}^n \to \mathbb{R}$ and two class $\mathscr{K}_\infty$ function κ_1 and κ_2 such that*

$$\kappa_1(|x(t)|) \leq V_{\sigma_i(t)}(x(t)) \leq \kappa_2(|x(t)|),$$
$$\dot{V}_{\sigma_i(t)}(x(t)) \leq -\alpha V_{\sigma_i(t)}(x(t)),$$

and $\forall(\sigma_i(t_l) = i, \sigma_i(t_l^-) = j) \in \mathbb{N} \times \mathbb{N}, i \neq j$,

$$V_i(x(t)) \leq \mu V_j(x(t)),$$

then the switched system $\dot{x}(t) = f_{\sigma_i(t)}(x(t))$ is GUAS for any switching signal with ADT

$$\tau_a > \tau_a^* = \frac{ln\mu}{\alpha}. \tag{7.11}$$

Remark 7.3 Lemma 7.2 is an important tool for studying switched systems. When the switched subsystem works, condition $\dot{V}_{\sigma_i(t)}(x(t)) \leq -\alpha V_{\sigma_i(t)}(x(t))$ gives the change trend of Lyapunov function, which α represents the attenuation rate of the Lyapunov function of the subsystem. When the system is at the switching time, condition $V_i(x(t)) \leq \mu V_j(x(t))$ gives the relationship between the Lyapunov functions of different subsystems before and after model switching, i.e., The Lyapunov function increased μ times. The larger the values of α and μ, the longer the average dwell time of switched system, so it is easier to achieve stability of the system. In addition, Lemma 7.2 does not define the type of switched systems, so it is applicable to linear or nonlinear systems and other types of switched systems.

7.1.3 Main Results

In this section, an H_∞/H_- FD observer is designed for continuous-time switched system (7.1), to make the generated residual $r(t)$ robust against the disturbances and sensitive to the faults.

7.1.3.1 Robust Performance Analysis

In this subsection, the sufficient conditions of H_∞ performance index are given.

If we set $f(t) = 0$, then the error system (7.8), is transformed to the following form:

$$\begin{cases} \dot{e}(t) = \sum_{i=1}^{N} \sigma_i(t)(N_i e(t) + \bar{D}_i d(t)) \\ r(t) = \sum_{i=1}^{N} \sigma_i(t)(C_i e(t) + D_{2i} d(t)) \end{cases} \tag{7.12}$$

In the following, the theorem is given to ensure that the error system (7.12), is GUAS and has H_∞ performance.

Theorem 7.1 *For given scalars $\alpha > 0$, $\mu \geq 1$ and $\gamma > 0$, the system (7.12) is GUAS with H_∞ performance index γ for any switching signal with ADT if the conditions (7.7), hold and there exist positive definite matrices P_i and P_j, $\forall i, j \in \mathbb{N}$, which satisfy*

$$P_i \leq \mu P_j, \tag{7.13}$$

$$\begin{bmatrix} N_i^T P_i + P_i N_i + \alpha P_i & P_i \bar{D}_i & C_i^T \\ * & -\gamma^2 I & D_{2i}^T \\ * & * & -I \end{bmatrix} < 0, \tag{7.14}$$

Proof Firstly, we establish the stability of system (7.12). Consider the following switched Lyapunov function for the error dynamic (7.12), as

$$V_i(t, e(t)) = e^T(t)\Big(\sum_{i=1}^{N} \sigma_i(t) P_i\Big) e(t). \tag{7.15}$$

Assume $d(t) = 0$, one has

$$\begin{aligned} &\dot{V}_i(t, e(t)) + \alpha V_i(t, e(t)) \\ &= \dot{e}^T(t)\Big(\sum_{i=1}^{N} \sigma_i(t) P_i\Big) e(t) + e^T(t)\Big(\sum_{i=1}^{N} \sigma_i(t) P_i\Big) \dot{e}(t) + e^T(t)\Big(\sum_{i=1}^{N} \sigma_i(t) \alpha P_i\Big) e(t) \\ &= \sum_{i=1}^{N} \sigma_i(t)\Big(e^T(t)(N_i^T P_i + P_i N_i + \alpha P_i) e(t)\Big). \end{aligned} \tag{7.16}$$

One can get from (7.14)

$$\dot{V}_i(t, e(t)) < -\alpha V_i(t, e(t)),$$

according to Lemma 7.2 and (7.13), one can get that the system (7.12), is GUAS.

Next, we give the proof of H_∞ performance γ as follows:

Consider the Lyapunov function as in (7.15), and set $\Gamma(t) = r^T(t)r(t) - \gamma^2 d^T(t)d(t)$, one has

$$\dot{V}_i(t, e(t)) + \alpha V_i(t, e(t)) + \Gamma(t) = \xi^T(t)\Theta\xi(t), \tag{7.17}$$

where

$$\xi(t) = \left[e^T(t)\ d^T(t) \right]^T,$$
$$\Theta = \begin{bmatrix} N_i^T P_i + P_i N_i + \alpha P_i + C_i^T C_i & P_i \bar{D}_i + C_i^T D_{2i} \\ * & -\gamma^2 I + D_{2i}^T D_{2i} \end{bmatrix}.$$

By (7.14), it follows that

$$\dot{V}_i(t, e(t)) \le -\alpha V_i(t, e(t)) - \Gamma(t). \tag{7.18}$$

From Lemma 7.1, one has

$$\begin{aligned}
V_i(t, e(t)) &\le e^{-\alpha(t-t_0)} V_i(t_0, e(t_0)) - \int_{t_0}^{t} e^{-\alpha(t-\tau)} \Gamma(\tau) d\tau \\
&\le \mu V_{\sigma_i(t_{k^-})}(t_{k^-}, e(t_{k^-})) e^{-\alpha(t-t_k)} - \int_{t_k}^{t} e^{-\alpha(t-\tau)} \Gamma(\tau) d\tau \\
&\le \mu [V_{\sigma_i(t_{k-1})}(t_{k-1}, e(t_{k-1})) e^{-\alpha(t_k - t_{k-1})} - \int_{t_{k-1}}^{t_k} e^{-\alpha(t_k-\tau)} \Gamma(\tau) d\tau] e^{-\alpha(t-t_k)} \\
&\quad - \int_{t_k}^{t} e^{-\alpha(t-\tau)} \Gamma(\tau) d\tau \\
&\le \cdots \le \mu^{N_{\sigma_i}(t_0,t)} V_{\sigma_i(t_0)}(t_0, e(t_0)) e^{-\alpha(t-t_0)} \\
&\quad - \mu^{N_{\sigma_i}(t_0,t)} \int_{t_0}^{t_1} e^{-\alpha(t-\tau)} \Gamma(\tau) d\tau - \cdots \\
&\quad - \mu^{N_{\sigma_i}(t_{k-1},t)} \int_{t_{k-1}}^{t} e^{-\alpha(t-\tau)} \Gamma(\tau) d\tau - \mu^{N_{\sigma_i}(t_k,t)} \int_{t_k}^{t} e^{-\alpha(t-\tau)} \Gamma(\tau) d\tau \\
&= e^{-\alpha(t-t_0) + N_{\sigma_i}(t_0,t) ln\mu} V_{\sigma_i}(t_0)(t_0, e(t_0)) \\
&\quad - \int_{t_0}^{t} e^{-\alpha(t-\tau) + N_{\sigma_i}(\tau,t) ln\mu} \Gamma(\tau) d\tau.
\end{aligned} \tag{7.19}$$

When $t_0 = 0$, and under the initial condition $e(0) = 0$, one has

$$\int_0^t e^{-\alpha(t-\tau)+N_{\sigma_i}(\tau,t)ln\mu}\Gamma(\tau)d\tau \le -V_i(t,e(t)). \tag{7.20}$$

Multiplying both sides of (7.20) by $e^{-N_{\sigma_i}(0,t)ln\mu}$, one has

$$\int_0^t e^{-\alpha(t-\tau)-N_{\sigma_i}(0,\tau)ln\mu}\Gamma(\tau)d\tau \le -e^{-N_{\sigma_i}(0,t)ln\mu}V_i(t,e(t)). \tag{7.21}$$

By the conditions (7.2) and (7.11), one has

$$-N_{\sigma_i}(0,t)ln\mu \ge -[N_0+\frac{t-0}{\tau_a}]ln\mu \ge -[(t-0)\frac{\alpha}{ln\mu}]ln\mu \quad \ge -\alpha t.$$

Then one can get

$$\begin{aligned} -e^{-\alpha t}V_i(t,e(t)) &\ge -e^{-N_{\sigma_i}(0,t)ln\mu}V_i(t,e(t)) \ge \int_0^t e^{-\alpha(t-\tau)-N_{\sigma_i}(0,\tau)ln\mu}\Gamma(\tau)d\tau \\ &\ge \int_0^t e^{-\alpha(t-\tau)-\alpha\tau}\Gamma(\tau)d\tau \ge \int_0^t e^{-\alpha t}\Gamma(\tau)d\tau \\ &\ge \int_0^t \Gamma(\tau)d\tau, \end{aligned} \tag{7.22}$$

which means

$$\int_0^t r^T(\tau)r(\tau)d\tau \le \gamma^2\int_0^t d^T(\tau)d(\tau)d\tau, \tag{7.23}$$

when $t\to\infty$, then

$$\int_0^\infty r^T(\tau)r(\tau)d\tau \le \gamma^2\int_0^\infty d^T(\tau)d(\tau)d\tau, \tag{7.24}$$

then one can conclude that the error system (7.12) has H_∞ performance γ. The proof is completed. (□)

7.1.3.2 Sensitive Performance Analysis

If we only consider the effect of the fault, i.e., we set $d(t)=0$, the error dynamic (7.8) is rewritten as

$$\begin{cases} \dot{e}(t)=\sum_{i=1}^N\sigma_i(t)(N_ie(t)+\bar{F}_if(t)) \\ r(t)=\sum_{i=1}^N\sigma_i(t)(C_ie(t)+F_{2i}f(t)) \end{cases} \tag{7.25}$$

The following theorem will provide the conditions for the error system (7.25), satisfying the H_- performance index.

Theorem 7.2 *For given scalars $\alpha > 0$, $\mu \geq 1$ and $\beta > 0$, under the ADT switching signal, the system (7.25) is GUAS with H_- performance index β if the conditions (7.7)–(7.7) hold and there exist positive definite matrices P_i and P_j, $\forall i, j \in \mathbb{N}$, which satisfy*

$$P_i \leq \mu P_j, \tag{7.26}$$

$$\begin{bmatrix} N_i^T P_i + P_i N_i + \alpha P_i - C_i^T C_i & P_i \bar{F}_i & 0 \\ * & -F_{2i}^T F_{2i} & \beta I \\ * & * & -I \end{bmatrix} < 0, \tag{7.27}$$

Proof By using Lemmas 7.1 and 7.2, and following the same proof lines as in Theorem 7.1, the condition (7.27), can be easily deduced. The detailed proof is thus omitted for the brevity. (□)

7.1.3.3 UIO Design

Theorem 7.3 *For given scalars $\alpha > 0$, $\mu \geq 1$ and $\gamma > 0$, the system (7.12), is GUAS with H_∞ performance index γ for any switching signal with ADT, if the conditions in (7.7), hold and there exist positive definite matrices P_i and P_j, matrices Q_i, $\forall i, j \in \mathbb{N}$, which satisfy*

$$P_i \leq \mu P_j, \tag{7.28}$$

$$E_i C_i D_{1i} = -D_{1i}, \tag{7.29}$$

$$\begin{bmatrix} \hat{A}_i^T P_i + P_i \hat{A}_i - C_i^T Q_i^T - Q_i C_i + \alpha P_i + I & P_i \bar{D}_i \\ * & -\gamma^2 I \end{bmatrix} < 0, \tag{7.30}$$

where $\hat{A}_i = T_i A_i$, $\bar{D}_i = T_i D_{1i} - L_i D_{2i}$, and $L_i = P_i^{-1} Q_i$.

Proof Substituting the conditions $\hat{A}_i$, $\bar{D}_i$, and L_i to (7.30), one can get that (7.30), is equivalent to (7.14). (□)

Theorem 7.4 *For given scalars $\alpha > 0$, $\mu \geq 1$ and $\beta > 0$, the system (7.25), is GUAS with H_- performance index β for any switching signal with ADT, if the conditions in (7.7), hold and there exist positive definite matrices P_i and P_j, matrices Q_i, $\forall i, j \in \mathbb{N}$, which satisfy*

$$P_i \leq \mu P_j, \tag{7.31}$$

$$E_i C_i D_{1i} = -D_{1i}, \tag{7.32}$$

$$\begin{bmatrix} \hat{A}_i^T P_i + P_i \hat{A}_i - C_i^T Q_i^T - Q_i C_i + \alpha P_i + I - C_i^T C_i & P_i \bar{F}_i \\ * & \beta^2 I - F_{2i}^T F_{2i} \end{bmatrix} < 0, \tag{7.33}$$

where $\hat{A}_i = T_i A_i$, $\bar{F}_i = T_i F_{1i} - L_i F_{2i}$, *and* $L_i = P_i^{-1} Q_i$.

Proof Substituting the conditions $\hat{A}_i$, $\bar{F}_i$ and L_i to (7.33), one can get that (7.33), is equivalent to (7.27). (□)

Theorem 7.5 *For given scalars* $\alpha > 0$, $\mu \geq 1$, $\beta > 0$, *and* $\gamma > 0$, *the system (7.25), is GUAS with* H_- *performance index* β *and* H_∞ *performance index* γ *for any switching signal with ADT, if the conditions in (7.7), hold and there exist positive definite matrices* P_i *and* P_j, *matrices* Q_i, $\forall i, j \in \mathbb{N}$, *which satisfy*

$$P_i \leq \mu P_j, \tag{7.34}$$

$$E_i C_i D_{1i} = -D_{1i}, \tag{7.35}$$

$$\begin{bmatrix} \hat{A}_i^T P_i + P_i \hat{A}_i - C_i^T Q_i^T - Q_i C_i + \alpha P_i + I & P_i \bar{D}_i \\ * & -\gamma^2 I \end{bmatrix} < 0, \tag{7.36}$$

$$\begin{bmatrix} \hat{A}_i^T P_i + P_i \hat{A}_i - C_i^T Q_i^T - Q_i C_i + \alpha P_i + I - C_i^T C_i & P_i \bar{F}_i \\ * & \beta^2 I - F_{2i}^T F_{2i} \end{bmatrix} < 0, \tag{7.37}$$

where $\bar{D}_i$, $\hat{A}_i$, *and* $\bar{F}_i$ *are defined as same as in Theorems 7.3 and 7.4, and* $L_i = P_i^{-1} Q_i$.

By (7.35), one can get

$$E_i = -D_{1i}(C_i D_{1i})^\dagger \tag{7.38}$$

where $(C_i D_{1i})^\dagger = ((C_i D_{1i})^T (C_i D_{1i}))^- (C_i D_{1i})^T$ denotes the pseudo inverse of $C_i D_{1i}$.

In the following algorithm, the procedure on how to get the gains of UIO (7.3) is given.

Algorithm 1:

Step1: Calculate the UIO gain E_i in (7.38), then G_i can be solved by (7.7);

Step2: Solve the matrices P_i and Q_i by Theorem 7.3, and calculate the gain $L_i = P_i^{-1} Q_i$;

Step3: Calculate the other gain N_i by (7.7).

7.1.4 FD Threshold Design

Following the construction of the residuals, the last step is to determine the residual evaluation function $J(r(t))$ and the threshold function J_{th}.

In this paper, the residual evaluation function $J(r(t))$ and the threshold function J_{th} can be selected as

$$J(r(t)) = \| r(t) \|_2 = \sqrt{\frac{1}{t}\sum_{s=t_0}^{t_0+t} r^T(t)r(t)}, \tag{7.39}$$

$$J_{th} = \sup_{d(t)\in l_2[0,\infty),\ f(t)=0} J(r(t)). \tag{7.40}$$

where t_0 devotes the initial evaluation time instant, and t is the evaluation time step.

Based on the above definitions, the occurrence of the fault can be detected by comparing $J(r(t))$ and J_{th} according to the following logical test:

$$J(r(t)) > J_{th} \Rightarrow \ with\ faults \tag{7.41}$$

$$J(r(t)) < J_{th} \Rightarrow \ no\ faults \tag{7.42}$$

7.1.5 Simulation Example

A liquid level control system investigated in the Sect. 1.2.8, is employed to prove the obtained results, which is described in the Fig. 1.10. The switching signal is set as in Fig. 7.1. The parameters of the system is given as follows:

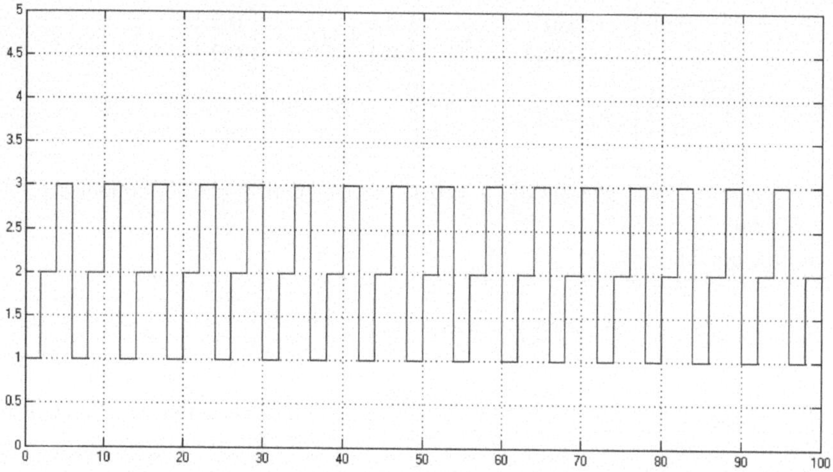

Fig. 7.1 The switching signal $\sigma_i(t)$

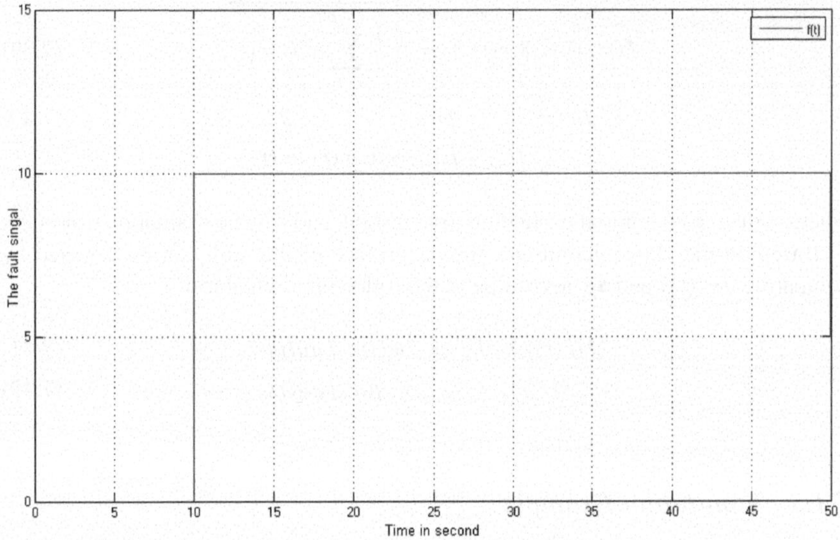

Fig. 7.2 The fault f(t)

$$A_1 = \begin{bmatrix} -0.0007 & 0.0007 \\ 0.0011 & -0.0011 \end{bmatrix}, A_2 = \begin{bmatrix} -0.0013 & 0.0007 \\ 0.0011 & -0.0011 \end{bmatrix}, B_i = \begin{bmatrix} 0.2 \\ 0.1 \end{bmatrix},$$
$$A_3 = \begin{bmatrix} -0.0007 & 0.0007 \\ 0.0011 & -0.0044 \end{bmatrix}, D_{1i} = \begin{bmatrix} 0.1 \\ 0.1 \end{bmatrix}, D_{2i} = \begin{bmatrix} 0.2 \\ 0.2 \end{bmatrix},$$
$$C_i = \begin{bmatrix} 0.01 & 0 \\ 0.01 & 0 \end{bmatrix}, F_{1i} = F_{2i} = \begin{bmatrix} 0.2 \\ 0.1 \end{bmatrix},$$

For given $\mu = 1.002$, $\alpha = 0.001$, $\gamma = 0.1$, $\varepsilon = 0.1$, and $\beta = 1.002$, one has $\tau_a^* = \frac{ln\mu}{\alpha} = 1.998$.

By the condition (7.38)

$$E_i = D_{1i}[(C_i D_{1i})^T (C_i D_{1i})]^{-1} (C_i D_{1i})^T = \begin{bmatrix} 100 & 0 \\ 50 & 0 \end{bmatrix}, i = 1, 2, 3.$$

By solving the conditions (7.36) and (7.37), a set of feasible solutions are obtained as follows:

$$P_1 = \begin{bmatrix} 3.9929 & -1.4933 \\ -1.4933 & 2.5559 \end{bmatrix}, \quad P_2 = \begin{bmatrix} 4.0583 & -1.4798 \\ -1.4798 & 2.6515 \end{bmatrix},$$
$$P_3 = \begin{bmatrix} 3.1591 & -0.4058 \\ -0.4058 & 4.1239 \end{bmatrix}, \quad Q_1 = \begin{bmatrix} 6.1546 & 0.6765 \\ 0.6765 & -2.2122 \end{bmatrix},$$
$$Q_2 = \begin{bmatrix} 6.2428 & 0.7888 \\ 0.7888 & -2.1929 \end{bmatrix}, \quad Q_3 = \begin{bmatrix} 5.7446 & 21.6199 \\ 21.6199 & -27.2793 \end{bmatrix}.$$

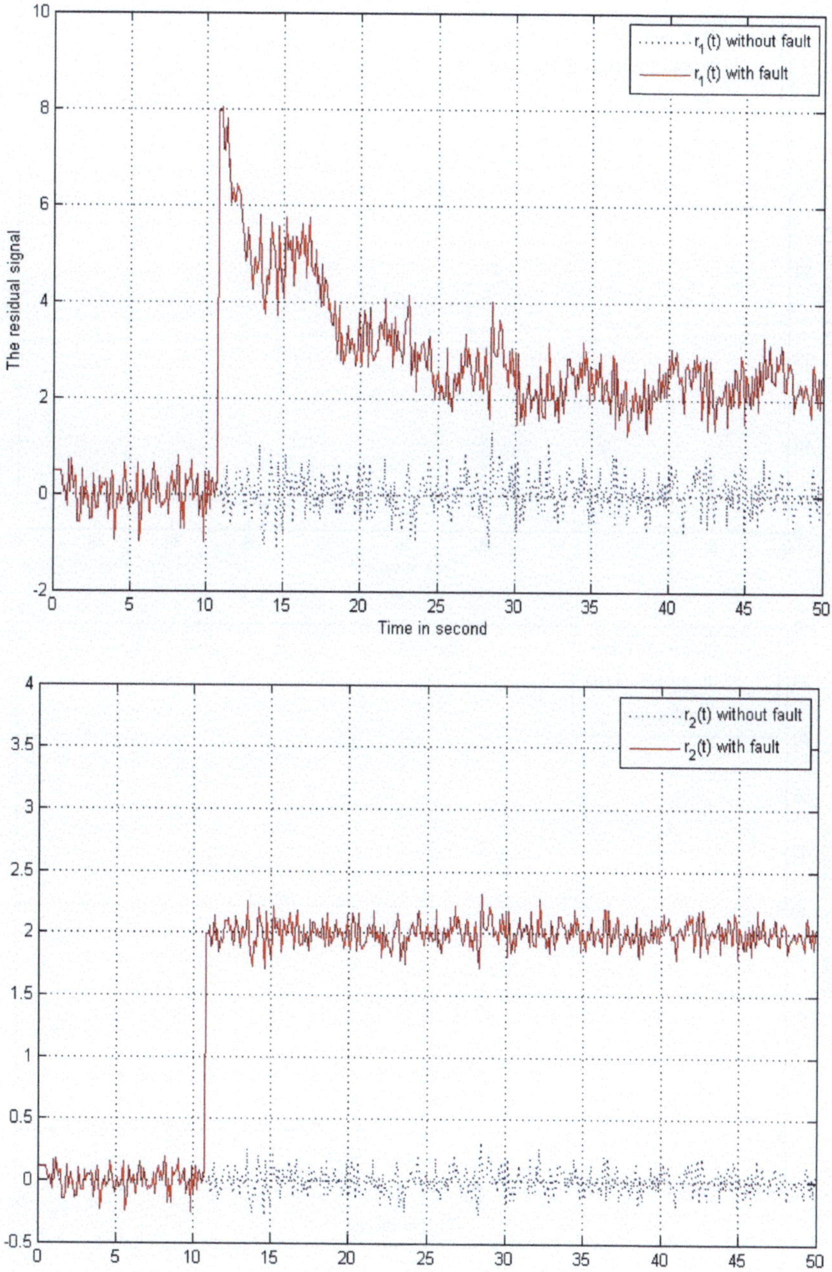

Fig. 7.3 The residual signal $r(t)$ with faulty and unfaulty cases

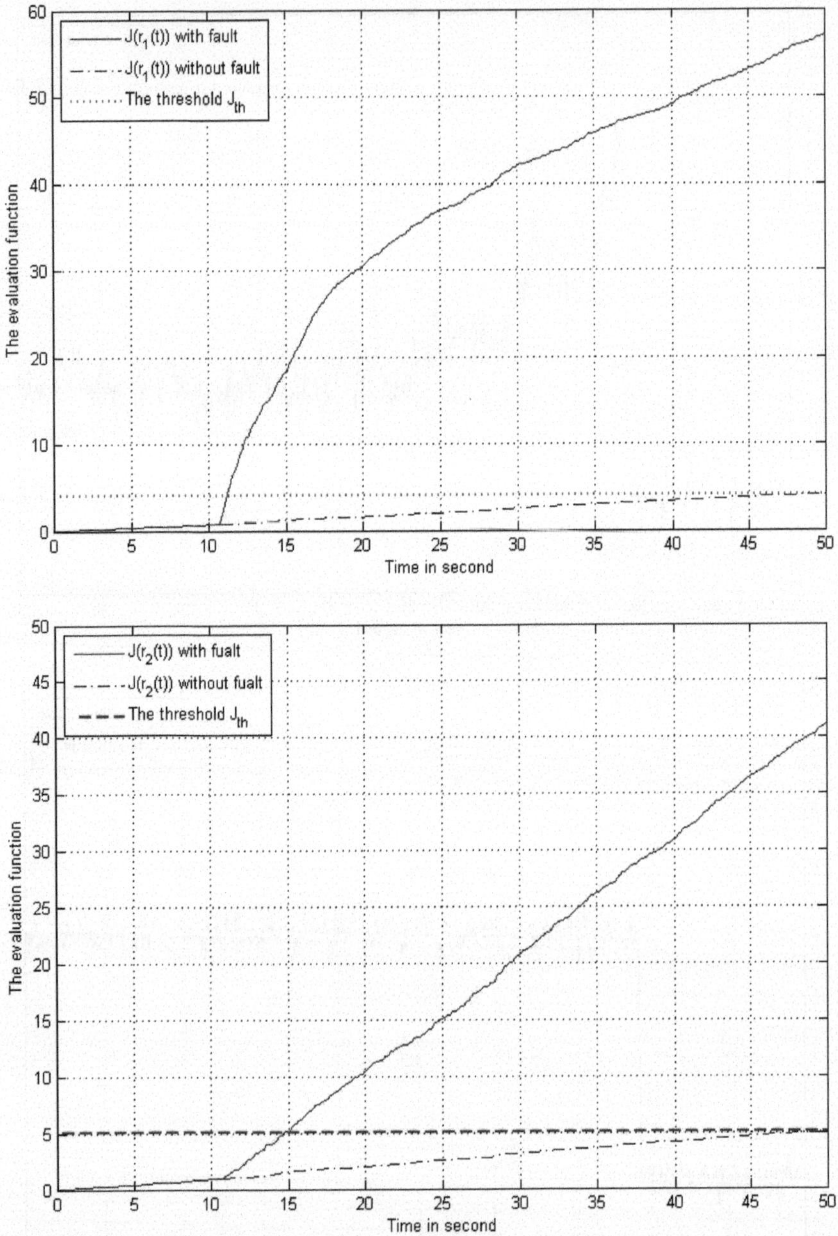

Fig. 7.4 The evaluation function $J(r(t))$

Then, one can obtain the other gains of the UIO (7.3) according to Algorithm 1:

$$N_1 = \begin{bmatrix} -0.0224 & 0.0014 \\ -0.0142 & -0.0008 \end{bmatrix}, \quad N_2 = \begin{bmatrix} -0.0233 & 0.0014 \\ -0.0141 & -0.0008 \end{bmatrix},$$
$$N_3 = \begin{bmatrix} -0.2462 & 0.0014 \\ -0.5318 & -0.0041 \end{bmatrix}, \quad L_1 = \begin{bmatrix} 2.0990 & -0.1974 \\ 1.4909 & -0.9809 \end{bmatrix},$$
$$L_2 = \begin{bmatrix} 2.0676 & -0.1346 \\ 1.4515 & -0.9022 \end{bmatrix}, \quad L_3 = \begin{bmatrix} 24.4761 & 58.8780 \\ 53.2539 & -58.3618 \end{bmatrix},$$
$$G_1 = G_2 = G_3 = \begin{bmatrix} 0.4000 \\ 0.2000 \end{bmatrix}.$$

In the simulation, we consider the following piecewise constant fault:

$$f_1(t) = \begin{cases} 0, & 0s \le t < 10s; \\ 10, & 10s \le t < 50s; \end{cases} \tag{7.43}$$

which is shown in Fig. 7.2. The residual responses for faulty and unfaulty cases are shown in Fig. 7.3. Figure 7.4 describes the residual evaluation function $J(r(t))$ with the faulty and unfaulty cases. By (7.41), the thresholds $J_{th}(r_1(t))$ and $J_{th}(r_2(t))$ can be chosen as 5.1094 and 5.0317. According to the logical rule in (7.41) and (7.42), and selecting $J_{th}(r(t)) = min\{J_{th}(r_1(t)), J_{th}(r_2(t))\}$, then the fault can be successfully detected.

7.1.6 Summary

In this paper, an UIO-based FD filter method was proposed for continuous-time switched systems. To make the generated residual sensitive to the fault and robust to the disturbance, we derived H_- fault sensitivity condition and H_∞ disturbance attenuation condition in the full frequency, respectively. Besides, the solutions of the observer were given in the formulation of LMIs. Finally, simulation results were obtained to demonstrate the performance of the proposed detection filter.

7.2 FE for Actuator and Sensor Faults

In this section, the problem of robust FE observer design for continuous-time switched systems subject to unknown input is investigated. Firstly, based on the ADT method and the switched Lyapunov function technique, a UIO is designed for the augmented switched system by eliminating the unknown input, which can guarantee that the error system is globally uniformly asymptotically stable with a prescribed H_∞ performance index. Next, the method is extended to the measurement disturbances case. Finally, an example is presented to illustrate the design and efficiency of the proposed FE observer technique.

7.2.1 Introduction

In the past decades, the issue of fault diagnosis has received considerable attention due to the increasing safety and reliability demands for various dynamic systems. Fruitful achievements can be found in several recent excellent sections and the references therein [25–28]. The process of fault diagnosis can be divided into two parts: FD and FE. FD focuses on detecting the fault as soon as possible. After that, the main task is to identify the position and the magnitude of the fault, which is called FE. Among all the approaches of online fault diagnosis, observer-based technique is regarded as one of the most popular methods [29]. By using the difference between the system output and the observer output as the so-called residual signal, generating an alarm and estimating the shape of the faults can be realized.

For the FE issue, many observer-based methods have been developed in recent years, such as adaptive observer [30], sliding-mode observer [31], fuzzy observer [32], etc. Adaptive observer-based FE technique can achieve asymptotical estimation for constant faults, while the sliding-mode observer-based FE technique can produce bounded estimation for constant faults. As for the case of time-varying fault, they both can realize bounded estimation. For example, the issues FE and FTC for fuzzy systems are separately investigated in [33] and [34], by assuming the fault in $Ł_2[0, \infty)$. However, it is well-known that the dynamic systems are inevitably affected by external disturbances. In order to reduce the influence of these disturbances, one approach is to carry out various optimization techniques such that the residual signal is sensitive to faults and robust to disturbances. Another solution is to decouple the residual from the disturbances such that their effects are removed. The UIO (UIO) technique has been proven to be an efficient method to achieve the disturbances decoupling [35]. For example, a robust fault diagnosis observer design for an autonomous spacecraft was considered in [36], by employing a nonlinear UIO. In [37], an actuator FE observer design for a wind turbine system was addressed by using the UIO technique. However, it should be noted that the UIO-based technique mentioned in the references is constrained by some existing restrictive conditions, which motivates us to search for a feasible technique to decouple the partial disturbances.

In another research direction, the control and synthesis of switched systems have caused more and more interests due to the large number of applications in practical engineering systems [38, 39]. Switching system is a class of hybrid systems, which consists of several subsystems and a switching signal specifying the switching sequence among them. According to the characterization of switched systems, the switching signal can be classified into two categories: arbitrary switching and constrained switching. ADT switching is one of the constrained switching signals and was firstly proposed in [15]. In [40], it has been proven that ADT switching is more general and less conservative than dwell-time switching and arbitrary switching. Furthermore, ADT technique has been widely used to deal with fault diagnosis and FTC for switched systems [41, 42].

In this section, we mainly focus on designing a UIO-based robust FE observer for continuous-time switched systems. The main contributions of this section can be summarized in the following aspects:

- an UIO-based robust FE observer is designed for switched systems. The FE is totally robust to the deterministic input disturbances, and the effect of the stochastic disturbances is minimized with respect to an H_∞ performance index;
- efficient conditions of the robust FE observer design for switched systems with measurement disturbances are obtained;
- the proposed FE algorithm is suitable to estimate all kinds of faults, such that $f^{(q)}(t) = 0, q \in \mathbb{N} = \{1, 2, \ldots\}$.

7.2.2 *Model Description and Preliminaries*

Consider the following continuous-time switched system:

$$\begin{cases} \dot{x}(t) = \sum_{i=1}^{N} \sigma_i(t)(A_i x(t) + B_i u(t) + D_{1i} d(t) + D_{2i}\omega(t) + F_{1i} f(t)) \\ y(t) = \sum_{i=1}^{N} \sigma_i(t)(C_i x(t) + F_{2i} f(t)) \end{cases} \tag{7.44}$$

where $x(t) \in \mathbb{R}^n$, $u(t) \in \mathbb{R}^u$, and $y(t) \in \mathbb{R}^m$ are, respectively, the state, the control input, and the measured output; $d(t) \in \mathbb{R}^d$ is the independent Gaussian white noise disturbance, and $\omega(t) \in \mathbb{R}^w$ is the bounded unknown input vector caused by either disturbances or modeling errors; $f(t) \in \mathbb{R}^l$ is the fault and a derivable signal up to an order q. The piecewise constant function

$$\sigma_i(t) : [0, \infty) \to \{0, 1\}, \quad \sum_{i=1}^{N} \sigma_i(t) = 1 \tag{7.45}$$

is the switching signal, which specifies which subsystem is activated at the switching instant. A_i, B_i, C_i, D_{1i}, D_{2i}, F_{1i}, and F_{2i} are constant real matrices with appropriate dimensions.

The following preliminaries are given:

Assumption 9 The matrix D_{1i} is full column rank.

Assumption 10 The dimension of the measured output is no smaller than the dimension of the unknown input, i.e., $m \geq d$.

7.2.3 Robust FE with Input Disturbances

Firstly, let us construct the augmented state vector as

$$\bar{x}(t) = \begin{bmatrix} x(t) \\ f(t) \\ \dot{f}(t) \\ \vdots \\ f^{(q-1)}(t) \end{bmatrix}, \; q \in \mathbb{N} \tag{7.46}$$

then the following augmented switched system can be derived from system (7.44):

$$\begin{cases} \dot{\bar{x}}(t) = \sum\limits_{i=1}^{N} \sigma_i(t)(\bar{A}_i\bar{x}(t) + \bar{B}_i u(t) + \bar{D}_{1i} d(t) + \bar{D}_{2i}\omega(t) + \bar{I} f^{(q)}(t)) \\ y(t) = \sum\limits_{i=1}^{N} \sigma_i(t)\bar{C}_i\bar{x}(t) \end{cases} \tag{7.47}$$

where

$$\bar{A}_i = \begin{bmatrix} A_i & F_{1i} & 0 & \cdots & 0 \\ 0 & 0 & I & \cdots & 0 \\ \vdots & \vdots & \vdots & \ddots & \vdots \\ 0 & 0 & 0 & \cdots & I \\ 0 & 0 & 0 & \cdots & 0 \end{bmatrix}, \quad \bar{B}_i = \begin{bmatrix} B_i \\ 0 \\ \vdots \\ 0 \\ 0 \end{bmatrix}, \quad \bar{C}_i = \begin{bmatrix} C_i^T \\ F_{2i}^T \\ 0 \\ \vdots \\ 0 \end{bmatrix}^T,$$

$$\bar{D}_{1i} = \begin{bmatrix} D_{1i} \\ 0 \\ \vdots \\ 0 \\ 0 \end{bmatrix}, \quad \bar{D}_{2i} = \begin{bmatrix} D_{2i} \\ 0 \\ \vdots \\ 0 \\ 0 \end{bmatrix}, \quad \bar{I} = \begin{bmatrix} 0 \\ 0 \\ \vdots \\ 0 \\ I \end{bmatrix}. \tag{7.48}$$

In this section, we assume that the q-order derivative of the fault is zero, i.e., $f^{(q)}(t) = 0, q \in \mathbb{N}$. The following UIO is employed to estimate the state of the augmented system (7.47):

$$\begin{cases} \dot{z}(t) = \sum\limits_{i=1}^{N} \sigma_i(t)(N_i z(t) + G_i u(t) + L_i y(t)) \\ \hat{\bar{x}}(t) = z(t) + \sum\limits_{i=1}^{N} \sigma_i(t) H_i y(t) \end{cases} \tag{7.49}$$

where $z(t)$ is the state of the observer, $\hat{\bar{x}}(t)$ is the state estimation of the augmented system (7.47). Matrices N_i, G_i, H_i, and $L_i = L_{1i} + L_{2i}$ are the UIO gains to be designed later.

Remark 7.4 As a result, one can conclude that if there exists a UIO (7.49), for the augmented system (7.47), then the fault vector of the system (7.44), can be estimated.

Define the state estimation error $e(t) = \bar{x}(t) - \hat{\bar{x}}(t)$, then one can get

$$e(t) = \bar{x}(t) - z(t) - \sum_{i=1}^{N} \sigma_i(t) H_i y(t) = \sum_{i=1}^{N} \sigma_i(t) T_i \bar{x}(t) - z(t)$$

where $T_i = I - H_i \bar{C}_i$.

From (7.47), (7.49), and (7.50), the derivative of the state error can be obtained

$$\begin{aligned}
\dot{e}(t) &= \sum_{i=1}^{N} \sigma_i(t) T_i \dot{\bar{x}}(t) - \dot{z}(t) \\
&= \sum_{i=1}^{N} \sigma_i(t)((T_i \bar{A}_i \bar{x}(t) + T_i \bar{B}_i u(t) + T_i \bar{D}_{1i} d(t) \\
&\quad + T_i \bar{D}_{2i} \omega(t) - N_i z(t) - G_i u(t) - L_i y(t)) \\
&= \sum_{i=1}^{N} \sigma_i(t)((T_i \bar{A}_i - L_{1i} \bar{C}_i) e(t) \quad + (T_i \bar{B}_i - G_i) u(t) \\
&\quad + (T_i \bar{A}_i - L_{1i} \bar{C}_i - N_i) z(t) \quad + ((T_i \bar{A}_i - L_{1i} \bar{C}_i) H_i - L_{2i}) y(t) \\
&\quad + T_i \bar{D}_{1i} d(t) + T_i \bar{D}_{2i} \omega(t))
\end{aligned} \tag{7.50}$$

If the following conditions hold:

$$T_i \bar{A}_i - L_{1i} \bar{C}_i - N_i = 0; \tag{7.51}$$

$$(T_i \bar{A}_i - L_{1i} \bar{C}_i) H_i - L_{2i} = 0; \tag{7.52}$$

$$T_i \bar{B}_i - G_i = 0; \tag{7.53}$$

$$T_i \bar{D}_{1i} = 0; \tag{7.54}$$

then the error system (7.50) reduces to

$$\dot{e}(t) = \sum_{i=1}^{N} \sigma_i(t)(N_i e(t) + T_i \bar{D}_{2i} \omega(t)) \tag{7.55}$$

Remark 7.5 Following the Assumptions 9 and 10, one can get that $rank(\bar{C}_i \bar{D}_{1i}) = rank(\bar{D}_{1i})$. By the condition (7.54), the matrices $H_i,\ (i = 1, 2, \ldots, N)$ of the UIO (7.49) can be calculated by

$$H_i = \bar{D}_{1i}[(\bar{C}_i \bar{D}_{1i})^T (\bar{C}_i \bar{D}_{1i})]^{-1} (\bar{C}_i \bar{D}_{1i})^T, \quad i = 1, 2, \ldots, N. \tag{7.56}$$

Remark 7.6 Condition (7.54), may be a strong constraint in some cases. If it is not possible to find a feasible solution T_i which fulfills condition (7.54), one can decouple partially the unknown input $d(t)$ from the system and integrate the non decoupled disturbances in the $\omega(t)$ vector.

Next, based on the error dynamic system (7.55), a UIO-based robust FE observer (7.49), is presented in the following theorem.

Theorem 7.6 *For given scalars $\alpha > 0$, $\mu > 1$, and $\gamma > 0$, if the conditions (7.51)–(7.54), hold and there exist positive definite matrices P_i and P_j, matrices Q_i, $\forall i, j \in \{1, 2, \ldots, N\}$, which satisfy*

$$P_i \leq \mu P_j, \tag{7.57}$$

$$\begin{bmatrix} \Psi_{11i} & P_i \hat{D}_{2i} \\ \hat{D}_{2i}^T P_i & -\gamma^2 I \end{bmatrix} < 0, \tag{7.58}$$

where $\Psi_{11i} = \hat{A}_i^T P_i + P_i \hat{A}_i - \bar{C}_i^T Q_i^T - Q_i \bar{C}_i + \alpha P_i + I$, $\hat{A}_i = T_i \bar{A}_i$, $\hat{D}_{2i} = T_i \bar{D}_{2i}$, and $L_{1i} = P_i^{-1} Q_i$, then, for the switching signal $\sigma_i(t)$ with ADT constraint (7.11), there exists a UIO (7.49), for the augmented switched system (7.47), such that the error system (7.55), is GUAS with H_∞ performance index γ.

Proof We first establish the stability of system (7.55). Let us set $\omega(t) = 0$ and construct a switched Lyapunov function for the error dynamics (7.55), as

$$V_i(t, e(t)) = e^T(t) \left(\sum_{i=1}^{N} \sigma_i(t) P_i \right) e(t) \tag{7.59}$$

One has

$$\begin{aligned} &\dot{V}_i(t, e(t)) + \alpha V_i(t, e(t)) \\ &= \dot{e}^T(t) \left(\sum_{i=1}^{N} \sigma_i(t) P_i \right) e(t) + e^T(t) \left(\sum_{i=1}^{N} \sigma_i(t) P_i \right) \dot{e}(t) + e^T(t) \left(\sum_{i=1}^{N} \sigma_i(t) \alpha P_i \right) e(t) \\ &= e^T(t) \left(\sum_{i=1}^{N} \sigma_i(t)(N_i^T P_i + P_i N_i + \alpha P_i) \right) e(t) \end{aligned} \tag{7.60}$$

From the conditions (7.51) and (7.58), we have $\dot{V}_i(t, e(t)) < -\alpha V_i(t, e(t))$. According to Lemma 7.2 and condition (7.57), one can get that the system (7.55), is GUAS.

In the following, the proof of the H_∞ performance γ is given.

Define

$$\Gamma = \int_0^T (e^T(t)e(t) - \gamma^2 \omega^T(t)\omega(t) + \alpha V_i(t, e(t)))dt \tag{7.61}$$

One has

$$\begin{aligned}\Gamma &= \int_0^T (e^T(t)e(t) - \gamma^2\omega^T(t)\omega(t) + \alpha V_i(t, e(t)) \\ &\quad + \dot{V}_i(t, e(t)))dt - \int_0^T \dot{V}_i(t, e(t))dt \\ &= \int_0^T \sum_{i=1}^N \sigma_i(t)(e^T(t)\Psi_{11i}e(t) + 2e^T(t)P_i\hat{D}_{2i}\omega(t) \\ &\quad - \gamma^2\omega^T(t)\omega(t)) - \int_0^T \dot{V}_i(t, e(t))dt\end{aligned} \tag{7.62}$$

Under the initial condition $e(0) = 0$, one has

$$\begin{aligned}\int_0^T \dot{V}_i(t, e(t))dt &= e^T(T)\left(\sum_{i=1}^N \sigma_i(t)P_i\right)e(T) - e^T(0)\left(\sum_{i=1}^N \sigma_i(t)P_i\right)e(0) \\ &= V(T, e(T)) > 0\end{aligned} \tag{7.63}$$

By (7.58), (7.62), and (7.63), we get

$$e^T(t)e(t) - \gamma^2\omega^T(t)\omega(t) + \dot{V}_i(t, e(t)) < -\alpha V_i(t, e(t)) \tag{7.64}$$

Then one can conclude that the error system (7.55), has H_∞ performance γ. The proof is concluded. (□)

Based on the obtained results in the Theorem 7.6, the UIO design procedure can be summarized as the following algorithm:

Algorithm 2:

Step 1: Construct the augmented system (7.47);

Step 2: Calculate the UIO gains H_i in (7.56);

Step 3: Solve the unknown matrices P_i and Q_i by the conditions (7.57) and (7.58), in the Theorem 7.6, and calculate the UIO gains L_{1i};

Step 4: Calculate the other UIO gains N_i, G_i, and L_{2i} by the conditions (7.51)–(7.53);

Step 5: Implement the UIO (7.49), to obtain the state estimation $\hat{\bar{x}}(t)$, and get the FE $\hat{f}(t)$ by the following equations:

$$\hat{f}(t) = \begin{cases} \begin{bmatrix} 0 & I \end{bmatrix}\hat{\bar{x}}(t); & \dot{f}(t) = 0 \\ \begin{bmatrix} 0 & I & 0 \end{bmatrix}\hat{\bar{x}}(t); & \ddot{f}(t) = 0 \\ \begin{bmatrix} 0 & I & 0 & \cdots & 0 \end{bmatrix}\hat{\bar{x}}(t); & f^{(k)} = 0 \ (k \geq 3) \end{cases} \tag{7.65}$$

7.2.4 Robust FE with Input and Measurement Disturbances

In this section, we assume that the switched system (7.44), is not only affected by process disturbances, but also affected by measurement disturbances. The system is now described by

$$\begin{cases} \dot{x}(t) = \sum_{i=1}^{N} \sigma_i(t)(A_i x(t) + B_i u(t) + D_{1i} d(t) + D_{2i}\omega(t) + F_{1i} f(t)) \\ y(t) = \sum_{i=1}^{N} \sigma_i(t)(C_i x(t) + E_i d_s(t) + F_{2i} f(t)) \end{cases} \tag{7.66}$$

where $d_s(t) \in \mathbb{R}^s$ is the measurement disturbance, E_i is a constant matrix with appropriate dimensions, and other symbols are defined as in the system (7.44).

Firstly, an augmented vector is defined as in (7.46), then one can get the augmented switched system as

$$\begin{cases} \dot{\bar{x}}(t) = \sum_{i=1}^{N} \sigma_i(t)(\bar{A}_i \bar{x}(t) + \bar{B}_i u(t) + \bar{D}_{1i} d(t) + \bar{D}_{2i}\omega(t) + \bar{I} f^{(q)}(t)) \\ y(t) = \sum_{i=1}^{N} \sigma_i(t)(\bar{C}_i \bar{x}(t) + E_i d_s(t)) \end{cases} \tag{7.67}$$

where the coefficient matrices are defined as in (7.48).

As previously, we assume that the q-order derivative of the fault is zero, i.e., $f^{(q)}(t) = 0, q \in \mathbb{N}$. Next, a UIO of the same form as in (7.49), is employed to estimate the state of the augmented switched system (7.67), then one can get

$$\begin{aligned} e(t) &= \bar{x}(t) - \hat{\bar{x}}(t) \\ &= \bar{x}(t) - z(t) - \sum_{i=1}^{N} \sigma_i(t) H_i y(t) \\ &= \sum_{i=1}^{N} \sigma_i(t)(T_i \bar{x}(t) - H_i E_i d_s(t)) - z(t) \end{aligned} \tag{7.68}$$

From (7.49), (7.67), and (7.69), the derivative of the state error can be derived as

$$\begin{aligned} \dot{e}(t) &= \sum_{i=1}^{N} \sigma_i(t)(T_i \dot{\bar{x}}(t) - H_i E_i \dot{d}_s(t)) - \dot{z}(t) \\ &= \sum_{i=1}^{N} \sigma_i(t)((T_i \bar{A}_i - L_{1i}\bar{C}_i)e(t) \quad + (T_i \bar{A}_i - L_{1i}\bar{C}_i - N_i)z(t) \\ &\quad + (T_i \bar{B}_i - G_i)u(t) \quad + ((T_i \bar{A}_i - L_{1i}\bar{C}_i)H_i - L_{2i})y(t) + T_i \bar{D}_{1i} d(t) \\ &\quad + T_i \bar{D}_{2i}\omega(t) - L_{1i} E_i d_s(t) - H_i E_i \dot{d}_s(t)) \end{aligned} \tag{7.69}$$

Under the same conditions (7.51)–(7.54), one can get

$$\dot{e}(t) = \sum_{i=1}^{N} \sigma_i(t)(N_i e(t) + M_i \upsilon(t)) \tag{7.70}$$

where $M_i = \begin{bmatrix} T_i \bar{D}_{2i} & -L_{1i}E_i & -H_i E_i \end{bmatrix}$, $\upsilon(t) = \begin{bmatrix} \omega^T(t) & d_s^T(t) & \dot{d}_s^T(t) \end{bmatrix}^T$. In the following theorem, the robust FE observer design of UIO (7.49), is developed for the switched system (7.44), affected by both input and measurement disturbances.

Theorem 7.7 *For given scalars $\alpha > 0$, $\mu > 1$, and $\gamma > 0$, if the conditions (7.51)–(7.54), hold and there exist positive definite matrices P_i and P_j, matrices Q_i, $\forall i, j \in \{1, 2, \ldots, N\}$, which satisfy*

$$P_i \leq \mu P_j, \tag{7.71}$$

$$\begin{bmatrix} \Psi_{11i} & P_i \hat{D}_{2i} & -Q_i E_i & -P_i \hat{E}_i \\ * & -\gamma^2 I & 0 & 0 \\ * & * & -\gamma^2 I & 0 \\ * & * & * & -\gamma^2 I \end{bmatrix} < 0, \tag{7.72}$$

where Ψ_{11i}, $\hat{A}_i$, and $\hat{D}_{2i}$ are the same as in Theorem 7.1, $\hat{E}_i = H_i E_i$, and $L_{1i} = P_i^{-1} Q_i$, then, for the switching signal $\sigma_i(t)$ with ADT constraint (7.11), there exists a UIO (7.49), for the augmented switched system (7.67), such that the error system (7.70), is GUAS with H_∞ performance index γ.

Proof We first establish the stability of system (7.70). If we set $\upsilon(t) = 0$ and choose a switched Lyapunov function as in (7.59), one has

$$\begin{aligned} &\dot{V}_i(t, e(t)) + \alpha V_i(t, e(t)) \\ &= \dot{e}^T(t) \left(\sum_{i=1}^{N} \sigma_i(t) P_i \right) e(t) + e^T(t) \left(\sum_{i=1}^{N} \sigma_i(t) P_i \right) \dot{e}(t) + e^T(t) \left(\sum_{i=1}^{N} \sigma_i(t) \alpha P_i \right) e(t) \\ &= e^T(t) \left(\sum_{i=1}^{N} \sigma_i(t)(N_i^T P_i + P_i N_i + \alpha P_i) \right) e(t) \end{aligned} \tag{7.73}$$

Conditions in (7.51) and (7.72), imply $\dot{V}_i(t, e(t)) < -\alpha V_i(t, e(t))$. According to Lemma 7.1 and the condition (7.57), one can get that the system (7.70), is GUAS.

In the following, the proof of the H_∞ performance γ is given. Define

$$\Gamma_s = \int_0^T (e^T(t)e(t) - \gamma^2 \upsilon^T(t)\upsilon(t) + \alpha V_i(t, e(t)))dt \tag{7.74}$$

one has

$$\begin{aligned}\Gamma_s &= \int_0^T (e^T(t)e(t) - \gamma^2 \upsilon^T(t)\upsilon(t) + \alpha V_i(t, e(t)) \\ &\quad + \dot{V}_i(t, e(t)))dt - \int_0^T \dot{V}_i(t, e(t))dt \\ &= \int_0^T \sum_{i=1}^{N} \sigma_i(t)(e^T(t)\Psi_{11i}e(t) + 2e^T(t)P_i\hat{D}_{2i}\upsilon(t) \\ &\quad - \gamma^2\upsilon^T(t)\upsilon(t)) - \int_0^T \dot{V}_i(t, e(t))dt \end{aligned} \tag{7.75}$$

Under the initial condition $e(t) = 0$, one has

$$\begin{aligned} &\int_0^T \dot{V}_i(t, e(t))dt \\ &= e^T(T)\left(\sum_{i=1}^{N} \sigma_i(t)P_i\right)e(T) - e^T(0)\left(\sum_{i=1}^{N} \sigma_i(t)P_i\right)e(0) \\ &= V(T, e(T)) > 0 \end{aligned} \tag{7.76}$$

From (7.72), (7.75), and (7.76), we obtain

$$e^T(t)e(t) - \gamma^2\upsilon^T(t)\upsilon(t) + \dot{V}_i(t, e(t)) < -\alpha V_i(t, e(t)) \tag{7.77}$$

Then one can conclude that the error system (7.70), has H_∞ performance index γ. The proof is completed. (□)

Remark 7.7 According to the same UIO design procedures as in Algorithm 1, the robust FE observer design for switched system (7.66), can be realized for the cases of $f^{(q)}(t) = 0, q \in \mathbb{N}$.

Remark 7.8 In this section, the actuator faults and measurement faults are simultaneously considered. Moreover, if only actuator faults or measurement faults are considered, the results remain suitable by setting $F_{2i} = 0$ or $F_{1i} = 0$.

7.2.5 An Illustrative Example

In this part, a liquid level control system in the Sect. 1.2.8, is employed to illustrate the obtained results, which is described in Fig. 1.10. The matrix parameters of the system (7.44), is defined as follows:

$$A_1 = \begin{bmatrix} -0.0007 & 0.0007 \\ 0.0011 & -0.0011 \end{bmatrix}, A_2 = \begin{bmatrix} -0.0013 & 0.0007 \\ 0.0011 & -0.0011 \end{bmatrix},$$
$$A_3 = \begin{bmatrix} -0.0007 & 0.0007 \\ 0.0011 & -0.0044 \end{bmatrix}, B_i = \begin{bmatrix} 0.2 \\ 0.1 \end{bmatrix}, D_{1i} = \begin{bmatrix} 0.1 \\ 0.1 \end{bmatrix},$$
$$D_{2i} = \begin{bmatrix} 0.2 \\ 0.2 \end{bmatrix}, C_i = \begin{bmatrix} 0.01 & 0 \\ 0.01 & 0 \end{bmatrix}, F_{1i} = F_{2i} = \begin{bmatrix} 0.2 \\ 0.1 \end{bmatrix}, i = 1, 2, 3.$$

For given $\mu = 1.002$, $\alpha = 0.001$, and $\gamma = 1.12$, one has $\tau_a^* = \frac{\ln \mu}{\alpha} = 1.9980$. The switching signal is described in Fig. 7.5.

Case I: constant faults. For the given switched system (7.44), if only constant faults are considered, i.e., $\dot{f}(t) = 0$, then the augmented state is $\bar{x}(t) = \begin{bmatrix} x^T(t) & f^T(t) \end{bmatrix}^T$, and the coefficient matrices in (7.48), are

$$\bar{A}_i = \begin{bmatrix} A_i & F_{1i} \\ 0 & 0 \end{bmatrix}, \bar{B}_i = \begin{bmatrix} B_i \\ 0 \end{bmatrix}, \bar{C}_i = \begin{bmatrix} C_i & F_{2i} \end{bmatrix},$$
$$\bar{D}_{1i} = \begin{bmatrix} D_{1i} \\ 0 \end{bmatrix}, \bar{D}_{2i} = \begin{bmatrix} D_{2i} \\ 0 \end{bmatrix}, \bar{I} = \begin{bmatrix} 0 \\ I \end{bmatrix}.$$

Under the conditions in the Assumptions 9 and 10, the parameters $H_i, i = 1, 2, 3$ in (7.56) can be calculated as follows:

$$H_i = \bar{D}_{1i}[(\bar{C}_i\bar{D}_{1i})^T(\bar{C}_i\bar{D}_{1i})]^{-1}(\bar{C}_i\bar{D}_{1i})^T = \begin{bmatrix} 500 & 500 \\ 500 & 500 \\ 0 & 0 \end{bmatrix}, \quad i = 1, 2, 3.$$

Solving the conditions (7.57) and (7.58), in Theorem 7.6, gives

$$P_1 = 10^5 \times \begin{bmatrix} 0.5422 & -0.0000 & 0.0000 \\ -0.0000 & 1.2607 & 0.0000 \\ 0.0000 & 0.0000 & 0.5422 \end{bmatrix},$$

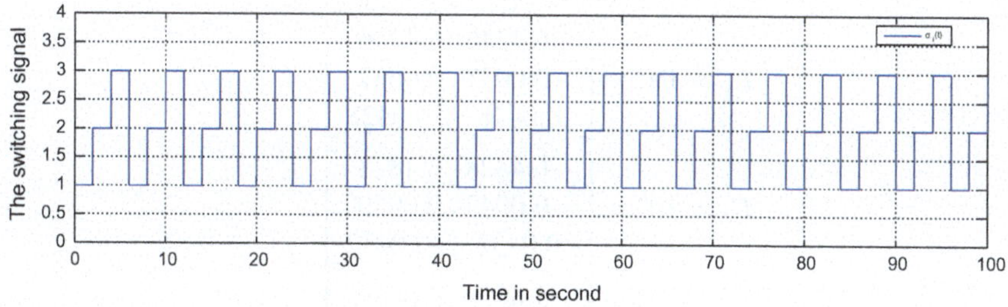

Fig. 7.5 The switching signal $\sigma_i(t)$

$$P_2 = 10^5 \times \begin{bmatrix} 0.5374 & -0.0000 & 0.0000 \\ -0.0000 & 1.2408 & 0.0000 \\ 0.0000 & 0.0000 & 0.5374 \end{bmatrix},$$

$$P_3 = 10^5 \times \begin{bmatrix} 0.7160 & -0.0000 & 0.0000 \\ -0.0000 & 1.2112 & 0.0000 \\ 0.0000 & 0.0000 & 0.7160 \end{bmatrix},$$

$$Q_1 = 10^8 \times \begin{bmatrix} -2.3225 & 4.6450 \\ -0.0035 & 0.0058 \\ 0.0232 & -0.0232 \end{bmatrix},$$

$$Q_2 = 10^8 \times \begin{bmatrix} -2.3225 & 4.6450 \\ -0.0042 & 0.0072 \\ 0.0232 & -0.0232 \end{bmatrix},$$

$$Q_3 = 10^8 \times \begin{bmatrix} -2.3226 & 4.6452 \\ -0.0034 & 0.0056 \\ 0.0232 & -0.0232 \end{bmatrix}.$$

Following the calculation procedures in Algorithm 2, one can obtain the other parameters of the UIO (7.49)

$$N_1 = \begin{bmatrix} -4.2831 & 0 & -0.0000 \\ -0.0000 & -0.0018 & 0.0000 \\ -0.0000 & 0 & -4.2832 \end{bmatrix},$$

$$N_2 = \begin{bmatrix} -4.3219 & 0 & -0.0000 \\ -0.0000 & -0.0018 & 0.0000 \\ -0.0000 & 0 & -4.3219 \end{bmatrix},$$

$$N_3 = \begin{bmatrix} -3.2439 & 0 & -0.0000 \\ -0.0000 & -0.0051 & 0.0000 \\ -0.0000 & 0 & -3.2440 \end{bmatrix},$$

$$G_1 = G_2 = G_3 = \begin{bmatrix} 0 \\ -0.1000 \\ 0 \end{bmatrix},$$

$$L_1 = 10^3 \times \begin{bmatrix} -6.4246 & 6.4246 \\ -0.0037 & 0.0037 \\ 0.0428 & -0.0428 \end{bmatrix},$$

$$L_2 = 10^3 \times \begin{bmatrix} -6.4828 & 6.4828 \\ -0.0043 & 0.0049 \\ 0.0432 & -0.0432 \end{bmatrix},$$

$$L_3 = 10^3 \times \begin{bmatrix} -4.8659 & 4.8659 \\ -0.0054 & 0.0021 \\ 0.0324 & -0.0324 \end{bmatrix}.$$

For the simulation, a piecewise constant fault is considered, which is defined as

$$f_1(t) = \begin{cases} 0, & 0s \le t < 20s; \\ 2, & 20s \le t < 50s; \\ -2, & 50s \le t < 100s; \end{cases}$$

The FE result is shown in Fig. 7.6. One can get that the FE $\hat{f}_1(t)$ accurately converges to the real fault $f_1(t)$.

Case II: slope faults. Firstly, for the augmented system (7.47), the augmented state is $\bar{x}(t) = \left[x^T(t)\ f^T(t)\ \dot{f}^T(t) \right]^T$ and the matrices in (7.48), are

$$\bar{A}_i = \begin{bmatrix} A_i & F_{1i} & 0 \\ 0 & 0 & I \\ 0 & 0 & 0 \end{bmatrix}, \bar{B}_i = \begin{bmatrix} B_i \\ 0 \\ 0 \end{bmatrix}, \bar{C}_i = \begin{bmatrix} C_i & F_{2i} & 0 \end{bmatrix},$$

$$\bar{D}_{1i} = \begin{bmatrix} D_{1i} \\ 0 \\ 0 \end{bmatrix}, \bar{D}_{2i} = \begin{bmatrix} D_{2i} \\ 0 \\ 0 \end{bmatrix}, \bar{I} = \begin{bmatrix} 0 \\ 0 \\ I \end{bmatrix}.$$

Then by substituting the matrices $\bar{C}_i$ and $\bar{D}_{1i}$ into (7.56), the gains of H_i $(i = 1, 2, 3)$ can be calculated as

$$H_i = \bar{D}_{1i}[(\bar{C}_i\bar{D}_{1i})^T(\bar{C}_i\bar{D}_{1i})]^{-1}(\bar{C}_i\bar{D}_{1i})^T = \begin{bmatrix} 500 & 500 \\ 500 & 500 \\ 0 & 0 \\ 0 & 0 \end{bmatrix}, \quad i = 1, 2, 3.$$

Solving the conditions in Theorem 7.6, gives

$$P_1 = 10^3 \times \begin{bmatrix} 3.1115 & -1.7385 & -0.0098 & 0.0363 \\ -1.7385 & 1.7386 & 0.0116 & 0.0107 \\ -0.0098 & 0.0116 & 1.6398 & -0.0002 \\ 0.0363 & 0.0107 & -0.0002 & 1.6404 \end{bmatrix},$$

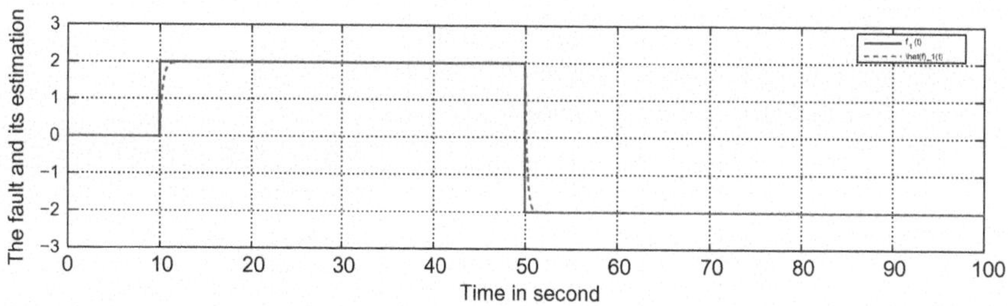

Fig. 7.6 The fault $f_1(t)$ and its estimation $\hat{f}_1(t)$

$$P_2 = 10^3 \times \begin{bmatrix} 3.1255 & -1.7190 & -0.0097 & 0.0370 \\ -1.7190 & 1.7191 & 0.0115 & 0.0100 \\ -0.0097 & 0.0115 & 1.6702 & -0.0002 \\ 0.0370 & 0.0100 & -0.0002 & 1.6706 \end{bmatrix},$$

$$P_3 = 10^3 \times \begin{bmatrix} 4.3136 & -2.0974 & -0.0117 & 0.0504 \\ -2.0974 & 2.0975 & 0.0140 & -0.0034 \\ -0.0117 & 0.0140 & 2.5574 & -0.0003 \\ 0.0504 & -0.0034 & -0.0003 & 2.5579 \end{bmatrix},$$

$$Q_1 = 10^8 \times \begin{bmatrix} -0.0705 & 0.1411 \\ -0.0001 & 0.0001 \\ 0.0007 & -0.0007 \\ 2.0596 & -4.1191 \end{bmatrix},$$

$$Q_2 = 10^8 \times \begin{bmatrix} -0.0705 & 0.1411 \\ -0.0001 & 0.0001 \\ 0.0007 & -0.0007 \\ 2.1099 & -4.2196 \end{bmatrix},$$

$$Q_3 = 10^8 \times \begin{bmatrix} -0.0705 & 0.1411 \\ -0.0001 & 0.0003 \\ 0.0007 & -0.0007 \\ 3.3245 & -6.6487 \end{bmatrix}.$$

Following the calculation procedures in Algorithm 2, the other gains of the UIO (7.49), can be solved as follows:

$$N_1 = \begin{bmatrix} -9.4478 & 0 & 0.0343 & -150 \\ -10.2220 & -0.0018 & 0.0681 & -150 \\ 0.0342 & 0 & -4.3053 & 1 \\ 125.8224 & 0 & -0.8398 & 0 \end{bmatrix},$$

$$N_2 = \begin{bmatrix} -9.2469 & 0 & 0.0335 & -150 \\ -9.9840 & -0.0018 & 0.0666 & -150 \\ 0.0334 & 0 & -4.2269 & 1 \\ 126.5488 & 0 & -0.8447 & 0 \end{bmatrix},$$

$$N_3 = \begin{bmatrix} -5.9474 & 0 & 0.0212 & -150 \\ -5.7385 & -0.0051 & 0.0383 & -150 \\ 0.0212 & 0 & -2.7607 & 1 \\ 130.0710 & 0 & -0.8683 & 0 \end{bmatrix},$$

$$G_1 = G_2 = G_3 = \begin{bmatrix} 0 \\ -0.1000 \\ 0 \\ 0 \end{bmatrix}, L_1 = 10^5 \times \begin{bmatrix} -0.1417 & 0.1417 \\ -0.1534 & 0.1534 \\ 0.0009 & -0.0009 \\ 1.8874 & -1.8874 \end{bmatrix},$$

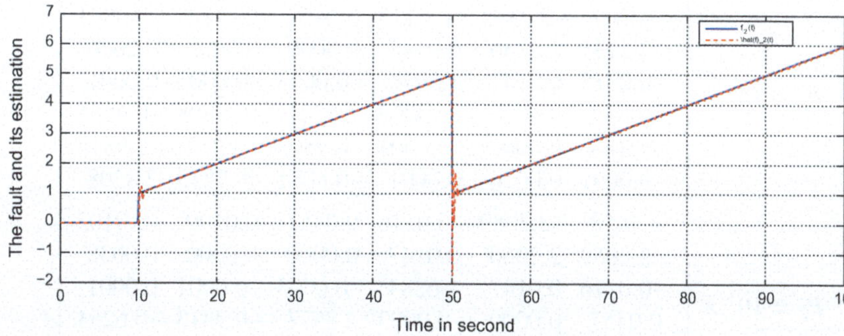

Fig. 7.7 The fault $f_2(t)$ and its estimation $\hat{f}_2(t)$

$$L_2 = 10^5 \times \begin{bmatrix} -0.1387 & 0.1387 \\ -0.1498 & 0.1498 \\ 0.0009 & -0.0009 \\ 1.8983 & -1.8983 \end{bmatrix}, L_3 = 10^5 \times \begin{bmatrix} -0.0892 & 0.0892 \\ -0.0861 & 0.0861 \\ 0.0006 & -0.0006 \\ 1.9512 & -1.9512 \end{bmatrix}.$$

For the simulation, a piecewise slope fault is considered, which is defined as

$$f_2(t) = \begin{cases} 0, & 0s \leq t < 10s; \\ 0.1t, & 10s \leq t < 50s; \\ 0.1(t-40), & 50s \leq t < 100s; \end{cases}$$

The FE result is presented in Fig. 7.7. It can be seen that the FE $\hat{f}_2(t)$ converges to the real fault $f_2(t)$ accurately.

Case III: $f^{(k)}(t) = 0$ $(k \geq 3)$. In this simulation part, the case of $f^{(4)}(t) = 0$ is considered. Following the same routine as in the cases I and II, the gains of $H_i, i = 1, 2, 3$ in (7.56) can be firstly computed as follows:

$$H_i = \bar{D}_{1i}[(\bar{C}_i\bar{D}_{1i})^T(\bar{C}_i\bar{D}_{1i})]^{-1}(\bar{C}_i\bar{D}_{1i})^T = \begin{bmatrix} 500 & 500 \\ 500 & 500 \\ 0 & 0 \\ 0 & 0 \\ 0 & 0 \\ 0 & 0 \end{bmatrix}, \quad i = 1, 2, 3.$$

Then solving the conditions in Theorem 7.6, gives

$$P_1 = 10^3 \times \begin{bmatrix} 3.3607 & -2.2135 & -0.0118 & 0.0457 & 0.0188 & -0.0091 \\ -2.2135 & 2.2136 & 0.0147 & 0.0000 & 0.0000 & 0.0026 \\ -0.0118 & 0.0147 & 1.5976 & -0.0003 & -0.0001 & 0.0001 \\ 0.0457 & 0.0000 & -0.0003 & 2.8221 & -0.9808 & -0.6081 \\ 0.0188 & 0.0000 & -0.0001 & -0.9808 & 1.6272 & -0.9818 \\ -0.0091 & 0.0026 & 0.0001 & -0.6081 & -0.9818 & 2.8208 \end{bmatrix},$$

$$P_2 = 10^3 \times \begin{bmatrix} 3.3658 & -2.2059 & -0.0116 & 0.0457 & 0.0190 & -0.0103 \\ -2.2059 & 2.2060 & 0.0147 & 0.0000 & 0.0000 & 0.0038 \\ -0.0116 & 0.0147 & 1.6284 & -0.0003 & -0.0001 & 0.0001 \\ 0.0457 & 0.0000 & -0.0003 & 2.8577 & -0.9813 & -0.6251 \\ 0.0190 & 0.0000 & -0.0001 & -0.9813 & 1.6447 & -0.9822 \\ -0.0103 & 0.0038 & 0.0001 & -0.6251 & -0.9822 & 2.8562 \end{bmatrix},$$

$$P_3 = 10^3 \times \begin{bmatrix} 4.4841 & -2.1943 & -0.0134 & 0.0459 & 0.0257 & -0.0082 \\ -2.1943 & 2.1944 & 0.0146 & -0.0001 & 0.0000 & 0.0006 \\ -0.0134 & 0.0146 & 2.4827 & -0.0003 & -0.0002 & 0.0001 \\ 0.0459 & -0.0001 & -0.0003 & 3.8534 & -1.1456 & -0.9429 \\ 0.0257 & 0.0000 & -0.0002 & -1.1456 & 2.1965 & -1.1446 \\ -0.0082 & 0.0006 & 0.0001 & -0.9429 & -1.1446 & 3.8490 \end{bmatrix},$$

$$Q_1 = 10^8 \times \begin{bmatrix} -0.0688 & 0.1377 \\ -0.0001 & 0.0001 \\ 0.0007 & -0.0007 \\ 1.7209 & -3.4416 \\ -0.0005 & 0.0009 \\ -0.0002 & 0.0004 \end{bmatrix},$$

$$Q_2 = 10^8 \times \begin{bmatrix} -0.0688 & 0.1377 \\ -0.0001 & 0.0002 \\ 0.0007 & -0.0007 \\ 1.7399 & -3.4797 \\ -0.0005 & 0.0009 \\ -0.0002 & 0.0004 \end{bmatrix},$$

$$Q_3 = 10^8 \times \begin{bmatrix} -0.0688 & 0.1377 \\ -0.0002 & 0.0003 \\ 0.0007 & -0.0007 \\ 3.4348 & -6.8694 \\ -0.0005 & 0.0009 \end{bmatrix}.$$

Following the calculation procedures in Algorithm 2, one can solve the other gains of the UIO (7.49), as follows:

$$N_1 = \begin{bmatrix} -11.6750 & 0 & 0.0491 & -150 & 0 & 0 \\ -11.7462 & -0.0018 & 0.0783 & -150 & 0 & 0 \\ 0.0490 & 0 & -4.3126 & 1 & 0 & 0 \\ 108.9852 & 0 & -0.7284 & 0 & 1 & 0 \\ 101.2109 & 0 & -0.6764 & 0 & 0 & 1 \\ 58.6896 & 0 & -0.3922 & 0 & 0 & 0 \end{bmatrix},$$

$$N_2 = \begin{bmatrix} -11.4652 & 0 & 0.0483 & -150 & 0 & 0 \\ -11.5652 & -0.0018 & 0.0771 & -150 & 0 & 0 \\ 0.0482 & 0 & -4.2310 & 1 & 0 & 0 \\ 107.3911 & 0 & -0.7178 & 0 & 1 & 0 \\ 98.4041 & 0 & -0.6577 & 0 & 0 & 1 \\ 57.3102 & 0 & -0.3831 & 0 & 0 & 0 \end{bmatrix},$$

$$N_3 = \begin{bmatrix} -6.6958 & 0 & 0.0262 & -150 & 0 & 0 \\ -6.7118 & -0.0051 & 0.0448 & -150 & 0 & 0 \\ 0.0260 & 0 & -2.7753 & 1 & 0 & 0 \\ 136.0602 & 0 & -0.9094 & 0 & 1 & 0 \\ 104.5792 & 0 & -0.6990 & 0 & 0 & 1 \\ 64.4097 & 0 & -0.4305 & 0 & 0 & 0 \end{bmatrix},$$

$$G_1 = G_2 = G_3 = \begin{bmatrix} 0 \\ -0.1000 \\ 0 \\ 0 \\ 0 \\ 0 \end{bmatrix},$$

$$L_1 = 10^5 \times \begin{bmatrix} -0.1751 & 0.1751 \\ -0.1762 & 0.1762 \\ 0.0012 & -0.0012 \\ 1.6349 & -1.6349 \\ 1.5182 & -1.5182 \\ 0.8804 & -0.8804 \end{bmatrix},$$

$$L_2 = 10^5 \times \begin{bmatrix} -0.1720 & 0.1720 \\ -0.1735 & 0.1735 \\ 0.0011 & -0.0011 \\ 1.6109 & -1.6109 \\ 1.4761 & -1.4761 \\ 0.8597 & -0.8597 \end{bmatrix},$$

$$L_3 = 10^5 \times \begin{bmatrix} -0.1004 & 0.1004 \\ -0.1007 & 0.1007 \\ 0.0007 & -0.0007 \\ 2.0410 & -2.0410 \\ 1.5688 & -1.5688 \\ 0.9662 & -0.9662 \end{bmatrix}.$$

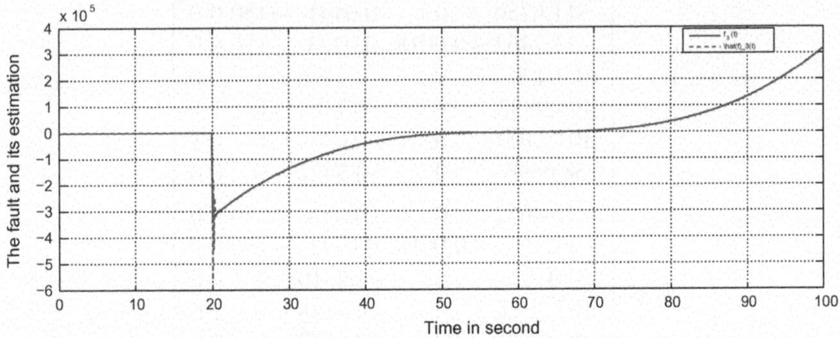

Fig. 7.8 The fault $f_3(t)$ and its estimation $\hat{f}_3(t)$

To illustrate the obtained techniques, a piecewise fault signal is defined as follows:

$$f_3(t) = \begin{cases} 0, & 0s \le t < 20s; \\ 5(t-60)^3, & 20s \le t < 100s; \end{cases}$$

The FE result is presented in Fig. 7.8. It can be seen that the FE $\hat{f}_3(t)$ converges to the real fault $f_3(t)$ accurately, which verifies the results for $f^{(k)}(t) = 0\ (k \ge 3)$ case.

7.2.6 *Summary*

In this section, a UIO-based robust FE observer is designed for continuous-time switched systems by decoupling the unknown input disturbances. By using the switched Lyapunov function method and the ADT technique, an FE algorithm is proposed to achieve robust FE. Furthermore, the results are extended to measurement disturbances case. Our design method is general in the sense that the same design procedure is applied for any fault vector such that $f^{(q)} = 0, q \in \mathbb{N}$. Finally, a two-tank system is borrowed to illustrate the developed techniques.

7.3 Nonlinear FEA with Actuator and Sensor Faults

This section investigates the problems of FE and fault accommodation for a class of nonlinear switched systems. First, an augmented switched system is constructed by forming an augmented state vector composed of the state vector and the fault vectors. Then, a UIO is designed for the augmented switched system to estimate the augmented state vector. With the assistance of the ADT method and the switched Lyapunov function technique, sufficient conditions are obtained to guarantee that the error system is GUAS with a prescribed H_∞ performance index. An algorithm is provided to show the procedures on how to design the UIO. Moreover, the results are extended to the measurement disturbances case. Based on the signal compensation principle, a dynamic output feedback controller is designed to ensure the asymptotical stability of the closed-loop system. Finally, simulation results are presented to demonstrate the proposed technique.

7.3.1 Introduction

With the quick development of modern techniques, the industrial systems are becoming more and more complex and expensive. Therefore, the demands of high reliability and security for such systems are increasing, which motivate the researches on fault diagnosis and FTC techniques. FD technique can be divided into three parts: FD, fault isolation, and FE [7]. FE is the final step of FD, which is mainly to estimate the magnitude of the faults.

Back into the past literature, observer-based FE is one of the most popular techniques. Until now, various observer-based approaches have been proposed. However, UIO is an effective observer to not only estimate the sensor but also estimate actuator faults, which can guarantee that is robust to unknown inputs and sensitive to faults. The problem of state estimation for nonlinear or linear system is solved by using UIO method in [43, 44]. The authors in [45], considered the problem of robust design by combining the UIO and FD filter.

Switched system is a kind of hybrid system, which is composed of several subsystems and a switching signal. ADT, one of the constrained switching signals, only needs to ensure that the average dwell time of the whole switched system is a constant, which is more general and less conservative than DT. In [42], FD for switched linear parameter-varying systems was proposed by using an ADT approach. In [46], this paper dealt with the problem of simultaneous finite-time control and FD for linear switched systems with state delay and parameter uncertainties. However, it should be noted that there still have no references to report the issues of FEA for nonlinear switched systems with actuator and sensor faults, and the estimated fault is generally a slope fault, which motivates us to try our best to investigate this issue.

The main contributions can be summarized as follows:

- UIO-based robust FE observer design is proposed for a class of nonlinear switched systems with the same term of actuator and sensor faults, which is capable of estimating all kinds of faults, i.e., $f^{(q)} = 0, q \in \mathbb{N} = \{1, 2, 3, \ldots\}$;
- the outside disturbances are considered as being partially decoupled, and the results are popularized into the case with measurement disturbances;
- a dynamic output feedback controller is designed to compensate the effect of the faults and ensure the asymptotical stability of the closed-loop system.

7.3.2 Model Description and Preliminaries

Consider the following nonlinear switched system:

$$\begin{cases} \dot{x}(t) = \sum_{i=1}^{N} \sigma_i(t)(A_i x(t) + B_i u(t) + D_{1i} d(t) + \\ \qquad D_{2i} w(t) + F_{1i} f(t) + \Phi_i(x(t), u(t))) \\ y(t) = \sum_{i=1}^{N} \sigma_i(t)(C_i x(t) + F_{2i} f(t)) \end{cases} \tag{7.78}$$

where $x(t) \in \mathbb{R}^n$, $u(t) \in \mathbb{R}^u$, $y(t) \in \mathbb{R}^m$ and $f(t) \in \mathbb{R}^l$ are the state vector, the control input vector, the measured output vector and the fault vector, respectively. $d(t) \in \mathbb{R}^d$ is the unknown input, and $w(t) \in \mathbb{R}^w$ is the independent Gaussian white noise disturbance, which can be easily decoupled. A_i, B_i, C_i, D_{1i}, D_{2i}, F_{1i}, and F_{2i} are known constant real matrices with appropriate dimensions. The piecewise constant function $\sigma_i(t) : [0, \infty) \to \{0, 1\}$ $(i \in \mathbb{N})$ is the switching signal satisfying $\sum_{i=1}^{N} \sigma_i(t) = 1$, which specifies which subsystem is activated at the switching instant. While $\Phi_i(x(t), u(t)) \in \mathbb{R}^n$ are real nonlinear vector functions with Lipschitz constant θ, namely,

$$\begin{aligned} &\| \Phi_i(\hat{x}(t), u(t)) - \Phi_i(x(t), u(t)) \| \leq \theta \| \hat{x}(t) - x(t) \|, \\ &\forall \{(\hat{x}(t), u(t)), (x(t), u(t))\} \in \mathbb{R} \times \mathbb{R}^n \times \mathbb{R}^u, \end{aligned} \tag{7.79}$$

where $\hat{x}(t)$ is the state estimation of $x(t)$.

For the section, the following preliminaries are given:

Assumption 11 The matrices D_{1i} are full column rank.

Assumption 12 The dimension of the measured output is no smaller than the dimension of the unknown input, i.e., $m \geq d$.

Assumption 13 $rank(C_i, D_{1i}) = rank(D_{1i})$.

Lemma 7.3 ([47]) *For any matrices $X \in \mathbb{R}^{s\times r}$, $Y \in \mathbb{R}^{r\times s}$, a time-varying matrix $F(t) \in \mathbb{R}^{r\times r}$ with $\| F(t) \| \leq 1$ and any scalar $\varepsilon > 0$, we have*

$$XF(t)Y + Y^T F^T(t) X^T \leq \varepsilon^{-1} X X^T + \varepsilon Y^T Y.$$

Lemma 7.4 (Schur complement) ([48]) *Let $S = \begin{bmatrix} S_{11} & S_{12} \\ S_{21} & S_{22} \end{bmatrix}$ to be a symmetric matrix, then the LMI $S < 0$ is equivalent to $S_{22} < 0$ and $S_{11} - S_{12} S_{22}^{-1} S_{12}^T < 0$.*

Lemma 7.5 *For a given positive constant β and positive definite matrices R_1, Q_1, if there exist matrices R_2, Q_2 and symmetric matrices R_3, Q_3 satisfying*

$$\begin{bmatrix} R_1 & R_2 \\ R_2^T & R_3 \end{bmatrix} > 0, \quad \begin{bmatrix} R_1 & R_2 \\ R_2^t & R_3 \end{bmatrix}^{-1} = \begin{bmatrix} Q_1 & Q_2 \\ Q_2^T & Q_3 \end{bmatrix},$$

if and only if

$$\begin{bmatrix} R_1 & I \\ I & Q_1 \end{bmatrix} \geq 0, rank\left(\begin{bmatrix} R_1 & I \\ I & Q_1 \end{bmatrix}\right) \leq n + min(n, \beta).$$

7.3.3 UIO Design

In this subsection, we mainly implement the design of UIO to achieve FE.

Firstly, we construct the augmented state vector as

$$\bar{x}(t) = \begin{bmatrix} x(t) \\ f(t) \\ \dot{f}(t) \\ \vdots \\ f^{(q-1)}(t) \end{bmatrix}, \tag{7.80}$$

we assume $f^{(q)} = 0, q \in \mathbb{N}$.

Then the following augmented system can be constructed:

$$\begin{cases} \dot{\bar{x}}(t) = \sum_{i=1}^{N} \sigma_i(t)(\bar{A}_i\bar{x}(t) + \bar{B}_i u(t) + \bar{D}_{1i} d(t) \\ \qquad + \bar{D}_{2i} w(t) + \bar{\Phi}_i(x(t), u(t)) + \bar{I} f^{(q)}(t)) \\ y(t) = \sum_{i=1}^{N} \sigma_i(t)\bar{C}_i\bar{x}(t) \end{cases}, \tag{7.81}$$

where

$$\bar{A}_i = \begin{bmatrix} A_i & F_{1i} & 0 & \cdots & 0 \\ 0 & 0 & I & \cdots & 0 \\ \vdots & \vdots & \vdots & \ddots & \vdots \\ 0 & 0 & 0 & \cdots & I \\ 0 & 0 & 0 & \cdots & 0 \end{bmatrix}, \quad \bar{B}_i = \begin{bmatrix} B_i \\ 0 \\ \vdots \\ 0 \\ 0 \end{bmatrix},$$

$$\bar{C}_i = \begin{bmatrix} C_i & F_{2i} & 0 & 0 & \cdots & 0 \end{bmatrix},$$

$$\bar{D}_{1i} = \begin{bmatrix} D_{1i} \\ 0 \\ \vdots \\ 0 \\ 0 \end{bmatrix}, \bar{D}_{2i} = \begin{bmatrix} D_{2i} \\ 0 \\ \vdots \\ 0 \\ 0 \end{bmatrix}, \bar{I} = \begin{bmatrix} 0 \\ 0 \\ \vdots \\ 0 \\ I \end{bmatrix},$$

$$\bar{\Phi}_i(x(t), u(t)) = \begin{bmatrix} \Phi_i(x(t), u(t)) \\ 0 \\ \vdots \\ 0 \\ 0 \end{bmatrix}. \tag{7.82}$$

Clearly, the state vector $x(t)$ and the fault vector $f(t)$ can be simultaneously estimated if an observer for the augmented system (7.81), can be constructed.

In order to estimate $\bar{x}(t)$, the following UIO is expressed by:

$$\begin{cases} \dot{z}(t) = \sum_{i=1}^{N} \sigma_i(t)(N_i z(t) + G_i u(t) + L_i y(t) + T_i \bar{\Phi}_i(\hat{x}(t), u(t))) \\ \hat{\bar{x}}(t) = z(t) + \sum_{i=1}^{N} \sigma_i(t) H_i y(t) \end{cases}, \tag{7.83}$$

where $z(t) \in \mathbb{R}^{\bar{n}}$ ($\bar{n} = n + ql$) is the state of the observer, $\hat{\bar{x}}(t) \in \mathbb{R}^{\bar{n}}$ represents the state estimation of $\bar{x}(t)$. Matrices N_i, G_i, H_i and L_i are the gains of UIO to be designed later. Further, $T_i = I_{\bar{n}} - H_i \bar{C}_i$, $L_i = L_{1i} + L_{2i}$, $L_{1i}, L_{2i} \in \bar{n} \times (\bar{n} - 1)$.

Define the state error estimation $e(t) = \bar{x}(t) - \hat{\bar{x}}(t)$, one has

$$e(t) = \bar{x}(t) - z(t) - \sum_{i=1}^{N} \sigma_i(t) H_i y(t) = \sum_{i=1}^{N} \sigma_i(t) T_i \bar{x}(t) - z(t). \tag{7.84}$$

From (7.81), (7.83), and (7.84), the derivative of the state error estimation can be calculated as

$$\begin{aligned} \dot{e}(t) &= \sum_{i=1}^{N} \sigma_i(t) T_i \dot{\bar{x}}(t) - \dot{z}(t) \\ &= \sum_{i=1}^{N} \sigma_i(t)((T_i \bar{A}_i - L_{1i} \bar{C}_i) e(t) + ((T_i \bar{A}_i - L_{1i} \bar{C}_i) H_i \\ &\quad - L_{2i}) y(t) + (T_i \bar{A}_i - N_i - L_{1i} \bar{C}_i) z(t) + (T_i \bar{B}_i - \\ &\quad G_i) u(t) + T_i \bar{D}_{1i} d(t) + T_i \bar{D}_{2i} w(t) + T_i \tilde{\Phi}), \end{aligned} \tag{7.85}$$

where $\tilde{\Phi} = \bar{\Phi}(x(t), u(t)) - \bar{\Phi}(\hat{x}(t), u(t))$.

If the following relationships hold:

$$(T_i \bar{A}_i - L_{1i} \bar{C}_i) H_i - L_{2i} = 0, \tag{7.86}$$

$$T_i \bar{A}_i - N_i - L_{1i} \bar{C}_i = 0, \tag{7.87}$$

$$T_i \bar{B}_i - G_i = 0, \tag{7.88}$$

$$T_i \bar{D}_{1i} = 0. \tag{7.89}$$

Then the state error estimation (7.85), reduces to

$$\dot{e}(t) = \sum_{i=1}^{N} \sigma_i(t)(N_i e(t) + T_i \bar{D}_{2i} w(t) + T_i \tilde{\Phi}). \tag{7.90}$$

Remark 7.9 From (7.89), the matrices $H_i (i \in \mathbb{N})$ can be calculated by

$$H_i = \bar{D}_{1i}[(\bar{C}_i \bar{D}_{1i})^T (\bar{C}_i \bar{D}_{1i})]^{-1} (\bar{C}_i \bar{D}_{1i})^T, \tag{7.91}$$

and the matrices G_i can be solved by (7.88) and (7.91).

Remark 7.10 If we have $rank(C_i, D_{1i}) = rank(D_{1i})$, one can get that $rank(\bar{C}_i, \bar{D}_{1i}) = rank(\bar{D}_{1i})$.

Next, based on the state error system (7.90), the following theorem provides how to design the UIO (7.83).

Theorem 7.8 *For given scalars $\alpha > 0$, $\mu > 1$, $\varepsilon > 0$ and $\gamma > 0$, if the conditions (7.86)–(7.89), hold and there exist positive definite matrices P_i and P_j, matrices Q_i, $\forall i, j \in \mathbb{N}$, which satisfy*

$$P_i \leq \mu P_j, \tag{7.92}$$

$$\begin{bmatrix} \Psi_{11i} & P_i \hat{D}_{2i} & P_i T_i \\ * & -\gamma^2 I_{l \times d} & 0 \\ * & * & -\varepsilon I_{\bar{n}} \end{bmatrix} < 0, \tag{7.93}$$

where $\Psi_{11i} = \hat{A}_i^T P_i + P_i \hat{A}_i - \bar{C}_i^T Q_i^T - Q_i \bar{C}_i + \alpha P_i + (\varepsilon\theta^2 + 1)I$, $\hat{A}_i^T = T_i \bar{A}_i$, $\hat{D}_{2i} = T_i \bar{D}_{2i}$, and $L_{1i} = P_i^{-1} Q_i$, then, for the switching signal $\sigma_i(t)$ with ADT constraint (7.11), there exists a UIO (7.83), for the augmented switched system (7.81), such that the error system (7.90), is GUAS with H_∞ performance index γ.

Proof Firstly, we establish the stability of system (7.90). If we assume $w(t) = 0$, consider the following switched Lyapunov function for the error dynamics (7.90), as

$$V_i(t, e(t)) = e^T(t)(\sum_{i=1}^{N} \sigma_i(t) P_i) e(t). \tag{7.94}$$

One has

$$\begin{aligned} &\dot{V}_i(t, e(t)) + \alpha V_i(t, e(t)) \\ &= \sum_{i=1}^{N} \sigma_i(t)(e^T(t)(N_i^T P_i + P_i N_i + \alpha P_i)e(t) + \tilde{\Phi}^T T_i^T P_i e(t) + e^T(t) P_i T_i \tilde{\Phi}. \end{aligned} \tag{7.95}$$

By using Lemma 7.3 and (7.79), one can get

$$
\begin{aligned}
&\dot{V}_i(t,e(t))+\alpha V_i(t,e(t))\\
&\leq \sum_{i=1}^{N}\sigma_i(t)(e^T(t)(N_i^T P_i + P_i N_i + \alpha P_i + \varepsilon^{-1}P_i T_i T_i^T P_i + \varepsilon\theta^2 I_{\bar{n}})e(t)). \quad (7.96)
\end{aligned}
$$

Lemma 7.4 and (7.93) imply

$$
\Psi_{11i} + \varepsilon^{-1}P_i T_i (P_i T_i)^T < 0, \tag{7.97}
$$

which leads to

$$
N_i^T P_i + P_i N_i + \alpha P_i + \varepsilon^{-1}P_i T_i T_i^T P_i + \varepsilon\theta^2 I_{\bar{n}} < 0. \tag{7.98}
$$

Apparently, we have $\dot{V}_i(t,e(t)) < -\alpha V_i(t,e(t))$. According to Lemma 7.2 and (7.92), one can get that the system (7.90), is GUAS.

Next, we give the proof of H_∞ performance γ as follows:

Let

$$
\Gamma = \int_0^T (e^T(t)e(t) - \gamma^2 w^T(t)w(t) + \alpha V_i(t,e(t))) \tag{7.99}
$$

One has

$$
\begin{aligned}
\Gamma &= \int_0^T (e^T(t)e(t) - \gamma^2 w^T(t)w(t) + \alpha V_i(t,e(t)) + \dot{V}_i(t,e(t)))dt - \int_0^T \dot{V}_i(t,e(t))dt\\
&= \int_0^T \sum_{i=1}^{N}\sigma_i(t)\begin{bmatrix} e^T(t) & w^T(T)\end{bmatrix}\Omega\begin{bmatrix} e(t)\\ w(t)\end{bmatrix}d(t) - \int_0^T \dot{V}_i(t,e(t))dt, \quad (7.100)
\end{aligned}
$$

where

$$
\Omega = \begin{bmatrix} \Psi_{11i} + \varepsilon^{-1}P_i T_i T_i^T P_i & P_i\hat{D}_{2i}\\ * & -\gamma^2 I_l\end{bmatrix} < 0. \tag{7.101}
$$

Under the initial condition $e(0) = 0$, one has

$$
\int_0^T \dot{V}_i(t,e(t))dt > 0 \tag{7.102}
$$

By (7.101) and (7.102), we get

$$
e^T(t)e(t) - \gamma^2 w^T(t)w(t) + \dot{V}_i(t,e(t)) < -\alpha V_i(t,e(t)). \tag{7.103}
$$

Then one can conclude that the error system (7.90), has H_∞ performance γ. The proof is completed. (□)

In the following algorithm, the procedure on how to design the UIO is given.

Algorithm 3:

Step1: Construct the augmented system (7.81);

Step2: Calculate the UIO gain H_i in (7.91), then G_i can be solved by (7.88);

Step3: Solve the matrices P_i and Q_i by Theorem 7.8, and calculate the gain $L_{1i} = P_i^{-1}Q_i$;

Step4: Calculate the other gains L_{2i} and N_i by (7.86) and (7.87);

Step5: Implement the robust UIO (7.83), to obtain the augmented state estimation $\hat{\bar{x}}(t)$, and the state estimation $\hat{x}(t)$ and FE $\hat{f}(t)$ can be got by the following equations:

$$\begin{aligned} \hat{x}(t) &= \begin{bmatrix} I_n \ \underbrace{0 \cdots 0}_{(q-1)0_{n\times 2l} matrices} \end{bmatrix} \hat{\bar{x}}(t), \\ \hat{f}(t) &= \begin{bmatrix} 0_{n\times(n+l)} \ I_l \ \underbrace{0 \cdots 0}_{(q-2)0_{l\times l} matrices} \end{bmatrix} \hat{\bar{x}}(t). \end{aligned} \tag{7.104}$$

7.3.4 UIO Design for Measurement Disturbances

In this section, a more general case is taken into consideration, i.e., the nonlinear switched system (7.78), is not only affected by process disturbances, but also corrupted by measurement disturbances, which is described by

$$\begin{cases} \dot{x}(t) = \sum_{i=1}^{N} \sigma_i(t)(A_i x(t) + B_i u(t) + D_{1i} d(t) + \\ \qquad D_{2i} w(t) + F_{1i} f(t) + \Phi_i(x(t), u(t))) \\ y(t) = \sum_{i=1}^{N} \sigma_i(t)(C_i x(t) + F_{2i} f(t) + E_i d_s(t)) \end{cases}, \tag{7.105}$$

where $d_s(t) \in \mathbb{R}^s$ stands for the distribution matrix of the measurement noise, E_i are constant matrices with appropriate dimensions, and the other symbols are the same as defined in the system (7.78).

Firstly, an augmented state vector is defined as in (7.80), then one can get the augmented switched system as

$$\begin{cases} \dot{\bar{x}}(t) = \sum_{i=1}^{N} \sigma_i(t)(\bar{A}_i \bar{x}(t) + \bar{B}_i u(t) + \bar{D}_{1i} d(t) + \\ \qquad \bar{D}_{2i} w(t) + \bar{\Phi}_i(x(t), u(t)) + \bar{I} f^{(q)}(t)) \\ y(t) = \sum_{i=1}^{N} \sigma_i(t)(\bar{C}_i \bar{x}(t) + E_i d_s(t)) \end{cases}, \tag{7.106}$$

where the coefficient matrices are defined as in (7.82), except for E_i and $d_s(t)$.

Next, a UIO is constructed as same form as in (7.83). Similarly, the error estimation $\bar{e}(t) = \bar{x}(t) - \hat{\bar{x}}(t)$ can be got.

By (7.83) and (7.106), and under the same conditions (7.86)–(7.89), the state error estimation can be calculated as

$$\dot{\bar{e}}(t) = \sum_{i=1}^{N} \sigma_i(t)(N_i\bar{e}(t) + M_i v(t) + T_i\tilde{\Phi}), \tag{7.107}$$

where $M_i = \begin{bmatrix} T_i\bar{D}_{2i} & -L_{1i}E_i & -H_iE_i \end{bmatrix}$ and $v(t) = \begin{bmatrix} w^T(t) & d_s^T(t) & \dot{d}_s^T(t) \end{bmatrix}^T$.

Remark 7.11 Similarly, from the condition (7.89), the gain matrices $H_i (i \in \mathbb{N})$ are the same as before.

The following theorem is given to design the parameters of observer (7.83).

Theorem 7.9 *For given scalars $\alpha > 0$, $\mu > 1$, $\varepsilon > 0$ and $\gamma > 0$, if the conditions (7.86)–(7.89), hold and there exist positive definite matrices P_i and P_j , matrices $Q_i, \forall i, j \in \mathbb{N}$, which satisfy*

$$P_i \leq \mu P_j, \tag{7.108}$$

$$\begin{bmatrix} \Psi_{11i} & P_i\hat{D}_{2i} & P_iT_i & -Q_iE_i & -P_i\hat{E}_i \\ * & -\gamma^2 I_{l\times d} & 0 & 0 & 0 \\ * & * & -\varepsilon I_{\bar{n}} & 0 & 0 \\ * & * & * & -\gamma^2 I_s & 0 \\ * & * & * & * & -\gamma^2 I_s \end{bmatrix} < 0, \tag{7.109}$$

where $\Psi_{11i} = \hat{A}_i^T P_i + P_i\hat{A}_i - \bar{C}_i^T Q_i^T - Q_i\bar{C}_i + \alpha P_i + (\varepsilon\theta^2 + 1)I$, $\hat{A}_i^T = T_i\bar{A}_i$, $\hat{D}_{2i} = T_i\bar{D}_{2i}$, $\hat{E}_{2i} = H_iE_i$ and $L_{1i} = P_i^{-1}Q(i)$, then, for the switching signal $\sigma_i(t)$ with ADT constraint (7.11), there exists a UIO (7.83), for the augmented switched system (7.106), such that the error system (7.107), is GUAS with H_∞ performance index γ.

Proof By using Lemmas 7.3–7.5, and following the same proof lines as in Theorem 7.8, the condition (7.109), can be easily deduced. Firstly, we establish the stability of system. If we assume $v(t) = 0$, then consider the same switched Lyapunov function (7.94), for the error dynamics (7.107), one has

$$\begin{aligned} &\dot{V}_i(t, \bar{e}(t)) + \alpha V_i(t, \bar{e}(t)) \\ &= \sum_{i=1}^{N} \sigma_i(t)(\bar{e}^T(t)(N_i^T P_i + P_iN_i + \alpha P_i)\bar{e}(t) + \\ &\quad \tilde{\Phi}^T T_i^T P_i\bar{e}(t) + \bar{e}^T(t)P_iT_i\tilde{\Phi}). \end{aligned} \tag{7.110}$$

By using Lemma 7.3 and (7.79), one can get

$$\begin{aligned} &\dot{V}_i(t, \bar{e}(t)) + \alpha V_i(t, \bar{e}(t)) \\ &\leq \sum_{i=1}^{N} \sigma_i(t)(\bar{e}^T(t)(N_i^T P_i + P_iN_i + \alpha P_i + \\ &\quad \varepsilon^{-1} P_iT_iT_i^T P_i + \varepsilon\theta^2 I_{\bar{n}})\bar{e}(t). \end{aligned} \tag{7.111}$$

Similar to Theorem 7.8, one can have $\dot{V}_i(t,\bar{e}(t)) < -\alpha V_i$, which means the system (7.107), is GUAS.

Next, the proof of the H_∞ performance γ is given.

Define

$$\Gamma = \int_0^T (\bar{e}^T(t)\bar{e}(t) - \gamma^2 v^T(t)v(t) + \alpha V_i(t,\bar{e}(t))). \tag{7.112}$$

One has

$$\begin{aligned}\Gamma &= \int_0^T (\bar{e}^T(t)\bar{e}(t) - \gamma^2 v^T(t)v(t) + \alpha V_i(t,\bar{e}(t)) + \dot{V}_i(t,\bar{e}(t)))dt - \int_0^T \dot{V}_i(t,\bar{e}(t))dt \\ &= \int_0^T \sum_{i=1}^N \sigma_i(t) \begin{bmatrix} \bar{e}(t) \\ v(t) \\ d_s(t) \\ \dot{d}_s(t) \end{bmatrix}^T \Theta \begin{bmatrix} \bar{e}(t) \\ v(t) \\ d_s(t) \\ \dot{d}_s(t) \end{bmatrix} d(t) - \int_0^T \dot{V}_i(t,\bar{e}(t))dt,\end{aligned} \tag{7.113}$$

where

$$\Theta = \begin{bmatrix} \Psi_{11i} + \varepsilon^{-1} P_i T_i T_i^T P_i & P_i \hat{D}_{2i} & -Q_i E_i & -Q_i \hat{E}_i \\ * & -\gamma^2 I_{l\times d} & 0 & 0 \\ * & * & -\gamma^2 I_s & 0 \\ * & * & * & -\gamma^2 I_s \end{bmatrix} < 0.$$

As the same as (7.102), under the initial condition $e(0) = 0$, one has $\int_0^T \dot{V}_i(t,\bar{e}(t))dt > 0$.

One can get

$$\bar{e}^T(t)\bar{e}(t) - \gamma^2 v^T(t)v(t) + \dot{V}_i(t,\bar{e}(t)) < -\alpha V_i(t,\bar{e}(t)). \tag{7.114}$$

Then one can conclude that the error system (7.107) has H_∞ performance γ. The proof is concluded. (□)

Remark 7.12 The procedures on how to design the UIO is similar to Algorithm 3.

7.3.5 Dynamic Output Feedback Controller Design

In this subsection, to guarantee the asymptotical stability of the closed-loop system, a dynamic output feedback controller is designed to compensating the faulty signals. The following dynamic output feedback controller is constructed:

$$\begin{cases} \dot{x}_c(t) = \sum_{i=1}^N \sigma_i(t)(A_{ci}x_c(t) + B_{ci}y_c(t)) \\ u(t) = \sum_{i=1}^N \sigma_i(t)(C_{ci}x_c(t) - B_i^\dagger F_{1i}\hat{f}(t)) \end{cases}, \tag{7.115}$$

where A_{ci}, B_{ci}, and C_{ci} are controller gains to be calculated later, $B_i^\dagger$ is the generalized inverse of matrices B_i. On the basis of the estimation for the augmented state vector $\hat{\bar{x}}(t)$, the considered faults can be constructed as

$$\hat{f}(t) = \begin{bmatrix} 0 \ I_l \ 0 \cdots 0 \end{bmatrix} \hat{\bar{x}}(t) = J\hat{\bar{x}}(t), \tag{7.116}$$

where $J = \begin{bmatrix} 0 \ I_l \ 0 \cdots 0 \end{bmatrix}$.

Then the measurement output can be compensated as follows:

$$y_c = y(t) - F_{2i}\hat{f}(t) = y(t) - F_{2i}J\hat{\bar{x}}(t). \tag{7.117}$$

Define $\tilde{x}(t) = \begin{bmatrix} x^T(t) \ x_c^T(t) \end{bmatrix}^T$, and substitute (7.115) to (7.78), one has

$$\dot{\tilde{x}}(t) = \sum_{i=1}^{N} \sigma_i(t)(\tilde{A}_i\tilde{x}(t) + \tilde{D}_i\psi(t) + \bar{I}\Xi_i), \tag{7.118}$$

where

$$\begin{aligned} \tilde{A}_i &= \begin{bmatrix} A_i & B_iC_{ci} \\ B_{ci}C_i & A_{ci} \end{bmatrix}, \ \Xi_i = \begin{bmatrix} \Phi_i(x(t), u(t)) \\ 0 \end{bmatrix}, \\ \psi(t) &= \begin{bmatrix} d(t) \\ w(t) \\ \tilde{f}(t) \end{bmatrix}, \ \tilde{D}_i = \begin{bmatrix} D_{1i} & D_{2i} & F_{1i} \\ 0 & 0 & B_{ci}F_{2i} \end{bmatrix}, \\ \bar{I} &= \begin{bmatrix} I_{n\times n} \ 0 \end{bmatrix}, \quad \tilde{f}(t) = f(t) - J\hat{\bar{x}}(t). \end{aligned} \tag{7.119}$$

In the following theorem, the design of the dynamic output feedback controller (7.115), for the descriptor switched system (7.78), is given.

Theorem 7.10 *For some given positive constants γ, $\tilde{\alpha}$, and $\tilde{\mu} > 1$, if there exist positive definite matrices $P_{11i} \in \mathbb{N}^{n\times n}$, $P_{22i} \in \mathbb{N}^{n\times n}$, $P_{11j} \in \mathbb{N}^{n\times n}$, $P_{22j} \in \mathbb{N}^{n\times n}$, $Q_{11i} \in \mathbb{N}^{n\times n}$, matrices $P_{12i} \in \mathbb{N}^{n\times n}$, $M_i \in \mathbb{N}^{n\times n}$, $N_i \in \mathbb{N}^{n\times m}$, and $W_i \in \mathbb{N}^{l\times n}$, for $\forall i, j \in \mathbb{N}$, which satisfy*

$$\begin{bmatrix} P_{11i} & I_n \\ I_n & Q_{11i} \end{bmatrix} > 0, \tag{7.120}$$

$$\begin{bmatrix} P_{11i} & P_{12i} \\ P_{12i}^T & P_{22i} \end{bmatrix} \leq \tilde{\mu} \begin{bmatrix} P_{11j} & P_{12j} \\ P_{12j}^T & P_{22j} \end{bmatrix}, \tag{7.121}$$

$$\begin{bmatrix} \Upsilon_{11i} & \Upsilon_{12i} & \Upsilon_{13i} & \Upsilon_{14i} \\ * & -\gamma I_{d+2l} & 0 & 0 \\ * & * & -I_n & 0 \\ * & * & * & \Upsilon_{14i} \end{bmatrix} < 0, \tag{7.122}$$

where

$$\Upsilon_{11i} = \begin{bmatrix} \mathscr{P} & A_i + M_i^T \\ A_i^T + M_i & \mathscr{Q} \end{bmatrix}, \Upsilon_{13i} = \begin{bmatrix} Q_{11i} \\ I_n \end{bmatrix}, \Upsilon_{14i} = \begin{bmatrix} \tilde{\alpha} I_n & 0 & \bar{I} \\ \tilde{\alpha} P_{11i} & \tilde{\alpha} P_{12i} & 0 \end{bmatrix},$$
$$\Upsilon_{12i} = \begin{bmatrix} D_{1i} & D_{2i} & F_{1i} & 0 \\ P_{11i} D_{1i} & P_{11i} D_{2i} & P_{11i} F_{1i} & N_i F_{2i} \end{bmatrix}, \Upsilon_{44i} = \begin{bmatrix} -\tilde{\alpha} P_{11i} & -\tilde{\alpha} P_{12i} & 0 \\ -\tilde{\alpha} P_{12i}^T & -\tilde{\alpha} P_{22i} & 0 \end{bmatrix}. \tag{7.123}$$

where $\mathscr{P} = A_i Q_{11i} + Q_{11i} A_i^T + B_i W_i + W_i^T B_i^T, \mathscr{Q} = P_{11i} A_i + A_i^T P_{11i} + N_i C_i + C_i^T N_i^T$.

Then, for the switching signal $\sigma_i(t)$ *with ADT constraint (7.11), there exists a dynamic output feedback controller (7.115), such that the system (7.78), is GUAS with* H_∞ *performance index* γ. *Under this case, the gains of the output feedback controller (7.115), are given by*

$$A_{ci} = P_{12i}^{-1}(M_i - P_{11i} A_i Q_{11i} - N_i C_i Q_{11i} - P_{11i} B_i W_i) Q_{12i}^{-T}, \tag{7.124}$$
$$B_{ci} = P_{12i}^{-1} N_i, \tag{7.125}$$
$$C_{ci} = W_i Q_{12i}^{-T}, \tag{7.126}$$

where the matrices P_{12i} *and* Q_{12i} *satisfy* $P_{11i} Q_{11i} + P_{12i} Q_{12i}^T = I_n$.

Proof The following switched Lyapunov function is constructed

$$V_i(t, \tilde{x}(t)) = \tilde{x}^T(t)(\sum_{i=1}^{N} \sigma_i(t) \tilde{P}_i) \tilde{x}(t), \tag{7.127}$$

where $\tilde{P}_i = \begin{bmatrix} P_{11i} & P_{12i} \\ P_{12i}^T & P_{22i} \end{bmatrix} > 0$. By using Eq. (7.118), one can get its derivative is

$$\dot{V}_i(t, \tilde{x}(t)) = \sum_{i=1}^{N} \sigma_i(t)(\tilde{x}^T(t)(\tilde{A}_i^T \tilde{P}_i + \tilde{P}_i \tilde{A}_i)\tilde{x}(t) + 2\tilde{x}^T(t)\tilde{P}_i$$
$$+ \tilde{D}_i \psi(t) + \Xi^T \bar{I}^T \tilde{P}_i \tilde{x}(t) + \tilde{x}^T(t) \tilde{P}_i \bar{I} \Xi). \tag{7.128}$$

Define

$$K = \int_0^T (\tilde{x}^T(t)\tilde{x}(t) - \gamma^2 \psi^T(t)\psi(t) + \alpha V_i(t, \tilde{x}(t)) + \dot{V}_i(t, \tilde{x}(t)))dt. \tag{7.129}$$

By submitting (7.127) and (7.128), to (7.129), one has

$$K = \sum_{i=1}^{N} \sigma_i(t) \int_0^T \xi^T(t) \Lambda_i \xi(t) dt, \tag{7.130}$$

where

$$\Lambda_i = \begin{bmatrix} \tilde{A}_i^T \tilde{P}_i + \tilde{P}_i \tilde{A}_i + \bar{I}^T \bar{I} + \tilde{\alpha} \tilde{P}_i & \tilde{P}_i \tilde{D}_i & \tilde{P}_i \bar{I} \\ \tilde{D}_i^T \tilde{P}_i - \gamma^2 I & 0 \\ \bar{I}^T \tilde{P}_i & 0 & 0 \end{bmatrix},$$

$$\xi(t) = \begin{bmatrix} \tilde{x}(t) \\ \psi(t) \\ \Xi \end{bmatrix}, \quad \bar{I} = \begin{bmatrix} I & 0 \end{bmatrix}. \tag{7.131}$$

If $\Lambda_i < 0$, based Lemma 7.4, which is equivalent to

$$\begin{bmatrix} \tilde{A}_i^T \tilde{P}_i + \tilde{P}_i \tilde{A}_i & \tilde{P}_i \tilde{D}_i & \bar{I}^T & \tilde{\alpha} \tilde{P}_i & \tilde{P}_i \bar{I} \\ * & -\gamma^2 I & 0 & 0 & 0 \\ * & * & -I & 0 & 0 \\ * & * & * & -\tilde{\alpha} \tilde{P}_i & 0 \\ * & * & * & * & 0 \end{bmatrix}. \tag{7.132}$$

Let

$$\tilde{P}_i = \begin{bmatrix} P_{11i} & P_{12i} \\ P_{12i}^T & P_{22i} \end{bmatrix}, \quad \tilde{P}_i^{-1} = \begin{bmatrix} Q_{11i} & Q_{12i} \\ Q_{12i}^T & Q_{22i} \end{bmatrix}, \tag{7.133}$$

which follows

$$\tilde{P}_i \begin{bmatrix} Q_{11i} \\ Q_{12i}^T \end{bmatrix} = \begin{bmatrix} I \\ 0 \end{bmatrix}. \tag{7.134}$$

Define

$$\Pi_i = \begin{bmatrix} Q_{11i} & I \\ Q_{12i}^T & 0 \end{bmatrix}, \tag{7.135}$$

and pre-multiply and post-multiply diag $\{\Pi_i^T, I, I, I, I\}$ and diag $\{\Pi_i, I, I, I, I\}$ to both sides of (7.132), which is equivalent to (7.122), with the definitions in (7.124)–(7.126). By combining the condition (7.121), it obtains that the augmented system (7.118), is asymptotically stable with H_∞ performance $\| x(t) \|_2 \leq \| g(t) \|_2$. By Lemma 7.5, the existence of matrices $\tilde{P}_i$ and $\tilde{P}_i^{-1}$ in the condition (7.120), is guaranteed. Then the proof is completed. (□)

Then, the following algorithm on how to design the dynamic output feedback controller is given.

Algorithm 4:

Step 1: Substitute dynamic output feedback controller (7.115), into system (7.78), one can get the augmented systems (7.118);

Step 2: Calculate the matrices P_{11i}, P_{12i}, Q_{11i}, M_i, N_i, W_i by using LMI toolbox to solving the condition (7.121) and (7.122), with the ADT constraint (7.11);

Step 3: According to the matrices P_{11i}, Q_{11i}, P_{12i}, the matrices Q_{12i} are calculated by using the equation P_{12i} and Q_{12i} satisfy $P_{11i} Q_{11i} + P_{12i} Q_{12i}^T = I_n$;

Step 4: Substitute the matrices P_{11i}, P_{12i}, Q_{11i}, Q_{12i}, M_i, N_i and W_i into (7.124)–(7.126), to calculate the controller gains A_{ci}, B_{ci} and C_{ci}.

7.3.6 Simulation Example

A liquid level control system in the Sect. 1.2.8, has been considered to test the effectiveness of the proposed UIO approach which is described in Fig. 1.10. The system parameters are set as:

$$
\begin{aligned}
&A_1 = \begin{bmatrix} -0.0007 & 0.0007 \\ 0.0011 & -0.0011 \end{bmatrix}, B_i = \begin{bmatrix} 0.2 \\ 0.1 \end{bmatrix},\\
&A_2 = \begin{bmatrix} -0.0013 & 0.0007 \\ 0.0011 & -0.0011 \end{bmatrix}, D_{1i} = \begin{bmatrix} 0.1 \\ 0.1 \end{bmatrix},\\
&A_3 = \begin{bmatrix} -0.0007 & 0.0007 \\ 0.0011 & -0.0044 \end{bmatrix}, D_{2i} = \begin{bmatrix} 0.2 \\ 0.2 \end{bmatrix},\\
&C_i = \begin{bmatrix} 0.01 & 0 \\ 0.01 & 0 \end{bmatrix}, F_{1i} = F_{2i} = \begin{bmatrix} 0.2 \\ 0.1 \end{bmatrix},\\
&\bar{\Phi}_i(x(t), u(t)) = \left[\Phi_i^T(x(t), u(t)) \; 0 \right]^T, i = 1, 2, 3.
\end{aligned}
$$

For given $\mu = 1.002$, $\alpha = 0.001$, $\gamma = 0.1$, $\varepsilon = 0.1$, and Lipschitz constant $\theta = 0.1$, one has $\tau_a^* = \frac{ln\mu}{\alpha} = 1.998$. We assume $\bar{\Phi}_i(x(t), u(t)) = sin(x(t))$. The switching signal is given in Fig. 7.9.

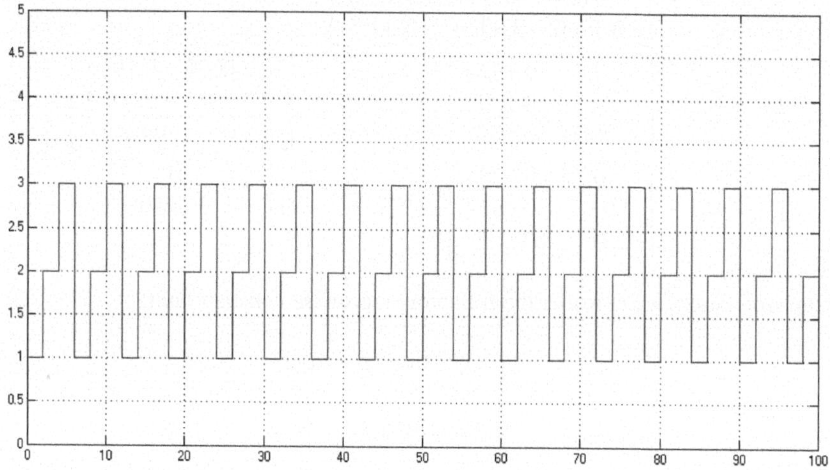

Fig. 7.9 The switching signal $\sigma_i(t)$

Case 1: Constant faults. For the given switched system (7.78), if only constant faults are considered, i.e., $\dot{f} = 0$. Then the augmented state is $\bar{x}(t) = \left[x^T(t)\ f^T(t)\right]^T$, and the coefficient matrices in 7.82, are

$$\bar{A}_i = \begin{bmatrix} A_i & F_{1i} \\ 0 & 0 \end{bmatrix}, \bar{C}_i = \begin{bmatrix} C_i & F_{2i} \end{bmatrix}, \bar{B}_i = \begin{bmatrix} B_i \\ 0 \end{bmatrix},$$

$$\bar{D}_{1i} = \begin{bmatrix} D_{1i} \\ 0 \end{bmatrix}, \quad \bar{D}_{2i} = \begin{bmatrix} D_{2i} \\ 0 \end{bmatrix}, \bar{I} = \begin{bmatrix} 0 \\ I \end{bmatrix}.$$

Then the observer gain $H_i (i = 1, 2, 3)$ can be calculated by Step 2 of Algorithm 3 as follows:

$$H_i = \bar{D}_{1i}[(\bar{C}_i\bar{D}_{1i})^T(\bar{C}_i\bar{D}_{1i})]^{-1}(\bar{C}_i\bar{D}_{1i})^T = \begin{bmatrix} 100 & 0 \\ 50 & 0 \\ 0 & 0 \end{bmatrix}, i = 1, 2, 3.$$

P_i and Q_i $(i = 1, 2, 3)$ in Theorem 7.8 can be solved by using Matlab tool. Then, one can obtain the other gains of the UIO (7.83), according to Algorithm 3

$$N_1 = 10^2 \times \begin{bmatrix} -1.4214 & 0.0000 & 0.1979 \\ -0.6916 & -0.1000 & 0.1000 \\ 0.0123 & 0.0000 & -0.3509 \end{bmatrix},$$

$$N_2 = 10^2 \times \begin{bmatrix} -1.4397 & 0.0000 & -0.1835 \\ -0.7049 & -0.1000 & -0.0734 \\ 0.1235 & 0.0000 & -0.3373 \end{bmatrix},$$

$$N_3 = 10^2 \times \begin{bmatrix} -1.2227 & 0.0000 & -0.4526 \\ -0.5834 & -0.3000 & -0.1585 \\ 0.1980 & 0.0000 & -0.3011 \end{bmatrix},$$

$$L_1 = 10^2 \times \begin{bmatrix} 0 & -0.1923 \\ 0 & -0.1003 \\ 0.0000 & 0.0040 \end{bmatrix}, L_2 = 10^2 \times \begin{bmatrix} -0.0000 & -0.1916 \\ -0.0023 & -0.0959 \\ 0 & 0.0188 \end{bmatrix},$$

$$L_3 = 10^2 \times \begin{bmatrix} 0 & -0.0185 \\ 0.0374 & -0.0859 \\ 0 & 0.0032 \end{bmatrix}, G_1 = G_2 = G_3 = \begin{bmatrix} 0 \\ 0.0050 \\ 0 \end{bmatrix}.$$

In the simulation, we consider the following piecewise constant fault:

$$f_1(t) = \begin{cases} 0, & 0s \le t < 10s; \\ 2, & 10s \le t < 50s; \\ -2, & 50s \le t < 100s. \end{cases}$$

The FE result is shown in Fig. 7.10.

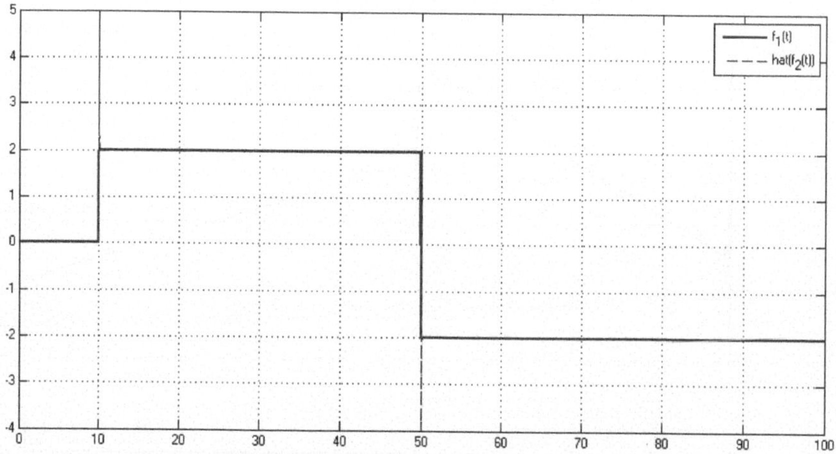

Fig. 7.10 The fault $f_1(t)$ (solid line) and its estimation $\hat{f}_1(t)$ (dotted line)

Case 2: Slope fault. For the given switched system (7.78), if slop faults are considered, i.e., $\ddot{f} = 0$. Then the augmented state is $\bar{x}(t) = \begin{bmatrix} x^T(t) & f^T(t) & \dot{f}^T(t) \end{bmatrix}^T$, and the coefficient matrices in (7.82) are

$$\bar{A}_i = \begin{bmatrix} A_i & F_{1i} & 0 \\ 0 & 0 & I \\ 0 & 0 & 0 \end{bmatrix}, \bar{B}_i = \begin{bmatrix} B_i \\ 0 \\ 0 \end{bmatrix}, \bar{D}_{1i} = \begin{bmatrix} D_{1i} \\ 0 \\ 0 \end{bmatrix},$$

$$\bar{D}_{2i} = \begin{bmatrix} D_{2i} \\ 0 \\ 0 \end{bmatrix}, \bar{I} = \begin{bmatrix} 0 \\ 0 \\ I \end{bmatrix}, \bar{C}_i = \begin{bmatrix} C_i & F_{2i} & 0 \end{bmatrix}. \tag{7.136}$$

Then by substituting the matrices $\bar{C}_i$ and $\bar{D}_{1i}$ into (7.91), the gains of $H_i (i = 1, 2, 3)$ can be calculated as

$$H_i = \bar{D}_{1i}[(\bar{C}_i\bar{D}_{1i})^T(\bar{C}_i\bar{D}_{1i})]^{-1}(\bar{C}_i\bar{D}_{1i})^T = \begin{bmatrix} 100 & 0 \\ 50 & 0 \\ 0 & 0 \\ 0 & 0 \end{bmatrix}, i = 1, 2, 3.$$

Similarly, by solving the conditions in Theorem 7.8, and following the calculation procedures in Algorithm 3, the other gains N_i, L_i, G_i of the UIO (7.83), can be obtained. The FE result is shown in Fig. 7.11.

Case 3: $f^{(q)} = 0, q \in \mathbb{N}$. Consider the case of $f^{(4)} = 0$ in this part. As the same routine as in the Case 1 and Case 2, the gains H_i, L_i, N_i, are obtained. To verify the accuracy of the UIO method, we define a piecewise fault signal as follows:

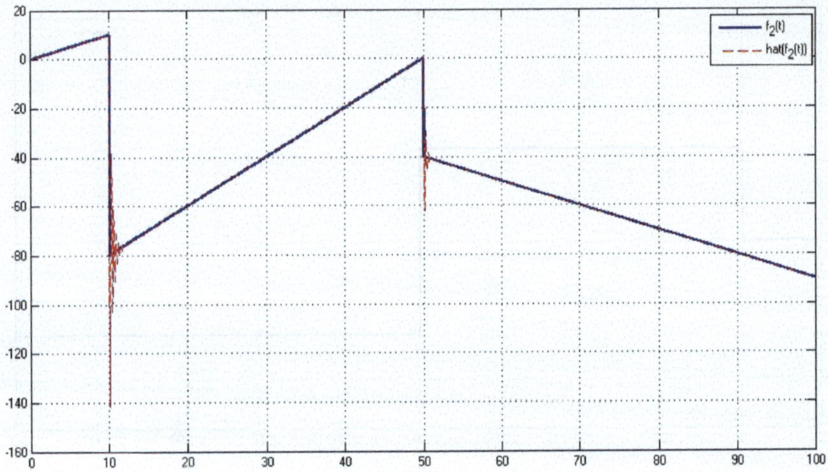

Fig. 7.11 The fault $f_2(t)$ (solid line) and its estimation $\hat{f}_2(t)$ (dotted line)

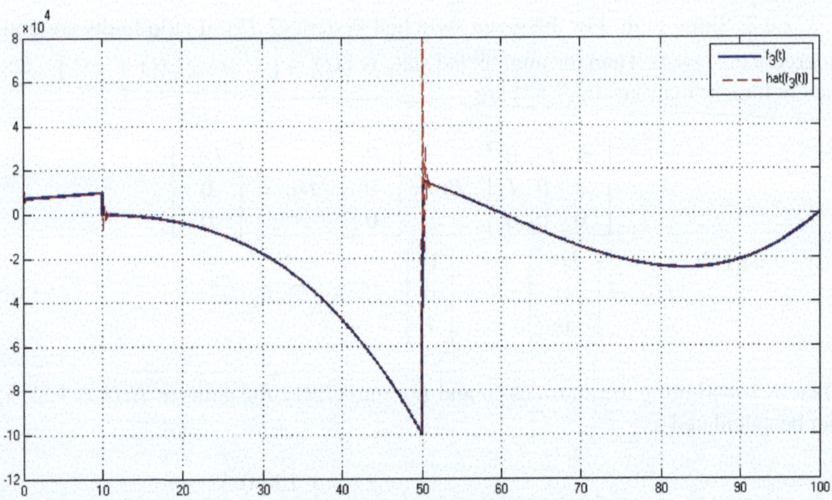

Fig. 7.12 The fault $f_3(t)$ (solid line) and its estimation $\hat{f}_3(t)$ (dotted line)

$$f_3(t) = \begin{cases} 2 \times (t+60)^2, & 0s \leq t < 10s; \\ t^2 \times (10-t), & 10s \leq t < 50s; \\ (t-20) \times (t-60) \times (t-100), & 50s \leq t < 100s. \end{cases}$$

The FE result is shown in Fig. 7.12.

7.3.7 Summary

This section deals with robust FE and fault accommodation problems for nonlinear switched systems. By using the switched Lyapunov function method and the ADT technique, a UIO is designed to be as robust as possible to the disturbances and as sensitive as possible to the faults. Furthermore, a dynamic output feedback controller is given to ensure the asymptotical stability of the closed-loop system. The results are also extended to measurement disturbances case. Finally, a liquid control system is provided to show the effectiveness of the presented scheme.

References

1. Chen, J., Patton, R.J.: Robust Model-Based Fault Diagnosis for Dynamic Systems. Kuwer Academic Publishers, Boston (1999)
2. Deutscher, J.: Fault detection for linear distributed-parameter systems using finite-dimensional functional observers. Int. J. Control **89**(3), 550–563 (2016)
3. Guo, F., Ren, X., Li, Z., Han, C.: Fault detection for linear discrete-time systems with output quantisation. Int. J. Control **90**(10), 2270–2283 (2017)
4. Huang, D., Duan, Z.: A new iterative approach for multi-objective fault detection observer design and its application to a hypersonic vehicle. Int. J. Control **91**(3), 554–570 (2018)
5. Patton, R.J., Frank, P.M., Clark, R.N.: Issues of Fault Diagnosis for Dynamic Systems. Springer (2000)
6. Qiu, A., Gu, J., Wen, C., Zhang, J.: Self-triggered fault estimation and FTC for networked control systems. Neurocomputing **272**, 629–637 (2018)
7. Hwang, I., Kim, Y.: A survey of fault detection, isolation, and reconfiguration methods. IEEE Trans. Control Syst. Technol. **18**(3), 636–653 (2010)
8. Li, J., Park, J.H., Ye, D.: Fault detection filter design for switched systems with quantisation effects and packet dropout. IET Control Theory Appl. **11**(2), 182–193 (2017)
9. Li, L., Ding, S.X., Qiu, J., Yang, Y., Zhang, Y.: Weighted fuzzy observer-based fault detection approach for discrete-time nonlinear systems via piecewise-fuzzy Lyapunov functions. IEEE Trans. Fuzzy Syst. **24**(6), 1320–1333 (2016)
10. Raoufi, R.C., Marquez, H.J., Zinober, A.S.: Sliding mode observer for uncertain nonlinear Lipschitz systems with fault estimation synthesis. Int. J. Robust Nonlinear Control **20**(16), 1785–1801 (2010)
11. Liu, C., Jiang, B., Zhang, K.: Distributed adaptive observers-based fault estimation for leader-following multi-agent linear uncertain systems with actuator faults. In: IEEE Chinese Guidance, Navigation and Control Conference, Nanjing, pp. 1162–1167 (2016)
12. Gritli, W., Gharsallaoui, H., Benrejeb, M.: Fault detection based on unknown input observers for switched discrete-time systems. In: 2017 International Conference on Advanced Systems and Electric Technologies, Hammamet, pp. 436–441 (2017)
13. Liu, L., Guo, R., Ma, S.: Input/output to state stability of switched nonlinear systems with an improved average dwell time approach. Int. J. Control Autom. Syst. **14**(2), 461–468 (2016)
14. Mahmoudi, A., Momeni, A., Aghdam, A.G., Gohari, P.: Switching between finite-time observers. Eur. J. Control **14**(4), 297–307 (2008)
15. Hespanha, J.P., Morse, A.S.: Stability of switched systems with average dwell time. In: IEEE Conference on Decision & Control, vol. 3(3), pp. 2655–2660 (2002)
16. Du, D., Xu, S.: Actuator fault detection for discrete-time switched linear systems with output disturbance. Int. J. Control Autom. Syst. **15**(6), 2590–2598 (2017)

17. Jiang, B., Du, D., Cocquempot, V.: Fault detection for discrete-time switched systems with interval time-varying delays. Int. J. Control Autom. Syst. **9**(2), 396–401 (2011)
18. Wang, D., Shi, P., Wang, W.: Robust fault detection for continuous-time switched delay systems: an linear matrix inequality approach. IET Control Theory Appl. **4**(1), 100–108 (2010)
19. Benzaouia, A., Eddoukali, Y.: Robust fault detection and control for continuous-time switched systems with average dwell time. Circuits Syst. Signal Process. **37**(6), 2357–2373 (2018)
20. Farhat, A., Koenig, D.: Robust fault detection for uncertain switched systems. IFAC-PapersOnLine **50**(1), 15265–15270 (2017)
21. Zhong, G., Yang, G.: Robust control and fault detection for continuous-time switched systems subject to a dwell time constraint. Int. J. Robust Nonlinear Control **25**(18), 3799–3817 (2016)
22. Gahinet, P., Apkarian, P.: An linear matrix inequality approach to H_∞ control. Int. J. Robust Nonlinear Control **4**, 422–448 (1994)
23. Tseng, C.S., Chen, B.S.: L_∞-gain fuzzy control for nonlinear dynamic systems with persistent bounded disturbance. In: Proceedings of the 13th IEEE International Conference on Fuzzy Systems, Budapest, Hungary, Luly, pp. 783–788 (2004)
24. Zhou, L., Ho, D.W.C., Zhai, G.: Stability analysis of switched linear singular systems. Automatica **49**, 1481–1487 (2013)
25. Long, Y., Park, J.H., Ye, D.: Transmission-dependent fault detection and isolation strategy for networked systems under finite capacity channels. IEEE Trans. Cybern. **47**(8), 2266–2278 (2017)
26. Marseglia, G.R., Raimondo, D.M.: Active fault diagnosis: a multi-parametric approach. Automatica **79**, 223–230 (2017)
27. Zhu, J., Yang, G., Wang, H., Wang, F.: Fault estimation for a class of nonlinear systems based on intermediate estimator. IEEE Trans. Autom. Control **61**(9), 2518–2524 (2016)
28. Yao, L., Qin, J., Wang, H., Jiang, B.: Design of new fault diagnosis and FTC scheme for non-Gaussian singular stochastic distribution systems. Automatica **48**(9), 2305–2313 (2012)
29. Zhong, M., Ding, S.X., Lam, J., Wang, H.: An LMI approach to design robust fault detection filter for uncertain LTI systems. Automatica **39**(3), 543–550 (2003)
30. Jiang, B., Staroswiecki, M., Cocquempot, V.: Fault accommodation for nonlinear dynamic systems. IEEE Trans. Autom. Control **51**(9), 1578–1583 (2006)
31. Yan, X., Edwards, C.: Robust decentralized actuator fault detection and estimation for large-scale systems using a sliding mode observer. Int. J. Control **81**(4), 591–606 (2007)
32. Zhang, K., Jiang, B., Shi, P., Xu, J.: Analysis and design of robust H_∞ fault estimation observer with finite-frequency specifications for discrete-time fuzzy systems. IEEE Trans. Fuzzy Syst. **45**(7), 1225–1235 (2015)
33. Huang, S., Yang, G.: FTCler design for T-S fuzzy systems with time-varying delay and actuator faults: a k-step fault-estimation approach. IEEE Trans. Fuzzy Syst. **22**(6), 1526–1540 (2014)
34. Zhang, K., Jiang, B., Staroswiecki, M.: Dynamic output feedback-FTCler design for Takagi-CSugeno fuzzy systems with actuator faults. IEEE Trans. Fuzzy Syst. **18**(1), 194–201 (2010)
35. Chen, J., Patton, R., Zhang, H.: Design of unknown input observers and robust fault detection filters. Int. J. Control **63**(1), 85–105 (1996)
36. Fonod, R., Henry, D., Charbonnely, C., Bornschleglz, E.: A class of nonlinear unknown input observer for fault diagnosis: application to fault tolerant control of an autonomous spacecraft. In: Proceedings of UKACC International Conference on Control, Loughborogh, U.K., pp. 19–24 (2014)
37. Sun, X., Patton, R.: Robust actuator multiplicative fault estimation with unknown input decoupling for a wind turbine system. In: Proceedings of the Conference on Control Fault-Tolerant Systems, Nice, France, pp. 263–268 (2013)
38. Johansson, M., Rantzer, A.: Computation of piecewise quadratic Lyapunov functions for hybrid systems. IEEE Trans. Autom. Control **43**(4), 555–559 (1998)
39. Sun, Z., Ge, S.: Analysis and synthesis of switched linear control systems. Automatica **41**, 181–195 (2005)
40. Hespanha, J.P.: Uniform stability of switched linear systems extensions of Lasalles invariance principle. IEEE Trans. Autom. Control **49**(4), 470–482 (2004)

41. Hu, H., Jiang, B., Yang, H.: Reliable guaranteed-cost control of delta operator switched systems with actuator faults: mode-dependent average dwell-time approach. IET Control Theory Appl. **10**(1), 17–23 (2016)
42. Li, J., Yang, G.: Fault detection for switched linear parameter-varying systems: an average dwell-time approach. IET Control Theory Appl. **7**(8), 1120–1130 (2013)
43. Liu, X., Gao, Z.W.: Unknown input observers for fault diagnosis in Lipschitz nonlinear systems. In: IEEE International Conference on Mechatronics and Automation, Beijing, pp. 1555–1560 (2015)
44. Chakrabarty, A., Corless, M.J., Buzzard, G.T., Żak, S.H., Rundell, A.E.: State and unknown input observers for nonlinear systems with bounded exogenous inputs. IEEE Trans. Autom. Control **62**(11), 5497–5510 (2017)
45. Li, G., Li, T., Zhao, Q.: Fault tolerant shape control for particulate process systems under simultaneous actuator and sensor faults. IET Control Theory Appl. **11**(15), 2448–2457 (2017)
46. Shokouhi-Nejad, H., Ghiasi, A.R., Badamchizadeh, M.A.: Robust simultaneous finite-time control and fault detection for uncertain linear switched systems with time-varying delay. IET Control Theory Appl. **11**(7), 1041–1052 (2017)
47. Li, L., Jia, Y., Du, J., Yuan, S.: Robust L_2/L_∞ control for uncertain singular systems with time-varying delay. Progress Nat. Sci. **18**(8), 1015–1021 (2008)
48. Boyd, S., Ghaoui, L.E., Feron, E., Balakrishnan, V.: Linear Matrix Inequalities in System and Control Theory. Philadelphia, PA, USA, Society for Industrial and Applied Mathematics (1994)

Chapter 8
UIO-Based Fault Diagnosis for Discrete-Time Switched Systems

8.1 UIO-Based Real-time-weighted Fault Detection

In this section, an observer-based fault detection method is proposed to deal with the issue of real-time-weighted fault detection for discrete-time switched systems. By employing the UIO technique and ADT method, the H_∞/H_2 and H_- performance levels are considered to ensure the robustness to the known and unknown input and the sensitivity to the fault. Finally, a practical example is given to demonstrate the effectiveness of the proposed methods.

8.1.1 Introduction

Driven by the considerably increasing demands for system reliability, the issue of fault diagnosis has attracted remarkable attention from both the application and research domains [1–5]. The fault diagnosis process is composed of fault detection and fault estimation. The purpose of fault detection is to detect faults as soon as possible, and then according to the process of identifying the size and type of the fault based on the detected fault is called fault estimation. The observer-based method is one of the most popular methods among all the fault diagnosis methods [6]. The residual signal can be obtained by using the difference between the output of the system and the output of the observer, if the residual is greater than the preset threshold, fault alarm can be realized.

For the issue of fault detection, many observer-based methods have been emerged in recent years, such as adaptive observer [7–9], sliding-mode observer [10], fuzzy observer [11–13], diagnostic observer [14], and unknown input observer [15], etc. However, many production processes can be divided into different modes in practice

D. Du et al., *Observer-Based Fault Diagnosis and Fault-Tolerant Control for Switched Systems*, Studies in Systems, Decision and Control 323,
https://doi.org/10.1007/978-981-15-9073-3_8

according to product requirements, such as chemical processes, power electronic systems, robotics, collaborative control, and other fields. Because of this, the engineering systems can usually be built into hybrid systems for theoretical research.

As is known to all, switched systems belong to hybrid systems, which consists of a set of continuous or discrete-time subsystems and a switching signal specifying the switching order among the subsystems [16]. According to the characterization of switched systems, the switching signal can be divided into arbitrary switching and constraint switching. For the past few decades, some switching signal techniques had been proposed to ensure that switching systems are stable, such as maximum or minimum dwell-time method [17], ADT [18], and PDT [19], etc. ADT switching is one of the constrained switching signals. Moreover, ADT switching has been shown to be more general and less conservative than dwell-time switching and arbitrary switching. In addition, ADT technique has been widely used in fault diagnosis and fault-tolerant control for switching systems [20, 21].

In this section, we devote to investigate the real-time-weighted fault detection issue for discrete-time switched systems. The objective of this work is to develop a UIO as a residual generator such that it is robust to the known and unknown input and sensitive to the fault. Moreover, by employing ADT and linear matrix inequality techniques, sufficient conditions for the existence of the fault detection are gained. Finally, the correctness of the designed approach is clarified by a numerical example. The main contributions of this paper can be summarized in the following aspects:

- for discrete-time switched systems, a real-time-weighted fault detection observer is designed. The weighted fault detection is completely robust to certain and uncertain inputs;
- the fault detection observer is designed to be sensitive to faults; (iii) the proposed fault detection algorithm is suitable to detect all kinds of faults.

8.1.2 The Preliminaries

Considering the following a class of discrete-time switched system formulated by:

$$\begin{cases} x(k+1) = A_{\sigma(k)}x(k) + B_{\sigma(k)}u(k) + D_{\sigma(k)}d(k) + F_{\sigma(k)}f(k) \\ y(k) = C_{\sigma(k)}x(k) + G_{\sigma(k)}f(k) \end{cases} \tag{8.1}$$

in which $x(t) \in \mathbb{R}^n$ is the state vector, $u(t) \in \mathbb{R}^m$ is the control input vector, and $y(t) \in \mathbb{R}^q$ is the measurable output vector. $d(t) \in \mathbb{R}^{n_d}$ and $f(t) \in \mathbb{R}^{n_f}$ are, respectively, the external disturbance and the detected fault, which belong to $l_2[0, \infty)$. $A_{\sigma(k)}$, $B_{\sigma(k)}$, $C_{\sigma(k)}$, $D_{\sigma(k)}$, $F_{\sigma(k)}$, and $G_{\sigma(k)}$ are the predefined constant matrices with appropriate dimensions. The piecewise function $\sigma(t) : [0, \infty) \to \mathscr{N} = \{1, 2, \ldots, N\}$ is the switching signal, which specifies which subsystem is activated at the instant t. When $\sigma(t) = i$, it means that the i-th subsystem is activated at the switching instant.

The switching signal is assumed to be a priori unknown, but its instantaneous value is available in real time.

For $\sigma(k) = i$, which means the i-th subsystem is activated. Then the original system can be simplified as

$$\begin{cases} x(k+1) = A_i x(k) + B_i u(k) + D_i d(k) + F_i f(k) \\ y(k) = C_i x(k) + G_i f(k) \end{cases} \tag{8.2}$$

where $i \in \mathcal{N}$.

Remark 8.1 Noted that the model described in (8.2), denotes a class of switched systems suffered with actuator and sensor faults. If the fault $f(t)$ is splitted as $f(t) = \begin{bmatrix} f_a^T(t) & f_s^T(t) \end{bmatrix}^T$, and $F_i = \begin{bmatrix} F_{ai} & O \end{bmatrix}$, $G_i = \begin{bmatrix} O & G_{si} \end{bmatrix}$, where $f_a(t)$ and $f_s(t)$ represent the actuator and sensor faults, F_{ai} and G_{si} denote the distribution matrices of them. Therefore, one can get that the considered issue is not loss of generality. In fact, similar models have been extensively considered in [22–24].

In order to obtain the objective of this paper, the following assumptions are given:

Assumption 14 The discrete-time switched system (8.2) is asymptotically stable.

Assumption 15 $(C_i, A_i), i \in \mathcal{N}$ is observable.

It is important to mention that the state variables of the process are not always fully measurable in practice. To deal with this issue, the Assumption 15, is given to guarantee the observability of the system. Then, a state observer is usually used to estimate the state and further generate the residual sinal. As a consequence, the objective of fault detection can be realized. For (8.2), we advise the following UIO as a residual signal generator:

$$\begin{cases} z(k+1) = S_i z(k) + M_i u(k) + L_i y(k) \\ \hat{x}(k) = z(k) - H_i y(k) \\ \hat{y}(k) = C_i \hat{x}(k) \\ r(k) = W_i (y(k) - \hat{y}(k)) \end{cases} \tag{8.3}$$

where $z(t) \in \mathbb{R}^n$ is the observer state, $\hat{x}(t) \in \mathbb{R}^n$ denotes the estimation of the state $x(t)$, $\hat{y}(t) \in \mathbb{R}^m$ is the observer output, $r(t) \in \mathbb{R}^m$ represents the residual signal. And matrices S_i, M_i, L_i, and H_i $(i \in \mathcal{N})$ are the observer parameters to be determined later, W_i $(i \in \mathcal{N})$ indicate the weighted matrices.

In this section, the main objective is to design a fault detection observer (8.3), such that the residual signal is robust to the known and unknown input and meanwhile sensitive to the fault, that is to say

- H_∞/H_2 robustness condition: for the case of $f(t) = 0$, one has

$$\sum_{k=k_0}^{\infty} (1-\alpha)^k r^T(k) r(k) \leq \gamma^2 \sum_{k=k_0}^{\infty} u^T(k) u(k) + \gamma^2 \sum_{k=k_0}^{\infty} d^T(k) d(k) \tag{8.4}$$

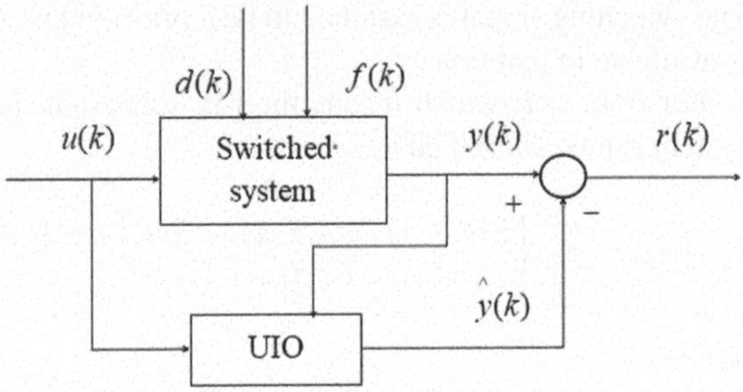

Fig. 8.1 The structure of fault detection

- H_- sensitivity condition: for the case of $d(t) = 0$, one has

$$\sum_{k=k_0}^{\infty}(1-\alpha)^k r^T(k)r(k) \geq \beta^2 \sum_{k=k_0}^{\infty} f^T(k)f(k) \tag{8.5}$$

where α, β and γ are given positive scalars.

In order to clarify the structure of this section, the structural diagram of the proposed approach is provided in the Fig. 8.1.

Assume the estimation error as $e(k) = x(k) - \hat{x}(k)$. For simplicity, we set $\sigma(k) = i$, then one has

$$e(k) = T_i x(k) - z(k) + H_i G_i f(k) \tag{8.6}$$

where $T_i = I + H_i C_i$.

Furthermore, the dynamic estimation error system can be rewritten as

$$\begin{aligned} e(k+1) = {} & S_i e(k) + (T_i A_i - S_i T_i - L_i C_i)x(k) + (T_i B_i - M_i)u(k) + T_i D_i d(k) \\ & + H_i G_i f(k+1) + (T_i F_i - S_i H_i G_i - L_i G_i) f(k) \end{aligned} \tag{8.7}$$

If we set

$$T_i A_i - S_i T_i - L_i C_i = 0 \tag{8.8}$$

$$T_i B_i - M_i = 0; \tag{8.9}$$

$$H_i G_i = 0. \tag{8.10}$$

one has

$$\begin{cases} e(k+1) = S_i e(k) + T_i D_i d(k) + (T_i F_i - S_i H_i G_i - L_i G_i) f(k) \\ r(k) = W_i (C_i e(k) + G_i f(k)) \end{cases} \tag{8.11}$$

Define

$$Z_i = S_i H_i + L_i \tag{8.12}$$

Combining (8.8) and (8.12), one has

$$S_i = T_i A_i - Z_i C_i \tag{8.13}$$

By introducing $\xi(k) = \left[e^T(k) \; x^T(k) \right]^T$, then an augmented system can be described as

$$\begin{cases} \xi(k+1) = \mathscr{A}_i \xi(k) + \mathscr{B}_i u(k) + \mathscr{D}_i d(k) + \mathscr{F}_i f(k) \\ r(k) = \mathscr{C}_i \xi(k) + \mathscr{G}_i f(k) \end{cases} \tag{8.14}$$

in which

$$\begin{aligned} \mathscr{A}_i &= \begin{bmatrix} S_i & 0 \\ 0 & A_i \end{bmatrix}, \mathscr{B}_i = \begin{bmatrix} 0 \\ B_i \end{bmatrix}, \mathscr{D}_i = \begin{bmatrix} T_i D_i \\ D_i \end{bmatrix}, \\ \mathscr{F}_i &= \begin{bmatrix} T_i F_i - Z_i G_i \\ F_i \end{bmatrix}, \mathscr{C}_i = \begin{bmatrix} W_i C_i \\ 0 \end{bmatrix}, \mathscr{G}_i = \begin{bmatrix} W_i G_i \\ 0 \end{bmatrix}. \end{aligned} \tag{8.15}$$

Before further investigating, we need to indroduce the following definition and lemma.

Definition 8.1 ([25]) For any switching signal $\sigma_i(t)$ and any $t_2 > t_1 > 0$, let $N_{\sigma_i(t)}(t_1, t_2)$ denote the number of switchings $\sigma_i(t)$ on an interval (t_1, t_2). If

$$N_{\sigma_i(t)}(t_1, t_2) \leq N_0 + \frac{t_2 - t_1}{\tau_a}$$

holds for a given $N_0 \geq 0$ and $\tau_a > 0$, then the constant τ_a is called the ADT and N_0 is the chattering bound.

Lemma 8.1 ([26]) *Consider the discrete-tiem switched system $x(k+1) = f_{\sigma(k)}(x(k))$ and let $0 < \alpha < 1$, $\mu > 1$ be given constants. Suppose that there exist a Lyapunov function $V_{\sigma(k)} : \mathbb{R}^n \to \mathbb{R}$ satisfying the following properties:*

- $\Delta V_i(x(k)) = V_i(x(k+1)) - V_i(x(k)) \leq -\alpha V_i(x(k)), \forall k \in [k_l, k_{l+1})$
- $V_i(x(k_l)) \leq V_j(x(k_l))$

then the system is globally exponentially stable for any switching signal with the average dwell time

$$\tau_a \geq \tau_a^* = -\frac{ln\mu}{ln(1-\alpha)} \tag{8.16}$$

8.1.3 Robust Condition

In this part, the H_∞/H_2 robustness condition is firstly addressed. Consider $f(t) = 0$, then the error dynamic system (8.12) is transformed as the following fault-free case:

$$\begin{cases} \xi(k+1) = \mathscr{A}_i\xi(k) + \mathscr{B}_i u(k) + \mathscr{D}_i d(k) \\ r(k) = \mathscr{C}_i x(k) \end{cases} \tag{8.17}$$

Theorem 8.1 *For given scalars* $0 < \alpha < 1$, $\mu \geq 1$ *and* $\gamma > 0$, *if there exist positive definite symmetric matrices* P_{1i}, P_{3i}, *matrices* P_{2i}, Ω_{1i}, Ω_{2i}, Ω_{3i}, Q_i, R_i, *and* W_i *such that*

$$\begin{bmatrix} P_{1i} & P_{2i} \\ * & P_{3i} \end{bmatrix} \leq \mu \begin{bmatrix} P_{1j} & P_{2j} \\ * & P_{3j} \end{bmatrix} \tag{8.18}$$

$$\begin{bmatrix} \Psi_{1i} & \Psi_{2i} & \Psi_{3i} \\ * & \Psi_{4i} & \Psi_{5i} \\ * & * & \Psi_{6i} \end{bmatrix} < 0 \tag{8.19}$$

where

$$\Psi_{1j} = \begin{bmatrix} P_{1j} - \Omega_{1j} - \Omega_{1j}^T & P_{2j} - \Omega_{2i} \\ * & P_{3j} - \Omega_{3i} - \Omega_{3i}^T \end{bmatrix}, \Psi_{2i} = \begin{bmatrix} \Omega_{1i}A_i + Q_iC_iA_i - R_iC_i & \Omega_{2i}A_i \\ 0 & \Omega_{3i}A_i \end{bmatrix},$$

$$\Psi_{3i} = \begin{bmatrix} \Omega_{2i}B_i & \Omega_{1i}D_i + Q_iC_iD_i + \Omega_{2i}D_i & 0 \\ \Omega_{3i}B_i & \Omega_{3i}D_i & 0 \end{bmatrix}, \Psi_{4i} = \begin{bmatrix} -(1-\alpha)P_{1i} & -(1-\alpha)P_{2i} \\ * & -(1-\alpha)P_{3i} \end{bmatrix},$$

$$\Psi_{5i} = \begin{bmatrix} 0 & 0 & W_iC_i \\ 0 & 0 & 0 \end{bmatrix}, \Psi_{6i} = \begin{bmatrix} -\gamma^2 I & 0 & 0 \\ * & -\gamma^2 I & 0 \\ * & * & -I \end{bmatrix}.$$

then, for the switching signal satisfying the ADT condition (8.16), the error dynamic system (8.14), is globally exponentially stable with H_∞/H_2 *performance index* γ. *Under this case, the observer parameters in (8.3), can be solved by* $H_i = \Omega_{1i}^{-1}Q_i$, $Z_i = \Omega_{1i}^{-1}R_i$.

Proof Consider the error dynamic system (8.17), we define the switched Lyapunov function as

$$V_{\sigma(k)} = \xi^T(k)P_{\sigma(k)}\xi(k) \tag{8.20}$$

Setting $\Delta V_i(\xi(k)) = V_i(\xi(k+1)) - V_i(\xi(k))$, then by Lemma 8.1, one has

$$\begin{aligned} &\Delta V_i(\xi(k)) + \alpha V_i(\xi(k)) + r^T(k)r(k) - \gamma^2(u^T(k)u(k) + d^T(k)d(k)) \\ &= \xi^T(k+1)P_{\sigma(k+1)}\xi(k+1) - (1-\alpha)\xi^T(k)P_{\sigma(k)}\xi(k) + r^T(k)r(k) \\ &\quad -\gamma^2(u^T(k)u(k) + d^T(k)d(k)) \\ &= \begin{bmatrix} \xi^T(k) & u^T(k) & d^T(k) \end{bmatrix} \left(\begin{bmatrix} \mathscr{A}_i^T \\ \mathscr{B}_i^T \\ \mathscr{D}_i^T \end{bmatrix} P_j \begin{bmatrix} \mathscr{A}_i & \mathscr{B}_i & \mathscr{D}_i \end{bmatrix} \right. \end{aligned}$$

$$
+\begin{bmatrix} -(1-\alpha)P_i+\mathscr{C}_i^T\mathscr{C}_i & 0 & 0 \\ 0 & -\gamma^2 I & 0 \\ 0 & 0 & -\gamma^2 I \end{bmatrix}\Bigg)\begin{bmatrix} \xi(k) \\ u(k) \\ d(k) \end{bmatrix}
$$
$$
= \eta^T(k)\Phi_i\eta(k) \tag{8.21}
$$

where $\eta(k) = \left[\xi^T(k)\ u^T(k)\ d^T(k)\right]^T$,

$$
\Phi_i = \begin{bmatrix} \mathscr{A}_i^T P_i\mathscr{A}_i-(1-\alpha)P_i+\mathscr{C}_i^T\mathscr{C}_i & \mathscr{A}_i^T P_i\mathscr{B}_i & \mathscr{A}_i^T P_i\mathscr{D}_i \\ * & \mathscr{B}_i^T P_i\mathscr{B}_i-\gamma^2 I & \mathscr{B}_i^T P_i\mathscr{D}_i \\ * & * & \mathscr{D}_i^T P_i\mathscr{D}_i-\gamma^2 I \end{bmatrix}. \tag{8.22}
$$

Thus, by the Schur Complement Lemma, the following inequality implies:

$$
\begin{bmatrix} -P_j^{-1} & \mathscr{A}_i & \mathscr{B}_i & \mathscr{D}_i \\ * & -(1-\alpha)P_i+\mathscr{C}_i^T\mathscr{C}_i & 0 & 0 \\ * & * & -\gamma^2 I & 0 \\ * & * & * & -\gamma^2 I \end{bmatrix} < 0. \tag{8.23}
$$

It can be observed that P_j^{-1} involved in (8.23), which leads to the coupling between P_j and the system matrices. To deal with the issue, a congruence transformation is performed to (8.23), by diag(Ω, I, I), which yields

$$
\begin{bmatrix} -\Omega_j P_j^{-1}\Omega_j^T & \Omega_i\mathscr{A}_i & \Omega_i\mathscr{B}_i & \Omega_i\mathscr{D}_i \\ * & -(1-\alpha)P_i+\mathscr{C}_i^T\mathscr{C}_i & 0 & 0 \\ * & * & -\gamma^2 I & 0 \\ * & * & * & -\gamma^2 I \end{bmatrix} < 0 \tag{8.24}
$$

It follows from

$$
(P_j-\Omega_j)P_j^{-1}(P_j-\Omega_j)^T = \Omega_j P_j^{-1}\Omega_j^T + P_j - \Omega_j - \Omega_j^T \geq 0 \tag{8.25}
$$

then

$$
-\Omega_j P_j^{-1}\Omega_j^T \leq P_j - \Omega_j - \Omega_j^T \tag{8.26}
$$

By (5.24), the following inequality implies $\Phi_i < 0$,

$$
\begin{bmatrix} P_j-\Omega_j-\Omega_j^T & \Omega_i\mathscr{A}_i & \Omega_i\mathscr{B}_i & \Omega_i\mathscr{D}_i \\ * & -(1-\alpha)P_i+\mathscr{C}_i^T\mathscr{C}_i & 0 & 0 \\ * & * & -\gamma^2 I & 0 \\ * & * & * & -\gamma^2 I \end{bmatrix} < 0 \tag{8.27}
$$

Define

$$P_i = \begin{bmatrix} P_{1i} & P_{2i} \\ * & P_{3i} \end{bmatrix}, \quad \Omega_i = \begin{bmatrix} \Omega_{1i} & \Omega_{2i} \\ 0 & \Omega_{3i} \end{bmatrix}, \quad Q_i = \Omega_{1i} H_i, \quad R_i = \Omega_{1i} Z_i, \tag{8.28}$$

and Schur complement, one can get (8.19), is equivalent to (8.27).

Then, one has

$$\Delta V_i(\xi(k)) + \alpha V_i(\xi(k)) + r^T(k)r(k) - \gamma^2(u^T(k)u(k) + d^T(k)d(k)) < 0 \tag{8.29}$$

Furthermore, one can get

$$\begin{aligned} V_i(\xi(k_1)) &< (1-\alpha)V_i(\xi(k_0)) - r^T(k_0)r(k_0) \\ &\quad + \gamma^2(u^T(k_0)u(k_0) + d^T(k_0)d(k_0)) \end{aligned} \tag{8.30}$$

Iterating (8.30), gives

$$V_i(\xi(k)) < (1-\alpha)^{k-k_0} V_i(\xi(k_0)) - \sum_{s=k_0}^{k-1} (1-\alpha)^{k-s-1} \Gamma(s) \tag{8.31}$$

where $\Gamma(s) = r^T(s)r(s) - \gamma^2(u^T(s)u(s) + d^T(s)d(s))$.

Define H_∞/H_2 performance index as

$$J_\infty = \sum_{k=k_0}^{\infty} (1-\alpha)^k r^T(k)r(k) - \gamma^2 \left(\sum_{k=k_0}^{\infty} u^T(k)u(k) + \sum_{k=k_0}^{\infty} d^T(k)d(k) \right) \tag{8.32}$$

From (8.18) and (8.29), one can obtain

$$\begin{aligned} V_{\sigma(k)}(\xi(k)) &< (1-\alpha)^{k-k_l} V_{\sigma(k)}(\xi(k_l)) - \sum_{s=k_l}^{k-1} (1-\alpha)^{k-s-1} \Gamma(s) \\ &\le (1-\alpha)^{k-k_l} \mu V_{\sigma(k_{l-1})}(\xi(k_l)) - \sum_{s=k_l}^{k-1} (1-\alpha)^{k-s-1} \Gamma(s) \\ &\le (1-\alpha)^{k-k_l} \mu \left[(1-\alpha)^{k_l-k_{l-1}} V_{\sigma(k_{l-1})}(\xi(k_{l-1})) - \sum_{s=k_{l-1}}^{k_l-1} (1-\alpha)^{k_l-s-1} \Gamma(s) \right] \\ &\quad - \sum_{s=k_l}^{k-1} (1-\alpha)^{k-s-1} \Gamma(s) \\ &\le \cdots\cdots\cdots\cdots \\ &\le (1-\alpha)^{k-k_0} \mu^{N_\sigma(k_0,k)} V_{\sigma(k_0)}(\xi(k_0)) - (1-\alpha)^{k-k_1} \mu^{N_\sigma(k_0,k)} \\ &\quad - \sum_{s=k_0}^{k_1-1} (1-\alpha)^{k_1-s-1} \Gamma(s) \cdots\cdots - \sum_{s=k_0}^{k-1} (1-\alpha)^{k-s-1} \Gamma(s) \end{aligned}$$

$$= (1-\alpha)^{k-k_0}\mu^{N_\sigma(k_0,k)}V_{\sigma(k_0)}(\xi(k_0)) - \sum_{s=k_0}^{k-1}\mu^{N_\sigma(s,k)}(1-\alpha)^{k-s-1}\Gamma(s) \tag{8.33}$$

Then, under zero initial condition, the above formula gives

$$\sum_{s=k_0}^{k-1}\mu^{N_\sigma(s,k)}(1-\alpha)^{k-s-1}\Gamma(s) \le 0 \tag{8.34}$$

Multiplying both sides of (8.34), by $\mu^{-N_\sigma(0,k)}$, one can obtain

$$\begin{aligned} &\mu^{-N_\sigma(0,k)}\sum_{s=k_0}^{k-1}\mu^{N_\sigma(s,k)}(1-\alpha)^{k-s-1}r^T(s)r(s) \\ \le&\mu^{-N_\sigma(0,k)}\sum_{s=k_0}^{k-1}\mu^{N_\sigma(s,k)}(1-\alpha)^{k-s-1}\gamma^2(u^T(s)u(s)+d^T(s)d(s)) \end{aligned} \tag{8.35}$$

which is equivalent to

$$\begin{aligned} &\sum_{s=k_0}^{k-1}\mu^{-N_\sigma(0,s)}(1-\alpha)^{k-s-1}r^T(s)r(s) \\ \le&\sum_{s=k_0}^{k-1}\mu^{-N_\sigma(0,s)}(1-\alpha)^{k-s-1}\gamma^2(u^T(s)u(s)+d^T(s)d(s)) \end{aligned} \tag{8.36}$$

Then, due to the fact that

$$N_\sigma(0,s) \le \frac{s}{\tau_a} \le \frac{-sln(1-\alpha)}{ln\mu} \tag{8.37}$$

we have

$$\begin{aligned} &\sum_{s=k_0}^{k-1}\mu^{\frac{sln(1-\alpha)}{ln\mu}}(1-\alpha)^{k-s-1}r^T(s)r(s) \\ \le&\sum_{s=k_0}^{k-1}\mu^{-N_\sigma(0,s)}(1-\alpha)^{k-s-1}\gamma^2(u^T(s)u(s)+d^T(s)d(s)) \\ \le&\sum_{s=k_0}^{k-1}(1-\alpha)^{k-s-1}\gamma^2(u^T(s)u(s)+d^T(s)d(s)) \end{aligned} \tag{8.38}$$

which means

$$\sum_{s=k_0}^{k-1}(1-\alpha)^s r^T(s)r(s) \le \sum_{s=k_0}^{k-1}\gamma^2(u^T(s)u(s)+d^T(s)d(s)) \tag{8.39}$$

Then the proof is concluded.

8.1.4 Sensitivity Condition

Next, the H_- sensitivity condition is investigated. By considering $d(t)=0$, the error dynamic system (8.14), is transformed into

$$\begin{cases} \xi(k+1) = \mathscr{A}_i\xi(k) + \mathscr{B}_i u(k) + \mathscr{F}_i f(k) \\ r(k) = \mathscr{C}_i\xi(k) + \mathscr{G}_i f(k) \end{cases} \tag{8.40}$$

Theorem 8.2 *For given scalars $0<\beta<1$ and $\mu\ge 1$, if there exist positive definite matrices P_{1i}, P_{3i}, matrices P_{2i}, Ω_{1i}, Ω_{2i}, Ω_{3i}, Q_i, R_i, W_i such that*

$$\begin{bmatrix} P_{1i} & P_{2i} \\ * & P_{3i} \end{bmatrix} \le \mu \begin{bmatrix} P_{1j} & P_{2j} \\ * & P_{3j} \end{bmatrix} \tag{8.41}$$

$$\begin{bmatrix} \Theta_{1i} & \Theta_{2i} & \Theta_{3i} \\ * & \Theta_{4i} & \Theta_{5i} \\ * & * & \Theta_{6i} \end{bmatrix} < 0 \tag{8.42}$$

where

$$\Theta_{1i} = \begin{bmatrix} P_{1j}-\Omega_{1j}-\Omega_{1j}^T & P_{2j}-\Omega_{2i} \\ * & P_{3j}-\Omega_{3i}-\Omega_{3i}^T \end{bmatrix}, \Theta_{2i} = \begin{bmatrix} \Omega_{1i}A_i+Q_iC_iA_i-R_iC_i & \Omega_{2i}A_i \\ 0 & \Omega_{3i}A_i \end{bmatrix},$$

$$\Theta_{3i} = \begin{bmatrix} \Omega_{1i}F_i+Q_iC_iF_i-R_iG_i+\Omega_{2i}F_i \\ \Omega_{3i}F_i \end{bmatrix}, \Theta_{4i} = \begin{bmatrix} -(1-\alpha)P_{1i} & -(1-\alpha)P_{2i} \\ * & -(1-\alpha)P_{3i} \end{bmatrix},$$

$$\Theta_{5i} = \begin{bmatrix} W_iC_i \\ 0 \end{bmatrix}, \Theta_{6i} = -\beta^2 I - G_i^T W_i^T W_i G_i$$

then, for the switching signal satisfying the ADT condition (8.16), the error dynamic system (8.40), is globally exponentially stable with H_- performance level β. Under this case, the observer parameters in (8.3), can be solved by $H_i=\Omega_{1i}^{-1}Q_i$, $Z_i=\Omega_{1i}^{-1}R_i$.

Proof For system (8.40), we define the switched Lyapunov function as

$$V_{\sigma(k)} = \xi^T(k)P_{\sigma(k)}\xi(k) \tag{8.43}$$

Setting $\Delta V_i(\xi(k)) = V_i(\xi(k+1)) - V_i(\xi(k))$, one has

$$
\begin{aligned}
&\Delta V_i(\xi(k)) + \alpha V_i(\xi(k)) - r^T(k)r(k) + \beta^2(f^T(k)f(k)) \\
&= V_i(\xi(k+1)) - (1-\alpha)V_i(\xi(k)) - r^T(k)r(k) + \beta^2(f^T(k)f(k)) \\
&= \xi^T(k)\mathscr{A}_i^T P_j \mathscr{A}_i \xi(k) + \xi^T(k)\mathscr{A}_i^T P_j \mathscr{F}_i f(k) + f^T(k)F_i^T P_j \mathscr{A}_i \xi(k) \\
&\quad + f^T(k)\mathscr{F}_i^T P_j \mathscr{F}_i f(k) - (1-\alpha)\xi^T(k)P_i\xi(k) - \xi^T(k)\mathscr{C}_i^T\mathscr{C}_i\xi(k) \\
&\quad - \xi^T(k)\mathscr{C}_i^T\mathscr{G}_i f(k) - f^T(k)\mathscr{G}_i^T\mathscr{C}_i\xi(k) - f^T(k)\mathscr{G}_i^T\mathscr{G}_i f(k) + \beta^2(f^T(k)f(k)) \\
&= \vartheta^T(k)\Xi_i\vartheta(k)
\end{aligned}
$$

where $\vartheta(k) = \begin{bmatrix} \xi^T(k) & f^T(k) \end{bmatrix}^T$,

$$
\Xi_i = \begin{bmatrix} \mathscr{A}_i^T P_j \mathscr{A}_i - (1-\alpha)P_i - \mathscr{C}_i^T\mathscr{C}_i & \mathscr{A}_i^T P_j \mathscr{F}_i - \mathscr{C}_i^T\mathscr{G}_i \\ * & \beta^2 I - \mathscr{G}_i^T\mathscr{G}_i + \mathscr{F}_i^T P_i \mathscr{F}_i \end{bmatrix}. \tag{8.44}
$$

thus, by the Schur Complement Lemma, the following inequality implies

$$
\begin{bmatrix} -P_j^{-1} & \mathscr{A}_i & \mathscr{F}_i \\ * & -(1-\alpha)P_i - \mathscr{C}_i^T\mathscr{C}_i & -\mathscr{C}_i^T\mathscr{G}_i \\ * & * & -\beta^2 I - \mathscr{G}_i^T\mathscr{G}_i \end{bmatrix} < 0. \tag{8.45}
$$

It can be observed that P_j^{-1} involved in (8.45), which leads to the coupling between P_j and the system matrices. To deal with the issue, a congruence transformation is performed to (8.45) by diag(Ω, I, I), which yields:

$$
\begin{bmatrix} -\Omega_j P_j^{-1}\Omega_j^T & \Omega_j\mathscr{A}_i & \Omega_j\mathscr{F}_i \\ * & -(1-\alpha)P_i - \mathscr{C}_i^T\mathscr{C}_i & -\mathscr{C}_i^T\mathscr{G}_i \\ * & * & -\beta^2 I - \mathscr{G}_i^T\mathscr{G}_i \end{bmatrix} < 0. \tag{8.46}
$$

It follows from

$$
(P_j - \Omega_j)P_j^{-1}(P_j - \Omega_j)^T = \Omega_j P_j^{-1}\Omega_j^T + P_j - \Omega_j - \Omega_j^T \geq 0 \tag{8.47}
$$

then

$$
-\Omega_j P_j^{-1}\Omega_j^T \leq P_j - \Omega_j - \Omega_j^T \tag{8.48}
$$

By (8.48), the following inequality implies $\Xi_i < 0$,

$$
\begin{bmatrix} P_j - \Omega_j - \Omega_j^T & \Omega_i\mathscr{A}_i & \Omega_i\mathscr{F}_i \\ * & -(1-\alpha)P_i - \mathscr{C}_i^T\mathscr{C}_i & -\mathscr{C}_i^T\mathscr{G}_i \\ * & * & \beta^2 I - \mathscr{G}_i^T\mathscr{G}_i \end{bmatrix} < 0 \tag{8.49}
$$

Define

$$P_i = \begin{bmatrix} P_{1i} & P_{2i} \\ * & P_{3i} \end{bmatrix}, \quad \Omega_i = \begin{bmatrix} \Omega_{1i} & \Omega_{2i} \\ 0 & \Omega_{3i} \end{bmatrix}, \quad Q_i = \Omega_{1i} H_i, \quad R_i = \Omega_{1i} Z_i, \tag{8.50}$$

and Schur complement, one can get (8.42) is equivalent to (8.49). Then

$$\Delta V_i(\xi(k)) + \alpha V_i(\xi(k)) - r^T(k) r(k) + \beta^2 (f^T(k) f(k)) > 0 \tag{8.51}$$

furthermore one has

$$V_i(\xi(k_1)) > (1-\alpha) V_i(\xi(k_0)) - r^T(k_0) r(k_0) + \beta^2 (f^T(k_0) f(k_0)) \tag{8.52}$$

Iterating (8.52) gives

$$V_i(\xi(k)) > (1-\alpha)^{k-k_0} V_i(\xi(k_0)) - \sum_{s=k_0}^{k-1} (1-\alpha)^{k-s-1} \Gamma(s) \tag{8.53}$$

where $\Gamma(s) = r^T(s) r(s) - \beta^2 (f^T(s) f(s))$. Define H_- performance index

$$J_\infty = \sum_{k=k_0}^{\infty} (1-\alpha)^k r^T(k) r(k) - \beta^2 \left(\sum_{k=k_0}^{\infty} f^T(k) f(k) \right) \tag{8.54}$$

From (8.41) and (8.52), one can obtain

$$\begin{aligned}
V_{\sigma(k)}(\xi(k)) &> (1-\alpha)^{k-k_l} V_{\sigma(k)}(\xi(k_l)) - \sum_{s=k_l}^{k-1} (1-\alpha)^{k-s-1} \Gamma(s) \\
&\geq (1-\alpha)^{k-k_l} \mu V_{\sigma(k_{l-1})}(\xi(k_l)) - \sum_{s=k_l}^{k-1} (1-\alpha)^{k-s-1} \Gamma(s) \\
&\geq (1-\alpha)^{k-k_l} \mu \left[(1-\alpha)^{k_l-k_{l-1}} V_{\sigma(k_{l-1})}(\xi(k_{l-1})) - \sum_{s=k_{l-1}}^{k_l-1} (1-\alpha)^{k_l-s-1} \Gamma(s) \right] \\
&\quad - \sum_{s=k_l}^{k-1} (1-\alpha)^{k-s-1} \Gamma(s) \\
&\geq \cdots\cdots\cdots\cdots \\
&\leq (1-\alpha)^{k-k_0} \mu^{N_\sigma(k_0,k)} V_{\sigma(k_0)}(\xi(k_0)) - (1-\alpha)^{k-k_1} \mu^{N_\sigma(k_0,k)} \\
&\quad - \sum_{s=k_0}^{k_1-1} (1-\alpha)^{k_1-s-1} \Gamma(s) \cdots\cdots - \sum_{s=k_0}^{k-1} (1-\alpha)^{k-s-1} \Gamma(s) \\
&= (1-\alpha)^{k-k_0} \mu^{N_\sigma(k_0,k)} V_{\sigma(k_0)}(\xi(k_0)) - \sum_{s=k_0}^{k-1} \mu^{N_\sigma(s,k)} (1-\alpha)^{k-s-1} \Gamma(s)
\end{aligned} \tag{8.55}$$

then, under zero initial condition, the above formula gives

$$\sum_{s=k_0}^{k-1} \mu^{N_\sigma(s,k)}(1-\alpha)^{k-s-1}\Gamma(s) \geq 0 \tag{8.56}$$

Multiplying both sides of (8.56) by $\mu^{-N_\sigma(0,k)}$, one can obtain

$$\begin{aligned} &\mu^{-N_\sigma(0,k)} \sum_{s=k_0}^{k-1} \mu^{N_\sigma(s,k)}(1-\alpha)^{k-s-1} r^T(s) r(s) \\ &\geq \mu^{-N_\sigma(0,k)} \sum_{s=k_0}^{k-1} \mu^{N_\sigma(s,k)}(1-\alpha)^{k-s-1} \beta^2(f^T(s) f(s)) \end{aligned} \tag{8.57}$$

which is equivalent to

$$\begin{aligned} &\sum_{s=k_0}^{k-1} \mu^{-N_\sigma(0,s)}(1-\alpha)^{k-s-1} r^T(s) r(s) \\ &\geq \sum_{s=k_0}^{k-1} \mu^{-N_\sigma(0,s)}(1-\alpha)^{k-s-1} \beta^2(f^T(s) f(s)) \end{aligned} \tag{8.58}$$

Then, due to the fact that

$$N_\sigma(0,s) \leq \frac{s}{\tau_a} \leq \frac{-sln(1-\alpha)}{ln\mu} \tag{8.59}$$

we have

$$\begin{aligned} &\sum_{s=k_0}^{k-1} \mu^{\frac{sln(1-\alpha)}{ln\mu}}(1-\alpha)^{k-s-1} r^T(s) r(s) \\ &\geq \sum_{s=k_0}^{k-1} \mu^{-N_\sigma(0,s)}(1-\alpha)^{k-s-1} \beta^2(f^T(s) f(s)) \\ &\geq \sum_{s=k_0}^{k-1} (1-\alpha)^{k-s-1} \beta^2(f^T(s) f(s)) \end{aligned} \tag{8.60}$$

which means

$$\sum_{s=k_0}^{k-1} (1-\alpha)^s r^T(s) r(s) \geq \sum_{s=k_0}^{k-1} \beta^2(f^T(s) f(s)) \tag{8.61}$$

Then the proof is concluded. (□)

As a conclusion, the design of the fault detection observer parameters in (8.3) can be summarized as the following algorithm.

Algorithm:

Step 1: By (8.10), to calculate H_i, then T_i can be solved by $T_i = I + H_i C_i$. Furthermore, the matrices M_i can be solved by (8.9);

Step 2: By solving the conditions in Theorem 1, the unknown matrices W_i and Z_i can be solved. Substituting Z_i into (8.13), then the matrices S_i can be computed;

Step 3: Substituting S_i and Z_i into (8.12), then the observer parameter L_i can be obtained.

8.1.5 *Residual Evaluation*

For the purpose of detecting the faults sensitively, it is necessary to set an appropriate threshold J_{th} and an evaluation function $J(r(k))$. In this paper, the residual evaluation function is defined as

$$J(r(k)) = \sum_{k=k_0}^{K}(1-\alpha)^k r^T(k)r(k) \tag{8.62}$$

where K means the terminal time of the total operating process. Set the threshold as

$$J_{th} = \max_{f(k)=0} J(r(k)) \tag{8.63}$$

Therefore, fault detection can be implemented through the following decision logic

$$\|r\| \geq J_{th} \Rightarrow withfaults \Rightarrow alarm \tag{8.64}$$

$$\|r\| < J_{th} \Rightarrow fault - free \tag{8.65}$$

Remark 8.2 Note that many approaches of selection proper residual evaluation function and fault detection threshold can be used, such as dynamic threshold method, root mean square function method, artificial experience method, and so on. Note that our main purpose is to design a fault detection observer. More details about the threshold selection can be found in [27, 28].

8.1.6 *Simulation Example*

In this section, a pulse width modulation (PWM) driven boost converter is shown in Fig. 8.2, to illustrate the validity and applicability of the proposed theoretical

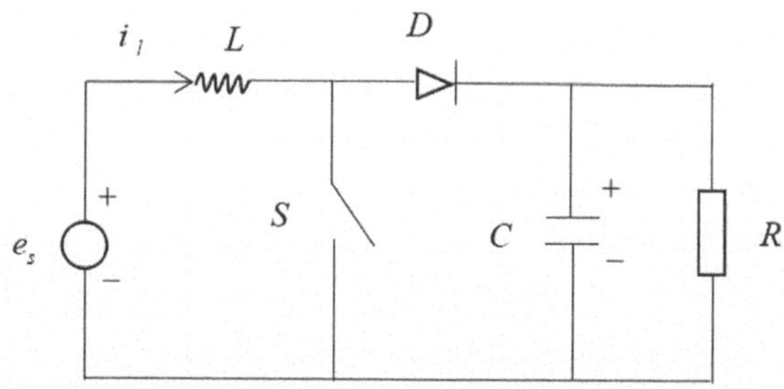

Fig. 8.2 Boost converter

results. The $e_s(t)$ in the figure is the power supply, the inductance is denoted by L, capacitance is expressed in terms of C, and R is the load resistance. The switch $s(t)$ is controlled by a PWM device and can switch at most once in each period T.

As a typical circuit system, the converter is used to transform the source voltage into a higher voltage. In recent years, such power converters have also been modeled as switching systems. The differential equation of the boost converter is as follows:

$$\dot{e}_c(\tau) = -\frac{1}{RC}e_c(\tau) + (1 - s(\tau))\frac{1}{C_1}i_L(\tau) \tag{8.66}$$

$$\dot{i}_L(\tau) = -(1 - s(\tau))\frac{1}{L_1}e_c(\tau) + s(\tau)\frac{1}{L_1} \tag{8.67}$$

where, $\tau = \frac{t}{T}$, $L_1 = \frac{L}{T}$ and $C_1 = \frac{C}{T}$. Then, we let $x = [e_c, i_l]^T$, so (8.66) and (8.67), can be formulated by

$$\dot{x}(t) = A_\sigma x(t) + B_\sigma u(t), \sigma \in (1, 2) \tag{8.68}$$

where

$$A_1 = \begin{bmatrix} -\frac{1}{RC_1} & \frac{1}{C_1} \\ -\frac{1}{L_1} & 0 \end{bmatrix}, A_2 = \begin{bmatrix} -\frac{1}{RC_1} & 0 \\ 0 & 0 \end{bmatrix}, B_1 = \begin{bmatrix} 0 \\ 0 \end{bmatrix}, B_2 = \begin{bmatrix} 0 \\ \frac{1}{L_1} \end{bmatrix}.$$

According to the normalization technique, the discretized matrices can be expressed as

$$A_1 = \begin{bmatrix} 0.0099 & 0.001 \\ -0.0001 & 0.01 \end{bmatrix}, A_2 = \begin{bmatrix} 0.0099 & 0 \\ 0 & 0.01 \end{bmatrix}, B_1 = \begin{bmatrix} 0 \\ 0 \end{bmatrix}, B_2 = \begin{bmatrix} 0 \\ 1 \end{bmatrix}.$$

Then, we assume that the other systems matrices to be

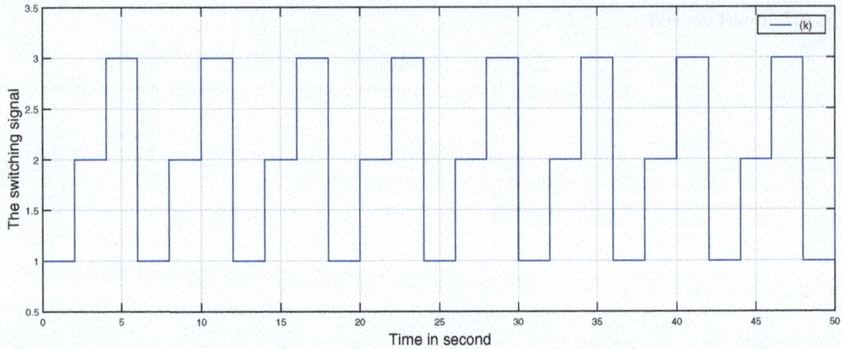

Fig. 8.3 The switching signal $\sigma(k)$

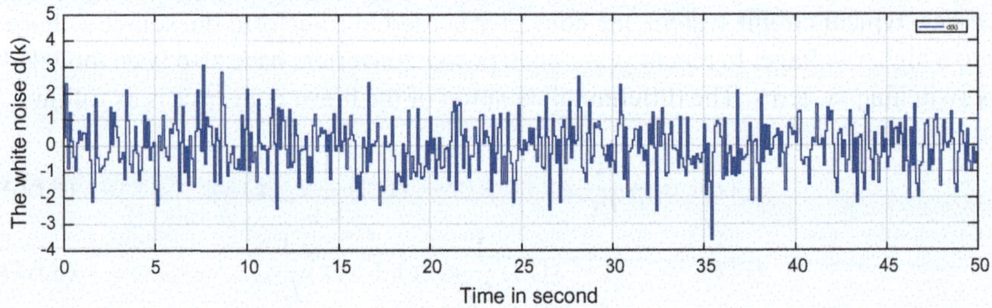

Fig. 8.4 The white noise $d(k)$

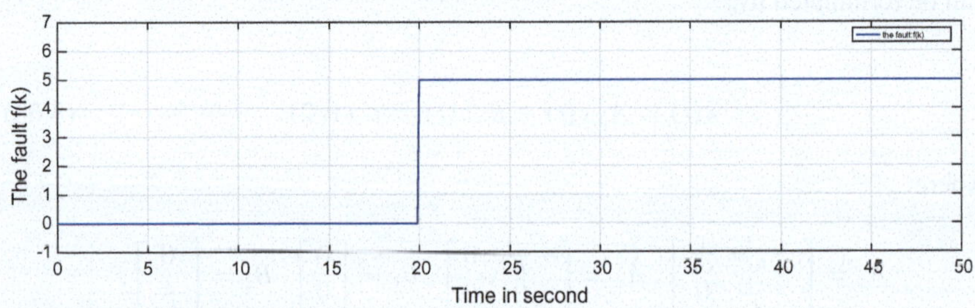

Fig. 8.5 The fault signal $f(k)$

$$D_1 = \begin{bmatrix} 0.003 \\ 0.005 \end{bmatrix}, D_2 = \begin{bmatrix} 0.004 \\ 0.007 \end{bmatrix}, F_1 = \begin{bmatrix} 0.001 \\ 0.002 \end{bmatrix}, F_2 = \begin{bmatrix} 0.001 \\ 0.001 \end{bmatrix},$$
$$C_1 = C_2 = \begin{bmatrix} 4.1 & 0.2 \\ 0.3 & -2 \end{bmatrix}, G_1 = G_2 = \begin{bmatrix} 0.1 \\ 0.2 \end{bmatrix}.$$

We set $\mu = 1.2$, $\alpha = 0.7$, $\beta = 0.4747$, by applying the results in Theorem 8.1, the parameters of the fault diagnosis observer are obtained as follows:

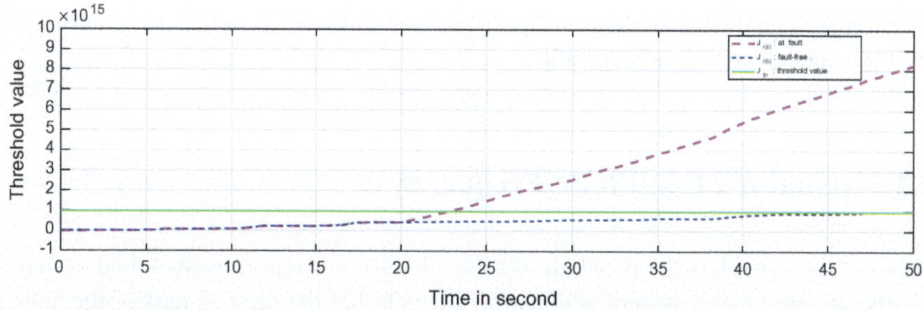

Fig. 8.6 Threshold value $J(r(k))$

$$
S_1 = \begin{bmatrix} -0.3405 & -0.2195 \\ 0.4371 & 0.1981 \end{bmatrix}, S_2 = \begin{bmatrix} -1.0376 & -0.1915 \\ 0.3780 & 0.4338 \end{bmatrix},
$$

$$
L_1 = \begin{bmatrix} 0.2854 & -0.1970 \\ -0.3871 & -0.2201 \end{bmatrix}, L_2 = \begin{bmatrix} 1.022 & -0.4504 \\ -0.2398 & 0.2676 \end{bmatrix},
$$

$$
R_1 = 1.0e^6 * \begin{bmatrix} 0.8375 & -1.2484 \\ -1.2484 & 0.5814 \end{bmatrix}, R_2 = 1.0e^6 * \begin{bmatrix} 4.3136 & 0.6191 \\ 0.6191 & 3.2310 \end{bmatrix},
$$

$$
W_1 = W_2 = 1.0e^7 * \begin{bmatrix} -5.9869 & -0.3566 \\ -0.3566 & 1.7762 \end{bmatrix},
$$

$$
M_1 = \begin{bmatrix} 0 \\ 0 \end{bmatrix}, M_2 = \begin{bmatrix} 0.96 \\ 0.52 \end{bmatrix}, H_1 = H_2 = \begin{bmatrix} 0.8 & -0.4 \\ -0.4 & 0.2 \end{bmatrix}.
$$

Furthermore, one can get the ADT value is $\tau_a^* = 0.1514$. The switching signal is described in Fig. 8.3, which satisfies the ADT condition in (8.16). The outside disturbance signal is set as a white noise, which is shown in Fig. 8.4. Then, a constant fault $f(k)$ is given by

$$
f(k) = \begin{cases} 5, & t \geq 20s \\ 0, & others \end{cases}
$$

which is shown in Figs. 8.5 and 8.6, gives the state response of the residual evaluation function $J_{r(k)}$ with fault and fault-free. By selecting a predefined threshold $J_{th} = 9.722e^{14}$, the simulation results shown that $J_{r(22.9)} > J_{th}$ which means the fault $f(k)$ can be detected quickly.

8.1.7 Summary

The main objective of this paper is to design a real-time-weighted fault detection observer for discrete-time switched systems. By using Lyapunov functions, ADT technique and linear matrix inequality technique, the observer parameters are solved successfully and the fault detection algorithm is obtained at the same time. Our design method can detect all kinds of faults.

8.2 Actuator FE in Finite-Frequency

This section considers the problem of robust FE for discrete-time switched system. Firstly, an augmented system is constructed such that the fault as part of the state. Then a UIO is employed to estimate the state of the augmented system. Based on the switched Lyapunov function and the generalized KYP lemma, a finite-frequency UIO-based FE observer is designed. Finally, the proposed technique is illustrated by a numerical example.

8.2.1 Introduction

It is a common phenomenon that the practical industrial plants are affected by the deviation of system parameters or unexpected faults. If the potential dangers can not be eliminated timely, they may cause great disastrous sequences. With the rapid development of fault diagnosis technique in recent decades, which makes it possible to avoid these unnecessary damages. Fault diagnosis technique mainly includes two parts: FD and FE. FD technique mainly focuses on judging whether a fault has occurred and giving a timely alarm. While FE technique can not only confirm when the fault occurs and the location of the fault, but also can provide the accurate shape of the fault.

Observer-based technique has been widely used to realize FE, such as adaptive observers [29], descriptor observers [30], and sliding-mode observers [31], etc. Among them, UIO-based technique is a popular method which can realize decouple process disturbance for robust FE [32, 33]. Therefore, to develop the UIO-based robust FE techniques for various control system with unknown inputs is a meaningful and challenging work.

In this section, the problem of robust FE for discrete-time switched systems is considered. Firstly, an augmented system is constructed such that the fault signal as part of the state. Then a UIO is employed to estimate the state of the augmented system. By using the switched Lyapunov function and the generalized KYP lemma, efficient conditions are obtained such that the error system is asymptotically stable with H_∞ performance. Finally, an example is provided to illustrate the proposed techniques.

Table 8.1 Three ranges of the fault frequencies

Frequency	Θ_f
LF	$\lvert\vartheta_f\rvert \leq \vartheta_{fl}$
MF	$\vartheta_{fm1} \leq \vartheta_f \leq \vartheta_{fm2}$
HF	$\lvert\vartheta_f\rvert \geq \vartheta_{fh}$

8.2.2 *Problem Formulation and Preliminaries*

Consider the following discrete-time switched systems:

$$\begin{cases} x(t+1) = \sum\limits_{i=1}^{N} \alpha_i(t)(A_i x(t) + B_i u(t) + D_i d(t) + F_i f(t)) \\ y(t) = \sum\limits_{i=1}^{N} \alpha_i(t) C_i x(t) \end{cases} \tag{8.69}$$

where $x(t) \in \mathbb{R}^{n_x}$ is the state vector, $u(t) \in \mathbb{R}^{n_u}$ is the control input vector, $y(t) \in \mathbb{R}^{n_y}$ is the measured output vector, $d(t) \in \mathbb{R}^{n_d}$ is the unknown disturbance input, and $f(t) \in \mathbb{R}^{n_f}$ is the fault. $\vartheta_f \in \Theta_f$ denotes the frequency of the fault, which is assumed to be finite and divided into three ranges, namely low-frequency (LF), middle-frequency (MF), and high-frequency (HF) as shown in the Table 8.1

where ϑ_{fl}, ϑ_{fm1}, ϑ_{fm2}, and ϑ_{fh} are all assumed to be known.

The piecewise function

$$\alpha_i(t) : [0, \infty) \rightarrow \{0, 1\}, \quad \sum_{i=1}^{N} \alpha_i(t) = 1 \tag{8.70}$$

is the switching signal, which specifies which subsystem is activated at the switching instant. The switching signal is assumed to be a priori unknown, but its instantaneous value is available in real time. A_i, B_i, C_i, D_i, and F_i are constant real matrices with appropriate dimensions.

Assumption 16 The matrices $C_i, i = 1, 2, \ldots, N$ is full row rank.

We rewrite the system (8.69), as the following form:

$$\begin{cases} \begin{bmatrix} x(t+1) \\ f(t+1) \end{bmatrix} = \sum\limits_{i=1}^{N} \alpha_i(t) \left(\begin{bmatrix} A_i & F_i \\ 0 & I \end{bmatrix} \begin{bmatrix} x(t) \\ f(t) \end{bmatrix} \right. \\ \left. + \begin{bmatrix} B_i \\ 0 \end{bmatrix} u(t) + \begin{bmatrix} D_i \\ 0 \end{bmatrix} d(t) + \begin{bmatrix} 0 \\ I \end{bmatrix} \Delta f(t) \right) \\ y(t) = \sum\limits_{i=1}^{N} \alpha_i(t) \begin{bmatrix} C_i & 0 \end{bmatrix} \begin{bmatrix} x(t) \\ f(t) \end{bmatrix} \end{cases} \tag{8.71}$$

where $\Delta f(t) = f(t+1) - f(t)$.

By defining the following denotations:

$$\bar{x}(t) = \begin{bmatrix} x(t) \\ f(t) \end{bmatrix}, \bar{A}_i = \begin{bmatrix} A_i & F_i \\ 0 & I \end{bmatrix}, \bar{B}_i = \begin{bmatrix} B_i \\ 0 \end{bmatrix},$$

$$\bar{D}_i = \begin{bmatrix} D_i \\ 0 \end{bmatrix}, \bar{I} = \begin{bmatrix} 0 \\ I \end{bmatrix}, \bar{C}_i = \begin{bmatrix} C_i & 0 \end{bmatrix}. \tag{8.72}$$

the system (8.71) reduces to

$$\begin{cases} \bar{x}(t) = \sum_{i=1}^{N} \alpha_i(t) \left(\bar{A}_i \bar{x}(t) + \bar{B}_i u(t) + \bar{D}_i d(t) + \bar{I} \Delta f(t) \right) \\ y(t) = \sum_{i=1}^{N} \alpha_i(t) \bar{C}_i \bar{x}(t) \end{cases} \tag{8.73}$$

The following UIO is employed to estimate the state of system (8.73):

$$\begin{cases} \xi(t+1) = \sum_{i=1}^{N} \alpha_i(t) \left(R_i \xi(t) + G_i u(t) + K_i y(t) \right) \\ \hat{\bar{x}}(t) = \sum_{i=1}^{N} \alpha_i(t) (\xi(t) - H_i y(t)) \end{cases} \tag{8.74}$$

in which $\xi(t)$ is the state of observer (8.74), $\hat{\bar{x}}(t)$ is the estimation of state $\bar{x}(t)$. R_i, G_i, K_i, and H_i are the gains to be determined later.

Define

$$e(t) = \hat{\bar{x}}(t) - \bar{x}(t), \quad e_f(t) = \hat{f}(t) - f(t) \tag{8.75}$$

then one can get

$$\begin{aligned} e(t) &= \sum_{i=1}^{N} \alpha_i(t)(\xi(t) - H_i y(t) - \bar{x}(t)) \\ &= \sum_{i=1}^{N} \alpha_i(t)(\xi(t) - (I + H_i \bar{C}_i)\bar{x}(t)) \\ &= \sum_{i=1}^{N} \alpha_i(t)(\xi(t) - T_i \bar{x}(t)) \end{aligned} \tag{8.76}$$

where $T_i = I + H_i \bar{C}_i$.

Then the error dynamic system can be deduced as follows:

$$\begin{aligned}
&e(t+1)\\
&=\sum_{i=1}^{N}\alpha_i(t)(\xi(t+1)-T_i\bar{x}(t+1))\\
&=\sum_{i=1}^{N}\alpha_i(t)(R_ie(t)+(R_iT_i+K_i\bar{C}_i-T_i\bar{A}_i)\bar{x}(t)\\
&+(G_i-T_i\bar{B}_i)u(t)-T_i\bar{D}_id(t)-T_i\bar{I}\Delta f(t))
\end{aligned}\tag{8.77}$$

If the following relationships hold:

$$R_iT_i+K_i\bar{C}_i-T_i\bar{A}_i=0 \tag{8.78}$$

$$G_i-T_i\bar{B}_i=0 \tag{8.79}$$

$$T_i\bar{D}_i=0 \tag{8.80}$$

then the error dynamic system (8.77) reduces to

$$\begin{cases} e(t+1)=\sum\limits_{i=1}^{N}\alpha_i(t)(R_ie(t)-T_i\bar{I}\Delta f(t))\\ e_f(t)=\bar{I}e(t)\end{cases}\tag{8.81}$$

Remark 8.3 By using UIO-based FE design, one can get the external disturbance $d(t)$ is eliminated completely, which means that the FE is not affected by the external disturbance $d(t)$.

In this section, we mainly focus on designing a finite-frequency robust FE observer for the augmented switched systems (8.73), such that the FE error is robust to the fault. Therefore, the problem of this section can be formulated as follows:

- for $\Delta f(t)=0$, the system (8.81) is asymptotically stable;
- the H_∞ performance from $\Delta f(t)$ to $e_f(t)$ is less than a given positive scalar γ.

The following lemmas are recalled:

Lemma 8.2 (Generalized KYP Lemma [34]) *Consider the following discrete-time switched system Σ_e:*

$$\begin{cases} x(t+1)=\sum\limits_{i=1}^{N}\alpha_i(t)(A_i(\lambda)x(t)+B_i(\lambda)f(t))\\ e(t)=\sum\limits_{i=1}^{N}\alpha_i(t)(C_i(\lambda)x(t)+D_i(\lambda)f(t))\end{cases}\tag{8.82}$$

where $A_i(\lambda)$, $B_i(\lambda)$, $C_i(\lambda)$, and $D_i(\lambda)$ are matrices with appropriate dimensions, which are defined as follows:

$$(A_i(\lambda), B_i(\lambda), C_i(\lambda), D_i(\lambda)) = \sum_{l=1}^{s} \lambda_l (A_{i,l}, B_{i,l}, C_{i,l}, D_{i,l}), \quad \sum_{l=1}^{s} \lambda_l = 1 \tag{8.83}$$

For a given symmetric matrix

$$\Pi = \begin{bmatrix} I & 0 \\ 0 & -\gamma^2 I \end{bmatrix} \tag{8.84}$$

the following statements are equivalent:
(i): The finite-frequency inequality

$$\rho_{max}(G^i_{fe}(e^{j\vartheta_f})) < \gamma, \tag{8.85}$$

where $\rho_{max}(G^i_{fe}(e^{j\vartheta_f}))$ *is the maximum singular value of the transfer function matrix* $G^i_{fe}(e^{j\vartheta_f})$ *for the* i*-th subsystem from the fault* $f(t)$ *to error output* $e(t)$*;*
(ii): There exist Hermitian matrix functions $P_i(\lambda)$*,* $Q_i(\lambda)$ *satisfying* $Q_i(\lambda) > 0$*, and*

$$\begin{bmatrix} A_i(\lambda) & B_i(\lambda) \\ I & 0 \end{bmatrix}^T \Xi_i(\lambda) \begin{bmatrix} A_i(\lambda) & B_i(\lambda) \\ I & 0 \end{bmatrix} + \begin{bmatrix} C_i(\lambda) & D_i(\lambda) \\ 0 & I \end{bmatrix}^T \Pi \begin{bmatrix} C_i(\lambda) & D_i(\lambda) \\ 0 & I \end{bmatrix} < 0 \tag{8.86}$$

where

$$\Xi_i(\lambda) = \begin{bmatrix} -P_i(\lambda) & Q_i(\lambda) \\ Q_i(\lambda) & P_i(\lambda) - 2cos(\vartheta_{fl}) Q_i(\lambda) \end{bmatrix} \tag{8.87}$$

for low-frequency range $|\vartheta_f| \leq \vartheta_{fl}$*,*

$$\Xi_i(\lambda) = \begin{bmatrix} -P_i(\lambda) & e^{j\vartheta_{fc}} Q_i(\lambda) \\ e^{-j\vartheta_{fc}} Q_i(\lambda) & P_i(\lambda) - 2cos(\vartheta_{fw}) Q_i(\lambda) \end{bmatrix} \tag{8.88}$$

for middle-frequency range $\vartheta_{fm1} < \vartheta_f < \vartheta_{fm2}$*, and*

$$\Xi_i(\lambda) = \begin{bmatrix} -P_i(\lambda) & -Q_i(\lambda) \\ -Q_i(\lambda) & P_i(\lambda) + 2cos(\vartheta_{fh}) Q_i(\lambda) \end{bmatrix} \tag{8.89}$$

for high-frequency range $|\vartheta_f| \geq \vartheta_{fh}$*, where* $\vartheta_{fc} = \frac{\vartheta_{fm1}+\vartheta_{fm2}}{2}$*,* $\vartheta_{fw} = \frac{\vartheta_{fm1}-\vartheta_{fm2}}{2}$*.*

Remark 8.4 Because $\rho_{max}(G^i_{fe}(e^{j\vartheta_f}))$ is the maximum singular value of the transfer function matrix $G^i_{fe}(e^{j\vartheta_f})$ for the i-th subsystem from the fault $f(t)$ to the error output $e(t)$, one can get that (8.85) is equivalent to minimize the error $e(t)$ over the energy bounded the fault $f(t)$, i.e.,

$$\min \sup_{f(t)\neq 0} \frac{\|e(t)\|_2}{\|f(t)\|_2} < \gamma$$

Remark 8.5 From the definitions in (8.83), if we set $A_{i,1} = A_{i,2} = \cdots = A_{i,s} = A_i$, $B_{i,1} = B_{i,2} = \cdots = B_{i,s} = B_i$, $C_{i,1} = C_{i,2} = \cdots = C_{i,s} = C_i$, $D_{i,1} = D_{i,2} = \cdots = D_{i,s} = D_i$, then the system Σ_e becomes a common linear switched system. Therefore, the parameters $P_i(\lambda)$ and $Q_i(\lambda)$ become P_i and Q_i in Lemma 8.2.

Lemma 8.3 (Projection Lemma [35]) *Given a symmetric matrix Σ and two matrices Γ, Λ of appropriate column dimensions, the problem*

$$\Sigma + \Gamma^T \Theta \Lambda + \Lambda^T \Theta^T \Gamma < 0 \tag{8.90}$$

is solvable with respect to the matrix Θ of compatible dimensions if and only if the following inequalities hold

$$\mathcal{N}_\Gamma^T \Sigma \mathcal{N}_\Gamma < 0, \quad \mathcal{N}_\Lambda^T \Sigma \mathcal{N}_\Lambda < 0, \tag{8.91}$$

where $\mathcal{N}_\Gamma$ and $\mathcal{N}_\Lambda$ are respectively any basis of the nullspace of Γ and Λ.

8.2.3 Main Results

In this section, the finite-frequency robust FE observer (8.74), will be designed for system (8.73), which guarantees that the error system (8.81), is asymptotically stable with H_∞ performance.

Theorem 8.3 *For a given positive scalar γ, the system (8.81), is asymptotically stable with H_∞ performance γ, if there exist positive definite matrices P_{1i}, P_{1j}, Hermitian matrices P_{2i}, $Q_i > 0$, and matrices S_i, M_i, Ω_i, for all $i, j \in \{1, 2, \ldots, N\}$, which satisfy*

$$\begin{bmatrix} P_{1j} - S_i - S_i^T & \Omega_i \\ * & -P_{1i} \end{bmatrix} < 0 \tag{8.92}$$

$$\begin{bmatrix} -P_{2i} & Q_i - S_i^T & 0 & 0 \\ * & \Phi_{1i} & -M_i \bar{I} & \bar{I} \\ * & * & -\gamma^2 I & 0 \\ * & * & * & -I \end{bmatrix} < 0 \tag{8.93}$$

for the low-frequency fault $|\vartheta_f| \leq \vartheta_{fl}$;

$$\begin{bmatrix} -P_{2i} & e^{j\vartheta_{fc}}Q_i - S_i^T & 0 & 0 \\ * & \Phi_{mi} & -M_i\bar{I} & \bar{I} \\ * & * & -\gamma^2 I & 0 \\ * & * & * & -I \end{bmatrix} < 0 \tag{8.94}$$

for the middle-frequency fault $\vartheta_{fm1} \leq \vartheta_f \leq \vartheta_{fm2}$*;*

$$\begin{bmatrix} -P_{2i} & -Q_i - S_i^T & 0 & 0 \\ * & \Phi_{hi} & -M_i\bar{I} & \bar{I} \\ * & * & -\gamma^2 I & 0 \\ * & * & * & -I \end{bmatrix} < 0 \tag{8.95}$$

for the high-frequency fault $|\vartheta_f| \geq \vartheta_{fh}$*;*
where

$$\begin{aligned} \Phi_{li} &= P_{2i} - 2cos(\vartheta_{fl})Q_i + \Omega_i + \Omega_i^T, \\ \Phi_{mi} &= P_{2i} - 2cos(\vartheta_{fw})Q_i + \Omega_i + \Omega_i^T, \\ \Phi_{hi} &= P_{2i} + 2cos(\vartheta_{fh})Q_i + \Omega_i + \Omega_i^T, \\ \vartheta_{fc} &= \frac{\vartheta_{fm1} + \vartheta_{fm2}}{2}, \ \vartheta_{fw} = \frac{\vartheta_{fm1} - \vartheta_{fm2}}{2}. \end{aligned}$$

Proof Firstly, we will establish the stability of system (8.81). If $\Delta f(t) = 0$, (8.81), becomes

$$e(t+1) = \sum_{i=1}^{N} \alpha_i(t) R_i e(t) \tag{8.96}$$

Define

$$\Omega_i = S_i R_i; \quad M_i = S_i T_i. \tag{8.97}$$

Substituting the condition (8.97) into (8.92), implies

$$\begin{bmatrix} \bar{P}_{1j} - \bar{S}_i - \bar{S}_i^T & S_i R_i \\ * & -\bar{P}_{1i} \end{bmatrix} < 0 \tag{8.98}$$

which is equivalent to

$$\begin{bmatrix} P_{1j} & 0 \\ 0 & -P_{1i} \end{bmatrix} + \begin{bmatrix} S_i \\ 0 \end{bmatrix} [-I \ R_i] + \begin{bmatrix} -I \\ R_i^T \end{bmatrix} [S_i^T \ 0] < 0 \tag{8.99}$$

By Lemma 8.3, if we define

$$\Sigma = \begin{bmatrix} P_{1j} & 0 \\ 0 & -P_{1i} \end{bmatrix}, \Lambda = \begin{bmatrix} -I & R_i \end{bmatrix}, \Theta = I, \Gamma = \begin{bmatrix} S_i^T & 0 \end{bmatrix}, \tag{8.100}$$

and choose $\mathcal{N}_\Lambda = \begin{bmatrix} R_i \\ I \end{bmatrix}$, then one can get from (8.99) that

$$\begin{aligned} &\mathcal{N}_\Lambda^T \Sigma \mathcal{N}_\Lambda \\ &= \begin{bmatrix} R_i^T & I \end{bmatrix} \begin{bmatrix} P_{1j} & 0 \\ 0 & -P_{1i} \end{bmatrix} \begin{bmatrix} R_i \\ I \end{bmatrix} \\ &= R_i^T P_{1j} R_i - P_{1i} < 0 \end{aligned} \tag{8.101}$$

Construct a switched Lyapunov function for dynamics (8.96), as

$$V(t, e(t)) = e^T(t) \left(\sum_{i=1}^{N} \alpha_i(t) P_{1i} \right) e(t) \tag{8.102}$$

From Lyapunov stability theory, if the following inequality

$$\begin{aligned} \Delta V(t) &= V(t+1, e(t+1)) - V(t, e(t)) \\ &= e^T(t+1) \left(\sum_{i=1}^{N} \alpha_i(t+1) P_{1i} \right) e(t+1) \\ &\quad - e^T(t) \left(\sum_{i=1}^{N} \alpha_i(t) P_{1i} \right) e(t) < 0 \end{aligned} \tag{8.103}$$

holds, then the system (8.96), is asymptotically stable. Then with the arbitrary switching instants $\alpha_i(t) = i, \alpha_i(t+1) = j$ and the condition (8.101), one can get that the system (8.96), is asymptotically stable.

In the following, we will consider the H_∞ performance of $\|e_f(t)\|_2 < \gamma \|\Delta f(t)\|_2$. By Schur complement lemma, (8.93), can be transformed into

$$\begin{bmatrix} -P_{2i} & Q_i - S_i^T & 0 \\ * & \Phi_{1i} + \bar{I}^T \bar{I} & -M_i \bar{I} \\ * & * & -\gamma^2 I \end{bmatrix} < 0 \tag{8.104}$$

(8.104) can be rewritten as

$$\begin{bmatrix} -P_{2i} & Q_i & 0 \\ * & P_{2i} - 2cos(\vartheta_{fl})Q_i + \bar{I}^T\bar{I} & 0 \\ * & * & -\gamma^2 I \end{bmatrix} + \begin{bmatrix} 0 \\ S_i \\ 0 \end{bmatrix} \begin{bmatrix} -I & R_i & -T_i\bar{I} \end{bmatrix} + \begin{bmatrix} -I \\ R_i^T \\ -\bar{I}^T T_i^T \end{bmatrix} \begin{bmatrix} 0 & S_i^T & 0 \end{bmatrix} < 0 \tag{8.105}$$

If we define

$$\begin{aligned} \Sigma &= \begin{bmatrix} -P_{2i} & Q_i & 0 \\ * & P_{2i} - 2cos(\vartheta_{fl})Q_i + \bar{I}^T\bar{I} & 0 \\ * & * & -\gamma^2 I \end{bmatrix}, \\ \Theta &= I, \Lambda = \begin{bmatrix} -I & R_i & -T_i\bar{I} \end{bmatrix}, \\ \Gamma &= \begin{bmatrix} 0 & \bar{S}_i^T & 0 \end{bmatrix}, \end{aligned} \tag{8.106}$$

and choose

$$\mathcal{N}_\Lambda = \begin{bmatrix} R_i & -T_i\bar{I} \\ I & 0 \\ 0 & I \end{bmatrix}, \tag{8.107}$$

then one can get from (8.98) that

$$\mathcal{N}_\Lambda^T \Sigma \mathcal{N}_\Lambda = \begin{bmatrix} \Pi_{11} & \Pi_{12} \\ \Pi_{12}^T & \Pi_{22} \end{bmatrix} < 0$$

where

$$\begin{aligned} \Omega_{11} &= -R_i^T P_{2i} R_i + Q_{1i} R_i + R_i^T Q_i - 2cos(\vartheta_{fl})Q_i + P_{2i} + \bar{I}\bar{I}^T; \\ \Omega_{12} &= R_i^T P_{2i} T_i \bar{I} - Q_{1i} T_i \bar{I}; \\ \Omega_{22} &= -(T_i\bar{I})^T P_{2i}(T_i\bar{I}) - \gamma^2 I. \end{aligned}$$

Therefore, if (8.93) holds, then we can deduce that (8.108) holds, which further implies that (8.86) holds. By Lemma 8.2, one can obtain that the system (8.81), satisfies the low-frequency H_∞ performance index γ.

The proofs of the conditions (8.94) and (8.95), can be directly deduced by setting

$$\Sigma = \begin{bmatrix} -P_{2i} & e^{j\vartheta_{fc}} Q_i & 0 \\ e^{-j\vartheta_{fc}} Q_i & P_{2i} - 2cos(\vartheta_{fm})Q_{1i} + \bar{I}\bar{I}^T & 0 \\ 0 & 0 & -\gamma^2 I \end{bmatrix}$$

and

$$\Sigma = \begin{bmatrix} -P_{2i} & -Q_i & 0 \\ -Q_{1i} & P_{2i} + 2cos(\vartheta_{fh})Q_i + \bar{I}\bar{I}^T & 0 \\ 0 & 0 & -\gamma^2 I \end{bmatrix}$$

According to the same proof lines, (8.93), (8.94), and (8.95) can be proven. (□)

In the following theorem, the design of FE UIO is given.

Theorem 8.4 *Consider the augmented switched system (8.73). For a given positive scalar γ, there exists a finite-frequency FE UIO (8.74), such that the error system (8.81), is asymptotically stable with H_∞ performances γ if there exist positive definite matrices P_{1i}, P_{1j}, Hermitian matrices P_{2i}, $Q_i > 0$, and matrices S_i, M_i, and Ω_i, for $\forall i, j \in \{1, 2, \ldots, N\}$, which satisfy the conditions in (8.92), (8.93) (or (8.94), or (8.95), and*

$$S_i^{-1} M_i \bar{D}_i = 0. \tag{8.108}$$

In this case, the parameters of the FD filter (8.71) are given by

$$R_i = S_i^{-1}\Omega_i; \tag{8.109}$$

$$G_i = S_i^{-1} M_i \bar{B}_i; \tag{8.110}$$

$$H_i = (S_i^{-1} M_i - I)\bar{C}_i^T (C_i C_i^T)^{-1}; \tag{8.111}$$

$$K_i = (S_i^{-1} M_i \bar{A}_i - S_i^{-1}\Omega_i S_i^{-1} M_i)\bar{C}_i^T (C_i C_i^T)^{-1}. \tag{8.112}$$

Proof (8.108) can be directly deduced by (8.80) and (8.97). From(8.97), one can get (8.109). By the condition (8.79), one can get the condition (8.110). From (8.97) and $T_i = I + H_i\bar{C}_i$, one can get that

$$H_i \bar{C}_i = S_i^{-1} M_i - I \tag{8.113}$$

Post-multiplying (8.113), by $\bar{C}_i^T$ with Assumption 11, one can get (8.111), directly. Similarly, (8.112), can be obtained by the condition (8.78). Then the proof is concluded. (□)

Remark 8.6 The FE UIO design problem can be concluded as following question: to minimize γ such that the conditions (8.92), (8.93) (or (8.94), or (8.95)) are satisfied.

Remark 8.7 After the UIO parameters have been calculated by solving the conditions in Theorem 8.2, we can construct the control input and the disturbance input for system (8.73). Then the state estimation $\hat{\bar{x}}(t)$ can be obtained by the UIO (8.74). The actuator FE $\hat{f}(t)$ can be direct computed by $\hat{f}(t) = \bar{I}^T\hat{\bar{x}}(t)$.

8.2.4 An Illustrate Example

Suppose that the discrete-time switched system (8.69), has two subsystems and the parameters are given by

$$A_1 = \begin{bmatrix} 0.23 & 0 \\ 2.51 & 0.98 \end{bmatrix}, B_1 = \begin{bmatrix} 0.1 \\ 0.1 \end{bmatrix}, D_1 = \begin{bmatrix} 1 \\ 1 \end{bmatrix}, C_1 = \begin{bmatrix} 1 & 1 \end{bmatrix}, F_1 = \begin{bmatrix} 0.5 \\ 0.5 \end{bmatrix};$$

$$A_2 = \begin{bmatrix} -0.68 & 0 \\ 1.74 & 0.27 \end{bmatrix}, B_2 = \begin{bmatrix} 0.2 \\ 0.2 \end{bmatrix}, D_2 = \begin{bmatrix} 0.5 \\ 0.5 \end{bmatrix}, C_2 = \begin{bmatrix} 0.5 & 0.5 \end{bmatrix}, F_2 = \begin{bmatrix} 0.2 \\ 0.2 \end{bmatrix}.$$

By solving the conditions in Theorem 8.4, one can get the minimum H_∞ performance index $\gamma = 0.8305$, and the UIO parameters can be calculated as follows:

$$R_1 = \begin{bmatrix} -0.3576 & -0.2918 & -0.2408 \\ -0.2918 & -0.3576 & -0.2408 \\ -0.2289 & -0.2289 & -0.4685 \end{bmatrix}, G_1 = \begin{bmatrix} -0.0005 \\ -0.0005 \\ 0.0076 \end{bmatrix},$$

$$H_1 = \begin{bmatrix} -0.5027 \\ -0.5027 \\ 0.0380 \end{bmatrix}, K_1 = \begin{bmatrix} 0.0147 \\ -0.0101 \\ 0.0895 \end{bmatrix},$$

$$R_2 = \begin{bmatrix} -0.0161 & 0.1557 & 0.1831 \\ 0.1557 & -0.0161 & 0.1831 \\ 0.1140 & 0.1140 & -0.1589 \end{bmatrix}, G_2 = \begin{bmatrix} -0.0055 \\ -0.0055 \\ 0.0227 \end{bmatrix},$$

$$H_2 = \begin{bmatrix} -1.0277 \\ -1.0277 \\ 0.1137 \end{bmatrix}, K_2 = \begin{bmatrix} -0.0145 \\ -0.0458 \\ 0.0784 \end{bmatrix}.$$

To illustrate the efficiency of the FE, the actuator fault is assumed to be in a low-frequency domain, i.e., $\vartheta_{fl} = \frac{\pi}{6}$, and it is set as follows:

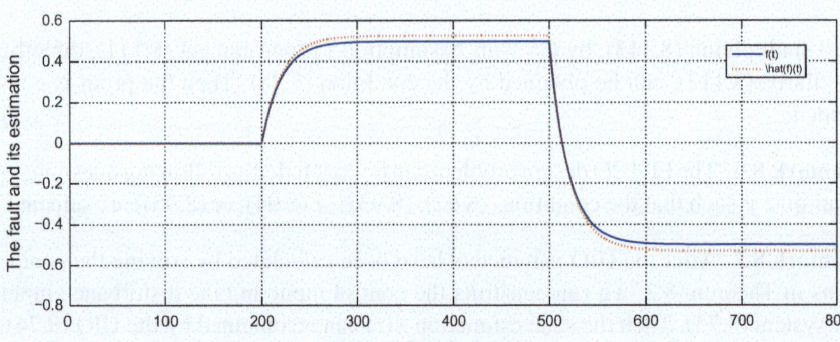

Fig. 8.7 The fault $f(t)$ and its estimation $\hat{f}(t)$

$$f(t) = \begin{cases} 0, & 0s \le t < 200s \\ 0.5(1 - e^{-0.05(t-200)}, & 200s \le t < 500s \\ -0.5 + e^{-0.05(t-500)}, & 500s \le t < 800s \end{cases}$$

The simulation result is shown in Fig. 8.7, one can get that the FE observer gives a fast estimation of the actuator fault.

8.2.5 Summary

In this section, the problem of UIO-based FE observer design is investigated for discrete-time switched systems with finite-frequency actuator fault. The stability and H_∞ performance index is ensured in the finite-frequency domain by using KYP lemma and the switched Lyapunov function. Sufficient conditions for the existence of the FE observer are obtained in terms of LMIs. Finally, a numerical example is provided to illustrate the proposed technique.

References

1. Xiong, J., Chang, X., Yi, X.: Design of robust nonfragile fault detection filter for uncertain dynamic systems with quantization. Appl. Math. Comput. **338**, 774–788 (2018)
2. Du, D.: Fault detection for discrete-time linear systems based on descriptor observer approach. Appl. Math. Comput. **293**, 575–585 (2017)
3. Mao, Z., Jiang, B., Shi, P.: Fault detection for a class of nonlinear networked control systems. Int. J. Adapt. Control Signal Process **24**, 610–622 (2010)
4. Li, T., Li, G., Zhao, Q.: Adaptive fault-tolerant stochastic shape control with application to particle distribution control. IEEE Trans. Syst, Man Cybern. Part A. Syst. **45**(12), 1592–1604 (2015)
5. Long, Y., Park, J.H., Ye, D.: Transmission-dependent fault detection and isolation strategy for networked systems under finite capacity channels. IEEE Trans. Cybern. **47**(8), 2266–2278 (2017)
6. Zhong, M., Ding, S.X., Lam, J., Wang, H.: An LMI approach to design robust fault detection filter for uncertain LTI systems. Automatica **39**(3), 543–550 (2003)
7. Jiang, B., Staroswiecki, M., Cocquempot, V.: Fault diagnosis based on adaptive observer for a class of non-linear systems with unknown parameters. Int. J. Control **77**, 415–426 (2004)
8. Zhang, Y., Sun, J., Liang, H., Li, H.: Event-triggered adaptive tracking control for multi-agent systems with unknown disturbances. IEEE Trans. Cybern. **53**(3), 890–901 (2020)
9. Zhang, Z., Lu, J., Xu, S.: Tuning functions-based robust adaptive tracking control of a class of nonlinear systems with time delays. Int. J. Robust Nonlinear Control **22**(14), 1631–1646 (2012)
10. Yan, X., Edwards, C.: Nonlinear robust fault reconstruction and estimation using a sliding mode observer. Automatica **43**(9), 1605–1614 (2007)
11. Dong, H., Wang, Z.D., Lam, J., Gao, H.: Fuzzy-model-based robust fault detection with stochastic mixed time delays and successive packet dropouts. IEEE Trans. Syst. Man Cybern. Part B. Cybern. **42**(2), 365–376 (2012)
12. Chadli, M., Karimi, H.R.: Robust observer design for unknown inputs Takagi-Sugeno models. IEEE Trans. Fuzzy Syst. **21**(1), 158–164 (2013)

13. Qiu, J., Tian, H., Lu, Q., Gao, H.: Non-synchronized robust filtering design for continuous-time T-S fuzzy affine dynamic systems based on piecewise Lyapunov functions. IEEE Trans. Cybern. **43**(6), 1755–1766 (2013)
14. Li, H., Pan, Y., Zhou, Q.: Filter design for interval type-2 fuzzy systems with D stability constraints under a unified frame. IEEE Trans. Fuzzy Syst. **23**(3), 719–725 (2015)
15. Huang, S., Yang, G.: Fault tolerant controller design for t-s fuzzy systems with time-varying delay and actuator faults: a k-step fault-estimation approach. IEEE Trans. Fuzzy Syst. **22**(6), 1526–1540 (2014)
16. Zhang, X., Polycarpou, M.M., Parisini, T.: Fault diagnosis of a class of nonlinear uncertain systems with Lipschitz nonlinearities using adaptive estimation. Automatica **46**(2), 290–299 (2010)
17. Xiao, X., Park, J.H., Zhou, L., Lu, G.: Event-triggered control of discrete-time switched linear systems with network transmission delays. Automatica **111** (2020)
18. Morse, A.S.: Supervisory control of families of linear setpoint controllers part I: exact matching. IEEE Trans. Automat. Control **41**(10), 1413–1431 (1996)
19. Zhang, L., Zhuang, S., Shi, P.: Non-weighted quasi-time-dependent H_∞ filtering for switched linear systems with persistent dwell-time. Automatica **54**, 201–209 (2015)
20. Hespanha, J.P.: Uniform stability of switched linear systems extensions of Lasalles invariance principle. IEEE Trans. Autom. Control **49**(4), 470–482 (2004)
21. Hu, H., Jiang, B., Yang, H.: Reliable guaranteed-cost control of delta operator switched systems with actuator faults: mode-dependent average dwell-time approach. IET Control Theory Appl. **10**(1), 17–23 (2016)
22. Liu, M., Shi, P.: Sensor fault estimation and tolerant control for Itô stochastic systems with a descriptor sliding mode approach. Automatica **49**, 1242–1250 (2013)
23. Zhang, K., Jiang, B., Cocquempot, V.: Adaptive observer-based fast fault estimation. Int. J. Control Autom. Syst. **6**(3), 320–326 (2008)
24. Fu, S., Qiu, J.: Adaptive fuzzy observer design for a class of switched nonlinear systems with actuator and sensor faults. IEEE Trans. Fuzzy Syst. **26**(6), 3730–3742 (2018)
25. Hespanha, J.P., Morse, A.S.: Stability of switched systems with average dwell time. In: Proceedings of 38th Conference on Decision Control, Phoenix, AZ, pp. 2655–2660 (1999)
26. Zhang, L., Shi, P.: $l_2 - l_\infty$ model reduction for switched systems with average dwell time. IEEE Trans. Automat. Control **53**(10), 2443–2448 (2008)
27. Chadli, M., Abdo, A., Ding, S.X.: H_-/H_∞ fault detection filter design for discrete-time Takagi-Sugeno fuzzy system. Automatica **49**, 1996–2005 (2013)
28. Li, L., Chadli, M., Steven, S.D., Qiu, J., Yang, Y.: Diagnostic observer design for T-S fuzzy systems: application to real-time-weighted fault-detection approach. IEEE Trans. Fuzzy Syst. **26**(2) (2018)
29. Wang, H., Huang, Z., Daley, S.: On the use of adaptive updating rules for actuator and sensor fault diagnosis. Automatica **33**(2), 217–225 (1997)
30. Gao, Z., Ding, S.X.: Actuator fault robust estimation and fault tolerant control for a class of nonlinear descriptor systems. Automatica **43**(5), 912–920 (2007)
31. Alwi, H., Edwards, C.: Robust fault reconstruction for linear parameter varying systems using sliding mode observers. Int. J. Robust Nonlinear Control **24**(14), 1947–1968 (2014)
32. Hrizi, O., Boussaid, B., Zouinkhi, A., Abdelkrim, M.: Robust unknown input observer based fast adaptive fault estimation: Application to unicycle robot. In: Proceedings of International Conference on Automation Control Engineering and Computer Science, Monastir, Tunisia, pp. 186–194 (2014)
33. Lee, K.S., Park, T.G.: New results on fault reconstruction using a finite-time converging unknown input observer. IET Control Theory Appl. **6**(9), 1258–1265 (2012)
34. Ding, D., Yang, G.: Finite frequency H_∞ filtering for uncertain discrete-time switched linear systems. Progress Nat. Sci. **19**, 1625–1633 (2009)
35. Gahinet, P., Apkarian, P.: A linear matrix inequality approach to H_∞ control. Int. J. Robust Nonlinear Control **4**(4), 421–448 (1994)

Chapter 9
Conclusions and Future Research Direction

In this book, some novel observer-based fault diagnosis and FTC techniques for switched systems are proposed. The FD filter design methods are summarized in Chap. 1. By using the residual signal generated in the filter, the fault can be successfully detected. Then, the fault detection, estimation and accommodation for actuator and sensor faults are respectively studied from Chaps. 2–8. In the present research, methods based on PIO, reduced-order observer, adaptive observer, and UIO have been developed. Several situations including finite-frequency, asynchronous, output disturbances, and nonlinear cases are considered. Based on ADT technique and switched Lyapunov function method, the sufficient conditions for the existence of various fault estimators are derived in terms of LMIs. The effectiveness of the proposed methodologies have been verified by some practical engineering systems.

Although fruitful literature have been reported about this topic, there are still many open problems to be further investigated. In the last, we want to provide some future research directions, which hope to be helpful to guide interested readers when exploring fault diagnosis and FTC for switched systems.

(1) Asynchronous fault estimation and accommodation for switched systems. Presently, some stability and robust control results for asynchronous switched systems have been proposed. How to use the existing theoretical results to improve the fault diagnosis theory of switched systems is still a worth studying topic. Especially for the observer-based fault estimation and accommodation methods need to be further exploited.

(2) Sliding-mode observer-based fault diagnosis for switched systems. Observer-based method in dealing with fault diagnosis issues has been proved to be efficient and effective. In the past, various approaches have been proposed, such as filter, adaptive observer, PIO, and UIO etc. However, few reports were focused on sliding-mode observer method. In truth, sliding-mode observer is

D. Du et al., *Observer-Based Fault Diagnosis and Fault-Tolerant Control for Switched Systems*, Studies in Systems, Decision and Control 323,
https://doi.org/10.1007/978-981-15-9073-3_9

an efficient method in dealing with fault diagnosis problems, especially for nonlinear cases. Therefore, it is necessary to employ the sliding-mode observer method to solve the fault diagnosis and FTC for switched systems.

(3) The practical application of the FTC methods for switched systems.
There exist two difficult aspects for the applications of the theoretical results. One is whether there is a feasible solution for the data collection. Another is whether the simulation results are correct for the obtained solutions. Therefore, the research on this issue is another arduous work.

The manufacturer's authorised representative in the EU is Springer Nature Customer Service Centre GmbH, Europaplatz 3, 69115 Heidelberg, Germany. If you have any concerns regarding our products, please contact ProductSafety@springernature.com

Printed and bound by CPI Group (UK) Ltd, Croydon, CR0 4YY
24/07/2026
02175961-0003